Pathomechanics of Common Foot Disorders

Douglas H. Richie Jr

Pathomechanics of Common Foot Disorders

Original Illustrations by Kevin B. Rosenbloom

Springer

Douglas H. Richie Jr
California School of Podiatric Medicine
at Samuel Merritt University
Oakland, CA
USA

ISBN 978-3-030-54203-0 ISBN 978-3-030-54201-6 (eBook)
https://doi.org/10.1007/978-3-030-54201-6

This Springer imprint is published by the registered company Springer Nature Switzerland AG
The registered company address is: Gewerbestrasse 11, 6330 Cham, Switzerland

To my father, Douglas H. Richie, who always believed in me, who supported me throughout my education, and encouraged me to write this book.

To my wife, Cathleen Richie, whose patience and continuous affirmation inspired me to complete this project.

This book would not be possible without the love and support from both of them.

Preface

Pathomechanics

(patho-mi-kan'i-ks)

Changes in the normal biomechanical function of a joint, an extremity, or the torso as the result of trauma or disease. Source: The Free Dictionary

This book is about common disorders of the human foot and ankle. The primary purpose is to explore the role of mechanical factors which contribute to the cause of foot pathologies. Reciprocally, the mechanics of the foot are in turn significantly altered by those same pathologies. Before studying foot disorders, the first chapters of this book will provide an understanding of how the human foot functions during its primary activities of standing and walking. It will be shown that the unique anatomic features of the human foot are the same structures which are vulnerable to injury and deformity.

Early in my career, I was influenced by certain theories of foot function which directed my philosophy in the evaluation and treatment of patients in clinical practice. Most notably, the teachers of biomechanics at the California School of Podiatric Medicine during the 1970s embraced the "Root Theory" as well as other explanations of foot function taught by Inman, Close, Efltman, Manter, and Hicks [1–6].

The pathomechanics of common foot disorders were taught primarily from the sentinel text "Normal and Abnormal Function of the Foot" written by Merton Root, John Weed and William Orien [7]. The conservative treatment options for most foot disorders focused mainly on foot orthotic therapy, which were also based upon the teachings of Root, Weed, and Orien. The surgical treatments I was taught combined a body of knowledge contributed by orthopedic surgeons and podiatric surgeons. In particular, forefoot surgery correcting digital deformities as well as hallux abducto valgus (HAV) focused heavily on procedures developed by podiatrists. The etiology of digital deformities and the correction of those deformities was based upon functional anatomy. At that time, podiatric physicians were surging far ahead of their orthopedic colleagues in developing innovative surgical interventions which were based upon biomechanical principles.

When I graduated from podiatric medical school in 1980, gait labs studying human movement were just being developed and were primarily supported by running shoe companies. Academic institutions such as Penn State University, the

University of Oregon, and the University of Calagary started offering graduate programs focusing on the study of the biomechanics of the lower extremity. These programs were headed by Peter Cavanaga, Barry Bates, and Benno Nigg.

Outcomes research testing the efficacy of treatment interventions for orthopedic conditions using randomized clinical trials was an unknown practice in the 1970s. Medical journals published papers describing orthopedic and podiatric treatment interventions that were mostly retrospective reviews or simply "expert opinion," which today would qualify as Level 4 or 5 evidence.

Shortly after I graduated from podiatric medical school, and for the next 30 years, a renaissance occurred in the world of lower extremity biomechanics and orthopedic surgery. An academic discipline developed by movement scientists and engineers known as "biomechanists" began studying lower extremity function during walking and running. New insights emerged regarding the kinematics, or movement of skeletal segments, as well as the kinetics, which studied the forces acting on those segments. Surgeons began performing research with biomechanists to study pathomechanics as well as evaluating procedures and their functional effects on the lower extremity. Surgical procedures took a quantum leap in technical sophistication thanks to arthroscopy as well as advances in hardware for internal fixation. Concomitantly, the academic community set standards for outcomes research and evidence based medicine to measure the reliability of treatment interventions. Certainly, the 1980s and 1990s were the most exciting times for anyone treating lower extremity pathologies as new understanding and new technologies for treatment underwent meteoric growth.

During this time period, some of what I had been taught in podiatry school was challenged, if not disproven, in terms of how the human foot functioned. Certainly, our concept of the ideal or "normal" foot was contradicted. As a result, the goals and strategies for foot orthotic therapy were heavily scrutinized.

Many of the surgical procedures which were "gold standard" in the correction of forefoot pathologies were abandoned and replaced by newer procedures. Outcomes research began showing which procedures worked better than others. A better understanding of pathomechanics as well as improved technologies produced more favorable and predictable outcomes.

As more insight became available about the pathomechanics of common foot and ankle problems, it became apparent that a knowledge of anatomy, physiology, and physics would be essential to understanding the etiology of musculoskeletal disorders. This posed a challenge to any podiatric physician in clinical practice and an even greater challenge to a teacher in the podiatric profession. I began teaching podiatric medical students almost immediately upon completion of my residency program in 1981, and within several years began lecturing at professional conferences on various topics of lower extremity biomechanics and surgery. I basically went back to school to learn from other disciplines and had to synthesize new information which could be clinically relevant to the ever-changing practice of podiatric medicine.

This interest in teaching started my own journey to better understand foot function. Early on, I focused solely on new research published in podiatric, orthopedic, and academic biomechanics journals. It was 20 years later in my career when I "discovered" the brilliant insight of anatomists and anthropologists who had a unique insight into function of the human foot. In most cases, this insight was based upon comparative anatomy, showing how and why the human foot functions in a superior fashion compared to all other primates.

The anatomists who provided the most influence for this book include Hicks, MacConaill, Lewis, Bojsen-Moller, and Sarrafian [5, 6, 9–17]. None of these researchers proposed a "theory" of foot function or foot treatment. Instead, they described the anatomy of the human foot in a manner different than what is seen in standard medical textbooks. In a sense, anatomy is a pure science and the descriptions of the component parts of the human foot cannot be distorted or be misrepresented. Conversely, describing functional anatomy and proposing how certain structures of the human foot function and differ from all other animals provides the potential for speculation and opinion. When these anatomists published their observations, there were no technologies available to validate their hypotheses. They often used the crude techniques of still photography and footprints. Notwithstanding, it will be shown that all five of these anatomists were nearly 100% correct in their initial assumptions as many subsequent scientific investigations validated their conclusions.

When evaluating current knowledge about function of the human foot as well as insights from the anatomists 50–100 years ago, there are 4 basic concepts which stand true:

1. The human foot is an osteoligamentous structure, twisted upon itself to provide stiffness as well as compliance. This spring-like structure of the human foot is capable of storing and releasing energy better than any other primate.
2. Pronation and supination of the forefoot is coupled with supination and pronation of the rearfoot, and all these motions are coupled with rotation of the talo-tibiofibular unit.
3. The lateral column of the foot is essential for providing medial transfer of load to the forefoot during terminal stance and pre-swing.
4. Pronation of the forefoot is essential to engage the windlass, stabilize the digits, and enable push off across the transverse axis of the 1st and 2nd metatarsophalangeal joints (MTPJ's).

As each foot disorder is studied, a common theme will surface where one or more of these four features has been disrupted. Recognizing the underlying mechanical aberration is an important prerequisite to implementing an effective treatment intervention. My goal in writing this book is to provide clinicians an understanding of pathomechanics which will ultimately improve treatment outcome for their patients.

Oakland, CA, USA

Douglas H. Richie Jr

References

1. Root ML, Orien WP, Weed JH. Clinical biomechanics: normal and abnormal function of the foot. Vd. 2. Los Angeles: Clinical Biomechanics Corp.; 1977, p. 138–39.
2. Close JR, Inman VT. The action of the ankle joint. Prosthetic devices research project. Institute of Engineering Research, University of California, Berkeley, Series 11, Issue 22, Berkeley: The Project; 1952.
3. Close JR, Inman VT, Poor PM, Todd FN. The function of the subtalar joint. Clin Orthop. 1967;50(1–2):159–79.
4. Close JR, Todd FN. The phasic activity of the muscles of the lower extremity and the effect of tendon transfer. J Bone Joint Surg. 1959;41A(2):189–208.
5. Inman VT, Ralston HJ, Todd FN. Human walking. Baltimore: Williams and Wilkins Company; 1981.
6. Inman VT. Human locomotion 1966. Clin Orthop Relat Res. 1993;288:3–9.
7. Elftman H, Manter J. The evolution of the human foot with special reference to the joints. J Anat. 1935;70:56–67.
8. Elftman H, Manter J. Chimpanzee and human feet in bipedal walking. Am J Phys Anthropol. 1935;20:69–79.
9. Hicks JH. The mechanics of the foot. Part I: the joints. J Anat. 1953;87(4):345–57.
10. Hicks JH. The mechanics of the foot: II. The plantar aponeurosis and the arch. J Anat. 1954;88:25–30.
11. Sarrafian SK. Functional characteristics of the foot and plantar aponeurosis under tibiotalar loading. Foot Ankle. 1987;8:4–18.
12. Sarrafian SK. Anatomy of the foot and ankle. 1st edition. Philadelphia: J.B. Lippincott Company; 1983.
13. Lewis OJ. Functional morphology of the evolving hand and foot. Oxford: Clarendon Press; 1989.
14. Bojsen-Møller F. Calcaneocuboid joint and stability of the longitudinal arch of the foot at high and low gear push off. J Anat. 1979;129:165.
15. Bojsen-Moller F, Flagstad KE. Plantar aponeurosis and internal architecture of the ball of the foot. J Anat. 1976;121:599–611.
16. Bojsen-Moller F, Lamoreux L. Significance of free dorsiflexion of the toes in walking. Acta Orthop Scand. 1979;50:471–9.
17. MacConaill MA. The postural mechanism of the human foot. Proc R Ir Acad. 1944–1945;50B:265–78.

Acknowledgments

I have had many extraordinary mentors, colleagues, and collaborators throughout my career whose influence and inspiration culminated in the writing of this book.

In my undergraduate education, two chemistry professors changed my life and made me realize my potential. A.J. Tharp, my inorganic chemistry professor, literally taught me how to study, how to learn, and how to believe in myself. Kenneth Marsi was a role model of the perfect teacher who loved his subject of organic chemistry and made learning fun. Both professors emulated the integrity and purity of the science they taught.

In podiatric medical school, I gravitated to my professors of biomechanics who uncovered the secrets of foot function and applied that knowledge to both surgical and non-operative treatment protocols. I was blessed to be taught and mentored by Merton Root, John Weed, Bill Orien, Chris Smith, Paul Scherer, Howard Marshall, and Ron Valmassey. Much of what they taught me appears throughout this book.

Early in my career, I conducted several research projects with colleagues which ultimately produced peer-reviewed publications in the medical literature. I am proud to have worked with Steve Kelso, Bill Olson, Kirk Herring, and Cliff Endo who collaborated with me on these projects. A highlight of my career was co-authoring a research paper with Herbert DeVries, Professor Emeritus of Exercise Physiology at the University of Southern California.

There are many other colleagues who have assisted me in the writing of this book. Among them, Daryl Phillips and Erin Ward stand out as being my "lifeline" in finding manuscripts, providing images, and offering expert insight. I am indebted to Kevin Rosenbloom who produced many of the original images in this book which convey the essential concepts in a unique and innovative format.

During my career, I have made lasting friendships with leaders of the foot orthotic industry, including Paul Paris, Jeff Root and the Marshall brothers (Kirk, Scott and Kent). They, along with other foot orthotic lab owners, had the vision to start the Prescription Foot Orthotics Laboratory Association (PFOLA) Scientific Meeting in 1998 and continued to organize the event for 12 years thereafter. Paul Scherer was the original Scientific Chair of this conference which brought together the best minds in lower extremity biomechanics in the world. I was privileged to serve as Scientific

Co-chair along with Chris MacLean the final two years of the PFOLA conferences. The many speakers at the PFOLA scientific conferences became lifetime friends and served as role models for their commitment to research and education. Their own works are cited many times in this book.

Contents

Comparative Anatomy and Introduction to the Twisted Plate Mechanism

1

"Man's foot is all his own. It is unlike any other foot. It is the most distinctly human part of his whole anatomical makeup. It is a human specialization and, whether he be proud of it or not, it is his hallmark and so long as Man has been Man and so long as he remains Man, it is by his feet that he will be known from all other members of the animal kingdom" [1].

Frederic Wood Jones, 20th century anatomist

Key Points

- The human foot contains certain anatomic features which enable efficient upright bipedal ambulation that are not seen in other primate feet.
- Failure of one or more of these unique structures accounts for most of the pathologies seen in the human foot today.
- The human foot resembles the chimpanzee foot during various phases of embryonic and fetal development until the lamina pedis or footplate "twists" into its uniquely human anatomic alignment.
- Humans are unique in possessing a lateral column with osseous locking at the calcaneocuboid joint which facilitates load transfer to the medial column during terminal stance.
- Push off across the transverse axis of the first and second MTPJs is another unique feature of humans compared to primates and is dependent upon pronation of the forefoot allowing pre-tensioning of the central band of the plantar aponeurosis.
- The human foot is not entirely rigid during midstance and terminal stance and actually undergoes more motion across the midfoot joints during pre-swing than what is seen in other primate feet.

D. H. Richie Jr, *Pathomechanics of Common Foot Disorders*,
https://doi.org/10.1007/978-3-030-54201-6_1

- The human foot does not possess an osseous locking mechanism across the midfoot joints but does rely on static and dynamic restraint provided by soft tissue structures which stiffen the foot prior to terminal stance and which provide the storage and release of energy for efficient ambulation.

The Specialization of the Human Foot

The unique function of the human foot is made possible by certain specialized anatomic features which are not found in any other primate species. Paradoxically, these unique structures are prone to failure and cause the most common disorders of the human foot. The trade-off for specialization of the human foot is a vulnerability to injuries which can ironically progress and significantly hamper the unique form of ambulation for which these structures are designed to enable.

The specialization of the human foot has enabled upright, bipedal ambulation. Human ambulation is not limited to upright bipedal walking. Certain features enable a human to participate in endurance running at a level of efficiency which exceeds all other mammals. Indeed, the human foot may be better suited for running than walking.

Teaching the specialization of the human foot requires comparison to other primate feet. Some observers would challenge this approach. For example, comparisons of the human foot to the ape foot are not valid from an evolutionary standpoint since there has never been identified a common ancestor shared by the two species. It is important to note that by teaching comparative anatomy, it does not suggest that humans descended from apes, as there is no supporting evidence of this theory [2]. Studying comparative anatomy teaches why the human foot performs better than the ape foot. *It also allows appreciation of the many unique features of the human foot which, when they fail, can lead to one or more significant pathologies.*

The specialized structures of the human foot include an enlarged calcaneus, well-developed longitudinal and transverse arches, short toes, and an adducted, non-opposable hallux. In comparative anatomy, the central plane of the human foot is the second metatarsal, according to French anatomic tradition. Thus, positioning of the hallux away from the midline of the body and toward the second metatarsal in this traditional anatomic tradition is called "adduction." That is why comparative anatomists describe the prehensile hallux and first ray of the primate foot as being "abducted." Prehension is the act of grasping or gripping. In all other discussion regarding the first ray in this book, the modern anatomic plane of reference is the midline of the body. Thus, the condition "hallux abductovalgus" refers to a movement of the hallux toward the second metatarsal and away from the midline.

The human midfoot has always been considered unique in its ability to become "rigid" during midstance, allowing heel rise and efficient propulsion via tension created by the tendo Achilles. More recent kinematic studies have revealed that stiffness or rigidity of the midfoot joints of humans is a misconception as there is significant motion or compliance across the midfoot joints when humans ambulate.

In fact, anthropologists have given better insight into the compliance of the human foot and have shown similarities with the ape foot, not previously discovered by any other academic disciplines. This will be explored in detail later in this chapter.

Bipedalism is not unique to humans. Many animals can ambulate while standing only on their hind feet, albeit for brief periods of time. Gibbons ambulate with a bipedal gait 10% of the time, yet have a mobile prehensile foot structure with an opposable hallux, just like all arboreal primates. What makes Gibbons unique compared to all other nonhuman primates is the presence of an Achilles tendon comparable in size to humans.

An adducted (non-opposable) hallux with a rectus, adducted (non-divergent) first ray is not unique to humans. Other animals such as squirrels, chipmunks, prairie dogs, beavers, and raccoons have non-divergent first rays in their hind feet which enable them to stand up on those feet for brief periods of time. The black bear has hind feet with tarsometatarsal joints in similar alignment as humans which allow this animal to also stand upright on two feet. Compared to humans, the first metatarsal of the black bear is the smallest in diameter, while the fifth metatarsal is the largest.

None of these animals, however, can run upright on two feet. Bramble and Lieberman discuss in their classic article that running, rather than walking, is the ultimate differentiating performance feature of the human foot [3]. They argue that the springlike mechanisms of the human foot, made possible by several muscle and ligamentous features, are activated far greater during running than walking. Further adaptations for long-distance or endurance running in humans are hypertrophy of the gluteal muscles and spinal extensor muscles which stabilize the trunk during running but not walking. Also, humans have a decoupled shoulder girdle which is essential for running. The development of a specialized system to dissipate body heat during running was a key advancement for the human species. With these features, Bramble and Lieberman point out that humans can outrun almost any animal on the planet over distances greater than 10 km or for periods of time exceeding 1 hour [3]. Endurance running rather than short sprints is what enables humans to be superior to all other primates in terms of mobility for hunting.

Within the foot itself are many specialized features which enable upright bipedal walking and running in a manner which cannot be achieved by any other primates. Among the apes, the chimpanzee foot most closely resembles the human foot. The chimpanzee foot contains the same number of bones as the human foot. At first glance, the bones of the human foot and chimpanzee foot look remarkably similar. However, subtle differences in shape of the bones as well as key soft tissue structures make the human foot unique among all other mammals (Fig. 1.1). The comparisons described are based upon the work of McNutt et al. [2].

In chimpanzees, the tibia is oriented in a varus alignment to the supportive surface. In addition, the ankle joint itself is angulated into varus due to a high lateral rim of the talus. The posterior facet of the subtalar joint in the chimpanzee is oriented almost vertical, or perpendicular, to the supportive surface, and the calcaneus is thus positioned lateral to the talus. The posterior calcaneal facet is highly convex in the chimpanzee, providing a high range of inversion and eversion for grasping. Therefore, chimpanzee foot can evert and achieve a plantigrade position despite the

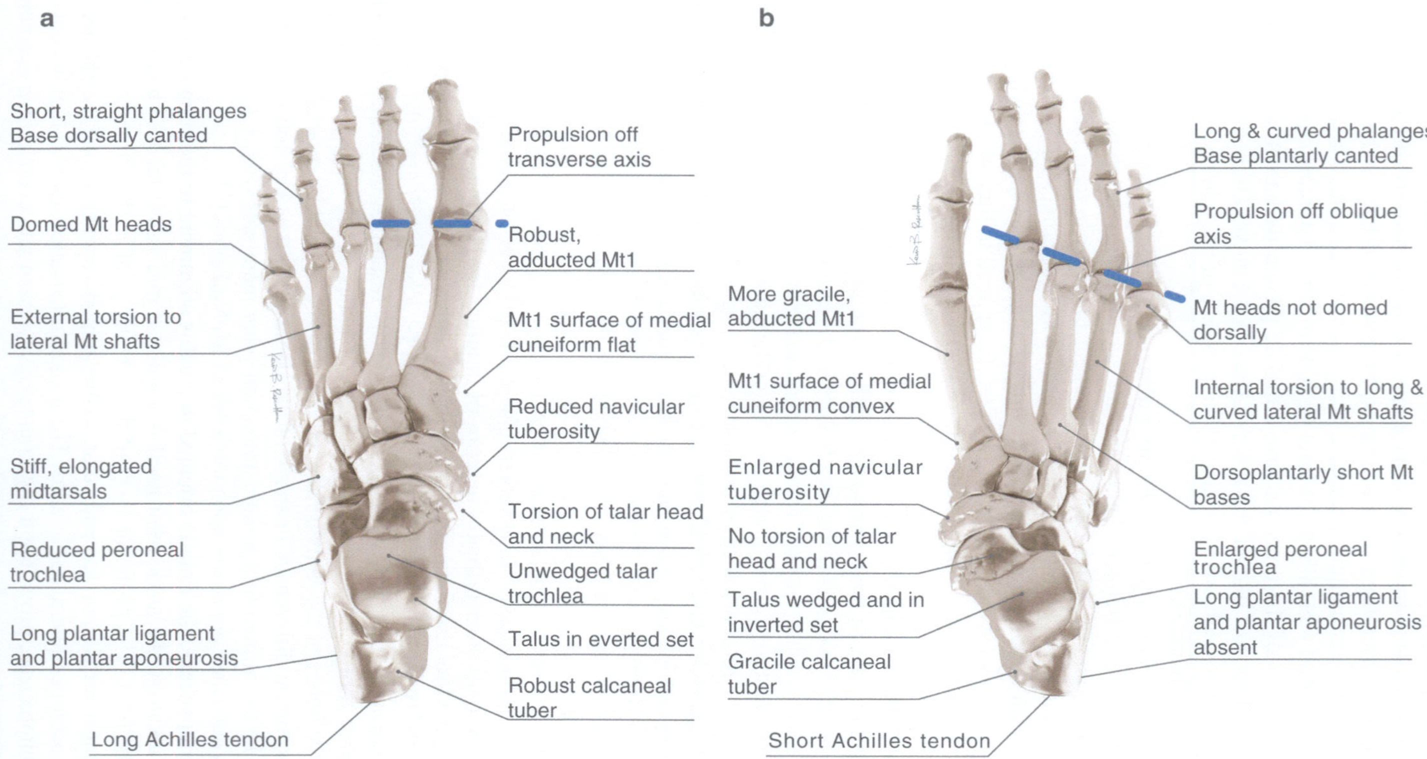

Fig. 1.1 Comparison of human (**a**) and chimpanzee (**b**) feet

varus alignment of the tibia and ankle because of the extreme range of motion available at the subtalar joint.

In the human ankle and hindfoot, the dorsal rim of the talus is parallel to the distal articular surface of the tibia and parallel to the supportive surface. Thus, the human foot is in a more everted position relative to the tibia and ground compared to the chimpanzee foot. The posterior facet of the subtalar joint is oriented parallel to the supportive surface but inclined or pitched dorsally from proximal to distal. Compared to the chimpanzee, the subtalar joint surface is flat, providing less mobility but more stability during upright bipedal gait.

The human calcaneus is unique in several ways. The entire posterior tuber of the calcaneus is three times the volume of the chimpanzee tuber [4]. Interestingly, the calcaneal tuber is wider in humans, but shorter than what is seen in the chimpanzee foot. A shorter calcaneus has a shorter moment arm for the Achilles acting at the ankle joint. Raichlen and co-workers theorize that the shorter calcaneal tuber in humans places more stretch on the Achilles which increases elastic energy storage [5]. They propose that storage of elastic energy by Achilles provides reduced energy cost during running, but not during walking.

Energy storage is also improved within the bones themselves in the lower extremity of humans. Latimer points out that while the human calcaneus is wider in girth than other primates, there is another critical difference in the ratio of cortical -to -cancellous bone [6, 7]. The human calcaneus has a thin cortical shell, and the primary contents are dominated by cancellous bone in a higher proportion than any other primate. This human adaptation is also seen in the distal tibia at the ankle and proximal tibia at the knee. Latimer shows that compared to cortical bone, cancellous bone has a lower elastic modulus and therefore is capable of storing energy better than cortical bone. Furthermore, the vast internal network of trabeculae in cancellous bone increases the overall skeletal surface area available for energy storage.

The human calcaneus has a lateral process which is plantarly positioned. This increases the weight-bearing surface of the calcaneus compared to the chimpanzee where the lateral process is dorsally positioned away from the supportive surface. The increased surface area and volume of the calcaneus allow the human foot to accept high-frequency loads during heel strike. The peroneal trochlea of the calcaneus is more robust and proximally positioned in the chimpanzee to allow a larger fulcrum and longer moment arm for the peroneus brevis and longus tendons to act on the divergent first ray for grasping. The peroneal trochlea is smaller and more dorsally positioned on the calcaneus of the human foot.

The articulation of the calcaneus and the cuboid is unique in humans with two osseous structures which facilitate "locking" of the joint into a close-packed position. A beak or proximally directed projection of bone is seen in the human cuboid at its plantar-medial proximal aspect (Fig. 1.2). This proximally directed beak of the cuboid articulates with the coronoid fossa on the inferior edge of the articular surface of the calcaneus. A dorsal "overhang" located in the superomedial corner of the human calcaneus locks over the dorsal surface of the cuboid. Both of these osseous structures, unique to the human calcaneus, provide restraint of motion in the sagittal plane at the calcaneocuboid joint when this joint is moved into a pronated, or close-packed position.

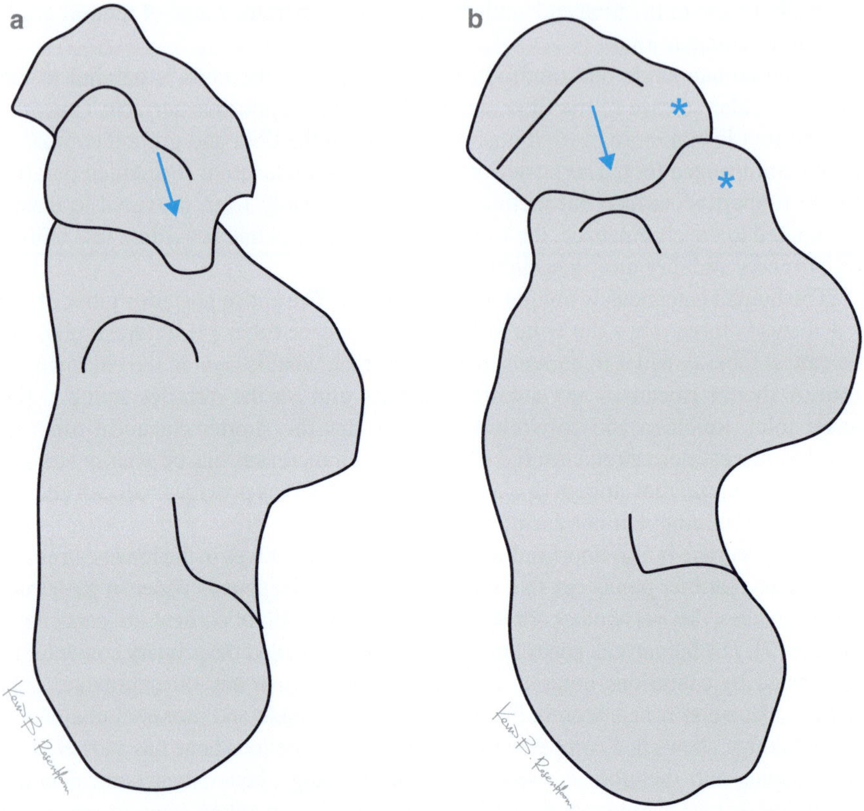

Fig. 1.2 Right calcaneocuboid joint from (**a**) man and (**b**) gorilla. Plantar aspect. The calcaneal process of the cuboid (arrow) extends posteriorly into a recess on the calcaneus. In the gorilla, the joint surfaces are extended medially (asterisks), while the calcaneal process is absent. (Adapted from: Bojsen-Møller [16])

There is an articular facet facilitating motion between the cuboid and navicular in chimpanzee feet, while human feet lack such a facet and instead have an interosseous ligament to restrict motion between these two bones. Thus, the navicular and cuboid are free to rotate upon each other in the chimpanzee foot. Overall, the midfoot joints of the human foot are adapted for stability at the sake of mobility.

The first ray of the human foot is adducted and parallel to the lesser metatarsals, while the first ray in the chimpanzee is divergent. The difference in alignment occurs at the articulation of the first metatarsal with the medial cuneiform. In the chimpanzee, the articular surface of the medial cuneiform is convex which enhances mobility. This articular surface of the first metatarsocuneiform joint is directed medially in the chimpanzee to facilitate grasping of the entire ray across the foot. In humans, this articulation is rotated laterally, bringing the first metatarsal parallel to the lesser metatarsals. Also, the first metatarsocuneiform joint in the human foot has a flat surface for stability and load bearing. The medial cuneiform of the human foot possesses a tubercle for insertion of the peroneus longus tendon, while this tendon primarily inserts on the first metatarsal in the chimpanzee to increase the lever arm for pulling the first ray across the foot in grasping.

The metatarsal heads of the human foot are designed for dorsiflexion of the digits, while in the chimpanzee foot, the metatarsal heads are adapted for plantarflexion of the digits. The articular surfaces of the heads of the metatarsals in the human foot are dorsally domed, while they are prolonged plantarly in the chimpanzee foot. The shafts of metatarsals 2–5 are inverted in the chimpanzee foot to enable opposition to the hallux for grasping. The lesser metatarsals are everted in the human compared to the chimpanzee, bringing the MTPJs and the toes out of a varus position and orienting them to the ground for weight bearing. The articulations of the tarsometatarsal joints are elongated from dorsal to plantar in humans, while they are flatter in the chimpanzee. Therefore, the midfoot joints are anatomically stiffer in humans.

The phalanges of the human foot are shorter and stouter than the chimpanzee. Shorter phalanges allow greater dorsiflexion range of the MTPJs due to reduced lever arm of the toe flexors. The base of the proximal phalanx is canted in humans to favor dorsiflexion, while in the chimpanzee foot, the cant of the proximal phalanx will favor plantarflexion and grasping.

Finally, there are unique ligament structures in the human foot which are essential for upright bipedal ambulation. In the human foot, the position of the calcaneus under the talus has moved it away from an articulation with the fibula, which is seen in the chimpanzee foot. While this position of the calcaneus in the human foot optimizes rearfoot alignment for a heel rocker and forward progression, it also creates a precarious situation for osseous stability in the frontal and transverse planes. As a result, in the human foot, the calcaneus relies upon ligamentous attachments to the bones located proximal in order to limit rotational moments from ground reaction forces. Adaptive features which address the lack of osseous stability laterally across the ankle and subtalar joints include the anterior talofibular ligament and calcaneofibular ligament which are well developed in the human ankle and poorly developed in the chimpanzee ankle [8].

There are several soft tissue structures unique to the human foot. Humans contain a long Achilles tendon designed to store and release elastic strain energy for ambulation. The establishment of the medial and lateral arches of the human foot requires four unique ligaments which are not well developed in the chimpanzee foot: the plantar aponeurosis, the long plantar ligament, the plantar cuboideonavicular ligament, and the bifurcated ligament [9]. The role of these ligaments in providing the unique bipedal gait in humans will be discussed further. First, it is important to show how the unique features of the human foot develop in the embryo and fetus.

Ontogeny of the Human Foot

Further insight into the comparative anatomy and specialization of the human foot can be gained by studying the embryonic development of the lower limb. While some subscribe to the notion that "ontogeny recapitulates phylogeny," it has been shown that such an assumption is filled with shortcomings. Just as in the proposed evolution of the human foot, there are gaps in understanding the uniqueness of the ontogeny relative to other species.

Does Ontogeny Recapitulate Phylogeny of the Human Foot?

Probably, the most unique feature of the human foot from an evolutionary standpoint is the alignment of the first ray and hallux to the long axis of the foot. There is a false, but well-accepted notion that the human foot evolved from an ape-like ancestor. A perceived monumental advancement allowing humans to assume an upright bipedal gait was the loss of the divergent first ray and prehensile hallux seen in all other primate creatures. Yet, at no time, during embryonic and fetal development does the human foot demonstrate a prehensile hallux [10]. Furthermore, there is no fossil evidence of a prehensile hallux in any of the "prehuman" hominoid species. A prehensile hallux is seen only in nonhuman primates.

Conversely, several "ape-like" foot structures are seen in the human embryo, which then differentiate into unique specialized features seen only in the adult human foot. We will explore the appearance of these features during the *embryonic stage*, which is the first 8 weeks after fertilization, while the *fetal stage* starts at 8 weeks after fertilization and culminates with birth. The designated weeks and months of the embryo and fetus refer to the time after fertilization. The descriptions are based upon the work of Sarrafian and Kelikian [11].

Ankle and Hindfoot Development: Twisting the Plate of Bones

In the ape species, which includes baboons, gorilla, and chimpanzee, the talus and the calcaneus are oriented side-by-side in the horizontal plane rather than stacked vertically as seen in the human foot. Therefore, in the ape foot, the talus and calcaneus lie adjacent in the transverse plane. This alignment orients the talus and calcaneus parallel to the weight-bearing surface, i.e., the *substrate*. With the talus lying next to the calcaneus, the articular surface of the subtalar joint is oriented in a vertical fashion, permitting wide ranges of transverse plane and frontal plane motion of the hindfoot.

The ankle joint of the ape foot is angulated medially due to the orientation of the distal tibia and the trochlea of the talus. This facilitates grasping for an arboreal existence. The tibia and fibula of the ape ankle are equidistant from the substrate, i.e., equally aligned from the ground. The fibula articulates with the calcaneus in most ape species and in many other mammalian species. In most ape feet, the articulation between the tibia and fibula, i.e., the ankle syndesmosis, seen in the human foot is either not well developed or may be nonexistent. Thus, the nonhuman primate ankle is wider than the human ankle and has weight bearing shared between the tibia and talus as well as the fibula which articulates with the calcaneus. The large range of frontal and transverse plane motion in the ankle and subtalar joints allows the ape to bring their feet plantigrade, despite a varus orientation of the tibia.

This parallel arrangement of the talus and calcaneus seen in the ape foot, oriented side-by-side in the transverse plane, is also seen clearly in the 7-week human embryo, or Horizon 20. At this stage, the calcaneus is in contact with the fibula, and

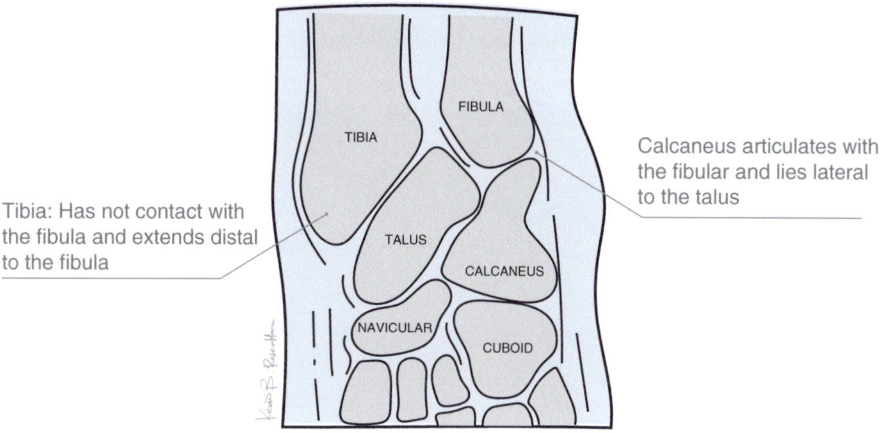

Fig. 1.3 The flat footplate: The 18-mm human embryonic foot

the distal margin of the tibia and fibula is on the same plane. All of these features are identical with a fully developed ape-like foot (Fig. 1.3).

A critical change occurs during the embryonic development of the human foot between Horizon 20 and 22 which is approximately the eighth embryonic week (Fig. 1.4). Rather quickly, a well-developed distal tibiofibular ankle articulation (syndesmosis) has formed providing cavitation for the ankle joint. The tibia has retracted proximally with the talus, while the fibula remains at its original distal location. Most importantly, the calcaneus has moved under, or inferior to the talus, and these two bones are now oriented vertically. This represents a vertical compression of the talus and calcaneus. It also partially establishes the "twisted plate" orientation of the bones of the human foot, unique to humans, where the rearfoot bones are compressed vertically and metatarsals are compressed horizontally. MacConnail was the first to describe the twisted plate arrangement of the bones of the human foot [10]. Saraffian expanded upon this observation to show how the "twist" of the plate of foot bones established the medial, lateral, and transverse arches of the human foot [11]. The osteoligamentous structure of the human foot, oriented as a twisted plate, will be explored in more detail and will be a common theme throughout this book.

The ankle joint has rapidly progressed between embryonic weeks 7 and 9 to align the bones for optimal dorsiflexion and plantarflexion, using one articulation between the tibia and the talus. This realignment sacrifices grasping, with reduced inversion and eversion of the ankle and hindfoot. With the need for grasping no longer necessary, the subtalar joint has rotated almost 90° to minimize frontal plane motion, perhaps an improvement for overall stability during bipedal gait.

The arches of the human foot cannot maintain their integrity without key supportive ligamentous structures which are not seen in most other primate feet. These

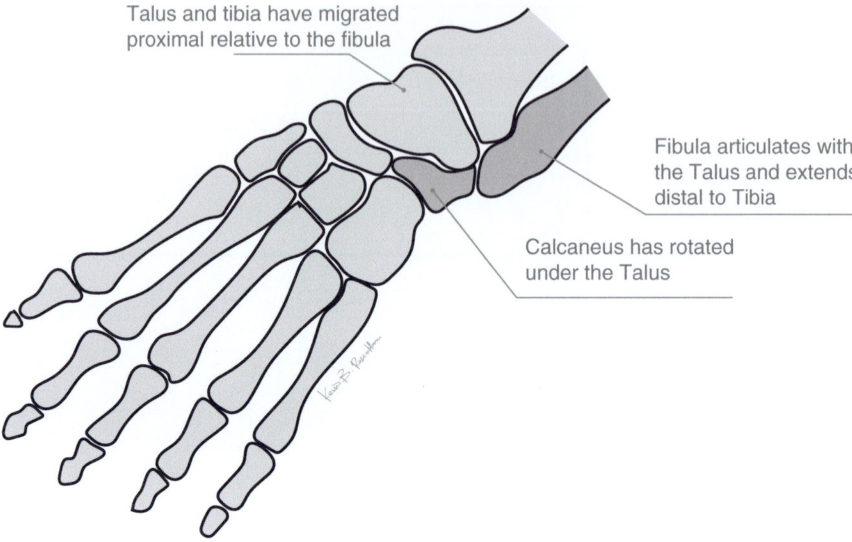

Fig. 1.4 The twisted footplate: 33-mm human embryonic foot

unique ligaments are the plantar fascia and the long and short plantar ligaments [9]. These structures do not develop until after the third month of fetal development and follow the appearance of important muscular structures.

Prior to the twisting events which positioned the calcaneus under the talus, key tendons have already developed. The insertions of the tibialis anterior and the Achilles are visible in the 7-week embryo. Some of the plantar intrinsic muscles are also visible including the quadratus plantae and abductor hallucis which appear in the 6-week embryo. The adductor hallucis and flexor hallucis brevis are not well developed at 6 weeks as the first metatarsal has not yet moved from its abducted position. This will not occur until after the eighth embryonic week.

Ontogeny of the Forefoot

When the metatarsals and digital rays first become visible in the mesenchymal stage, or the fifth and sixth week of embryonic development, the metatarsals are spread apart. The angle of divergence of the first and second metatarsals in the 2-month embryo is 32° which decreases to 9° at 9 months and 6° in the adult. The first metatarsal has a lateral, (valgus) twist or torsion starting at 25°valgus at 3 months, ending up at 13° valgus in the adult. This valgus torsion in the adult first metatarsal is often confused to be a result of hallux abductovalgus deformity, yet it is a normal finding. The second metatarsal has a medial (varus) twist of 13° at 3 months reducing to 5° varus in the adult. The medial torsion of the second metatarsal and lateral (pronatory) torsion of the first metatarsal are remnants of the grasping function of the foot seen in other primates whose twisting arrangement of the metatarsals is much more exaggerated [3].

The hallux and first metatarsal are deviated medially in the transverse plane in the 7-week embryo. This is partly attributed to the angle of the neck of the talus. However, the more important contribution to alignment of the first metatarsal is the distal articular surface of the medial cuneiform which is angulated 45° medial in the 7-week embryo. This angle diminishes to almost zero by 12 weeks. The adducted, or medially deviated, first metatarsal suggests an embryonic portrayal of a prehensile hallux, or opposable first ray and hallux. However, while the first ray is medially deviated in the 7-week human embryo, it never assumes the appearance of an independent, opposable structure or a true prehensile hallux [10]. When the deep transverse metatarsal ligament becomes visible at 12 weeks, the bridge or connection between the first and second metatarsal is evident. There is no true independence of the first ray from the rest of the foot. *Never is there a time when this soft tissue bridge between metatarsals 1 and 2 is not evident* [10]. When the deep transverse ligament develops, it bridges across all five metatarsals. Conversely, a prehensile thumb is fully developed in the hand of the 7-week human embryo with no bridging of a transverse metatarsal ligament between metacarpals 1 and 2.

The 12-week fetus demonstrates an adducted hallux and adducted first metatarsal. Based on current anatomic terms, this means that the hallux and first metatarsal are *abducted away from the midline* of the body and, according to traditional French anatomic terms, *adducted toward the second metatarsal*. This realignment coincides with the appearance of the insertion of the peroneus longus tendon on the base of the first metatarsal and medial cuneiform. The oblique course of the peroneus longus tendon across the midfoot to insert on both the first metatarsal and medial cuneiform is a unique human feature. In the ape foot, the peroneus longus follows more along the long axis of the foot and inserts only to the first metatarsal.

In the 12-week fetus, the deep transverse metatarsal ligament has wrapped fully around the head of the first metatarsal to maintain its new adducted alignment closer to the second metatarsal. The sesamoids also chondrify at 12 weeks which is clearly a milestone of fetal development of the first ray and supportive structures of the first MTPJ. The sesamoids are now anchored to the deep transverse metatarsal ligament which is a unique human feature. It is also a critical anatomic relationship which allows the progression of hallux abductovalgus deformity.

Another key differentiating feature of the human foot is decreased length of the digits. This is thought to be an advantage for bipedal gait and progression of the center of mass over the toes during forward propulsion. Longer digits have a longer lever arm for the soft tissue restraints, i.e., the long digital flexor tendons to resist dorsiflexion at the metatarsophalangeal joints. In the fetus, the lesser digits 2–5 are longer than in the adult. The gradual shortening of the digits which occurs during fetal development to the adult occurs mainly at distal phalanges. The unique human characteristic of a "great toe," or hallux, which is wider than any of the other toes, is fully evident by the tenth week of fetal development. Conversely, the first metatarsal is shorter than the second metatarsal at 3 months (ratio 0.73) and then grows faster than the second metatarsal ending up with a ratio of 0.33 at birth. Thus, digital length as well as metatarsal length demonstrates an ape-like configuration in the human embryo, progressing to a unique human-like relationship by the end of fetal development. The increased length of metatarsals 1 and 2 increases the lever arm for push off by the triceps during terminal stance and offers a distinct advantage over the nonhuman primate foot where the first metatarsal is the shortest of all five metatarsals.

Rotation of Segments

At the seventh embryonic week, the feet are in full equinus, with no dorsal crease at the ankle, and the lower extremities are positioned in extreme external rotation. In the eighth week, the thigh and lower leg begin internally rotating. At 12 weeks, the foot is now positioned external relative to the leg yet is still positioned in equinus or plantarflexion at the ankle. The magnitude of equinus reduces during the third fetal month, but the foot remains inverted or supinated. The first metatarsal has already adducted or moved toward the second metatarsal at 12 weeks. By the fourth month, the foot pronates out of its supinated position, but still ends up slightly supinated, while equinus positioning at the ankle fully reduces. A slight metatarsus varus remains at 12 weeks. The pronation rotation of the entire foot seen at month 4 continues thru fetal development and is not yet complete in the newborn.

During fetal development, the calcaneus increases in length, particularly within the main body of this bone. An angle of torsion exists between the long axis of the calcaneal tuber and tibial diaphysis at 12 weeks. The calcaneus is rotated into a supination-varus angle of 36° at 3 months which reduces to 26° at 9 months. The calcaneus continues to rotate in a valgus direction in the newborn to finally reach an angulation of 3° varus in the adult.

At the end of the eighth embryonic week, the skeletal components of the foot are composed of cartilage and are fully formed. At the ninth week, the talus begins changing into a "human" form, compared to a previous ape-like shape during the embryonic stage. The declination angle of the talar neck has increased to 25° (i.e., the head and neck of the talus plantarflex after the eighth embryonic week). The lateral process of the talus is well developed and articulates with the fibula. This coincides with the unique position of the talus which has retracted with the tibia proximally, placing it on top of the calcaneus. The proximal migration of the tibia and talus leaves the fibula behind which is now positioned distal to the tibia and the talus (Fig. 1.4).

The next major change in the talus occurs between 5 months and 9 months of fetal development when the angle of the talar neck, relative to the trochlea, moves laterally, into abduction, from a 32-degree angle to a 25-degree angle (Fig. 1.5). With respect to the calcaneus, the neck of the talus rotates lateral. The talar neck/calcaneal angle measures 36° at 5 months, reducing to 30° at 9 months and 23° in adult. The trochlear/calcaneal angle measures 6° at 5 months and 1° at 9 months and adult. The net result is adduction (lateral displacement) of the head and neck of the talus to align over the calcaneus, closing down the talocalcaneal angle in the transverse plane.

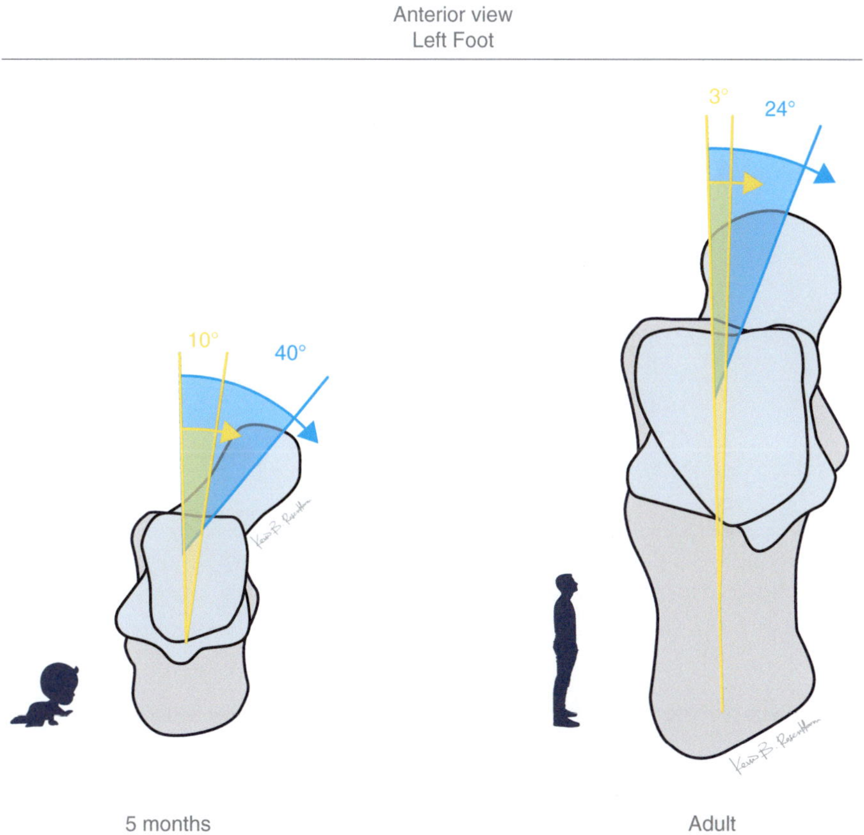

Fig. 1.5 Transverse plane rotation of the trochlea (yellow) and neck of talus (blue) relative to calcaneus

The head of the talus undergoes lateral or valgus rotation measuring 19° valgus or pronated at 5 months, increasing to 26° valgus at 9 months and finally achieving 37° of valgus or lateral rotation in the adult (Fig. 1.6). The biplane changes of shape of the talus during fetal development are a critical and unique specialization of the human foot compared to other primates. These changes in the talus initiate a twisting of the forefoot relative to the rearfoot in the direction of pronation and also carry the first ray into adduction toward the second metatarsal. Ultimately, this places the plane of all of the metatarsals flat on the ground (Fig. 1.7).

It takes 9 months to complete the twisted plate configuration of the osseous structures of the human foot during embryonic and fetal development. This begins at the eighth embryonic week, with rotation of the calcaneus medially from its lateral position, to become positioned inferior to the talus. The twisting of the entire

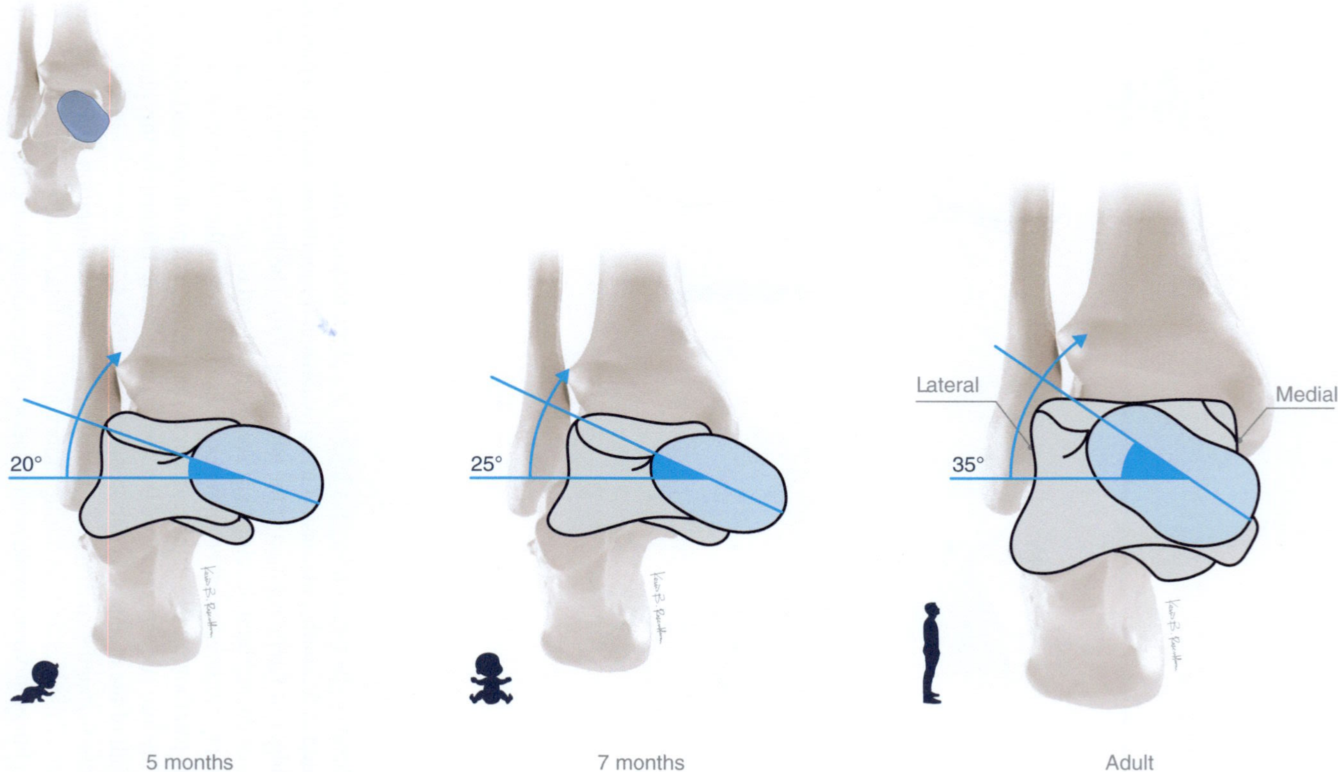

Fig. 1.6 Coronal view of the head of the left talus. Progressive valgus twist from newborn to adult

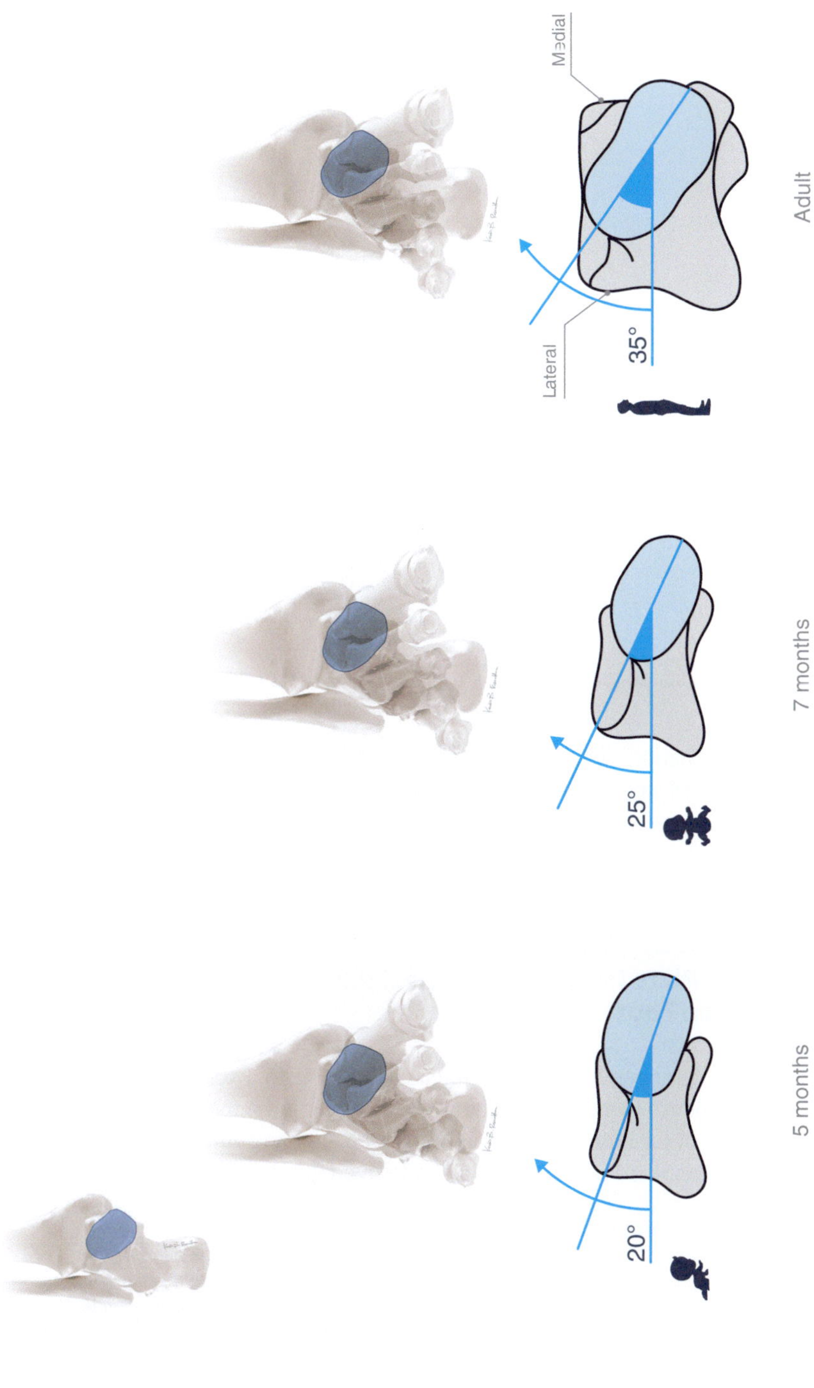

Fig. 1.7 This is a combined twisting of the head and neck of the talus and metatarsals. Progressive valgus twist from newborn to adult

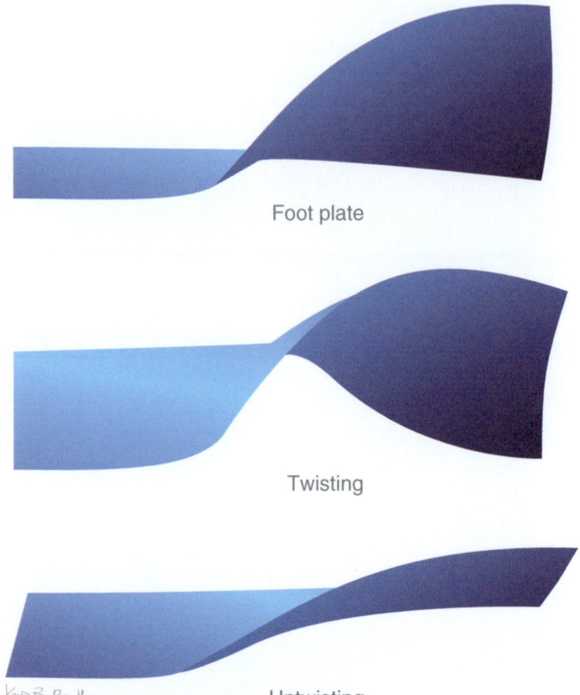

Fig. 1.8 Twisting a plate of bones establishes the medial, lateral, and transverse arches of the foot

foot out of supination, which occurs between the fourth and ninth fetal months, occurs within the shape of multiple bones, as well as at the articulations between the rearfoot and the forefoot. Further valgus twisting of the forefoot on the rearfoot after birth and until adulthood establishes the medial, lateral, and transverse arches of the foot (Fig. 1.8).

The Twisted Plate Provides the Specialized Function of the Human Foot

The horizontal arrangement of the metatarsals relative to a vertical position of the talus and calcaneus is what most anatomists consider to be the final and most specialized feature of the human foot. This "twisted plate" orientation of the foot bones was elegantly described by MacConaill, an Irish anatomist: [12]

> *The bones and ligaments of the human foot form a twisted osteo-fibrous plate of irregular thickness; the degree of torsion of the tarsometatarsal part of this plate is alterable. The torsion of the foot skeleton is brought about in intra-uterine life. By comparison with the upper limb; the whole lower limb is defined to be permanently pronated; the twist of the foot skeleton is a twist in the same sense as that of the whole limb. Accordingly, any movement,*

or set of movements, which brings about a lessening of the torsion of the foot skeleton will be called supinative; and any movement, or set of movements, which brings about an increase of the torsion will be called pronative. The foot will be said to be supinated when it has been untwisted as much as is normally possible; and as being pronated when it has been twisted as much as is normally possible.

MacConaill uses the term "lamina pedis" to describe the bones of the foot, absent the talus. He, like many anatomists, considers the talus to be part of the leg, not the foot. The twisted plate model proposed by MacConaill describes medial rotation of the calcaneus to align vertically under the talus. The forefoot rotates into eversion, moving away from a grasping position to become plantigrade on the ground. Twisting of the lamina pedis during embryonic development results in a vertical "compression" of the bones of the rearfoot (calcaneus and talus), while the bones of the forefoot are compressed horizontally (Fig. 1.9). This leaves the narrow calcaneus as the only bone in the rearfoot to remain in contact with the supportive surface. Therefore, ground reaction forces have a limited surface area to act upon in the rearfoot. The narrow weight-bearing surface of the calcaneus limits the length of the lever arm of ground reaction forces to apply torque on the subtalar joint. At the same time, with placement of the calcaneus beneath the talus, the Achilles has an increased length of lever arm acting at the ankle and subtalar joints. The posterior facet of the subtalar joint is now oriented perpendicular to the line of action of the Achilles which further facilitates the lever arm. The longer, perpendicular lever arm of the Achilles has significant influence on frontal plane movement of the rearfoot at the subtalar joint in the human foot.

In the forefoot, the wide flat plane of the metatarsals will allow ground reaction forces to provide a cantilever effect on the narrow calcaneus supporting the rearfoot. From an anatomical standpoint, it is easy to see how ground reaction forces

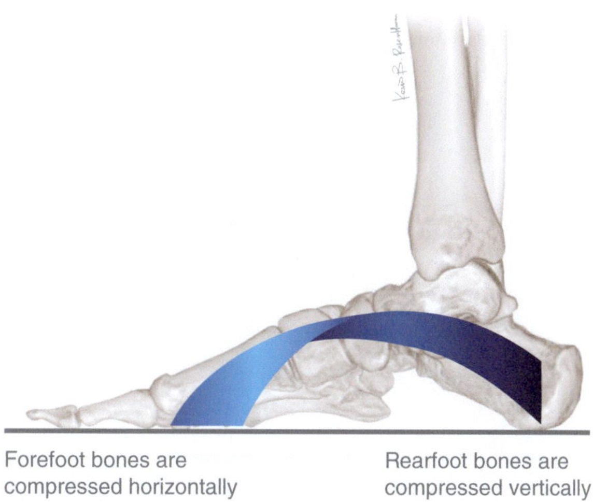

Fig. 1.9 The twisted plate arrangement of the bones of the human foot

acting on the weight-bearing surface of the metatarsal heads will have an extended lever arm to apply torque on the midtarsal joint and the subtalar joint. This mechanism is entirely dependent upon the stiffness of the numerous joints which lie between the metatarsal heads and the calcaneus. Conversely, the narrow weight-bearing surface of the rearfoot has reduced capacity to influence alignment of the forefoot.

When the forefoot is supinated, the lamina pedis is "untwisted" and the human foot is stiff due to maximal tensioning of the plantar ligaments. MacConaill proposes that this intrinsic stiffening is an essential mechanism for the human foot to achieve upright bipedal gait. According to MacConaill, the untwisting of the lamina pedis or footplate is a response to rearfoot position: as the rearfoot pronates, the forefoot supinates (Fig. 1.10). This notion validated a previous observation from Steindler who proposed that all five metatarsals will rotate on their proximal joints to remain on the ground in response to rotation of the rearfoot [13]. Finally, MacConaill was the first to describe how twisting of the forefoot into pronation established the medial, lateral, and transverse arches of the foot (Fig. 1.8). "Twisting" of the forefoot into pronation and the rearfoot into supination will relax the plantar aponeurosis, the long plantar ligament, and the spring ligament (Figs. 1.10 and 1.11). These assertions were later validated by Saraffian [14].

It is interesting to note the contrast of opinion about the "ideal" position for the function of the human foot. MacConaill asserted that the "untwisted" or supinated position of the forefoot on the rearfoot provided increased tension on the plantar

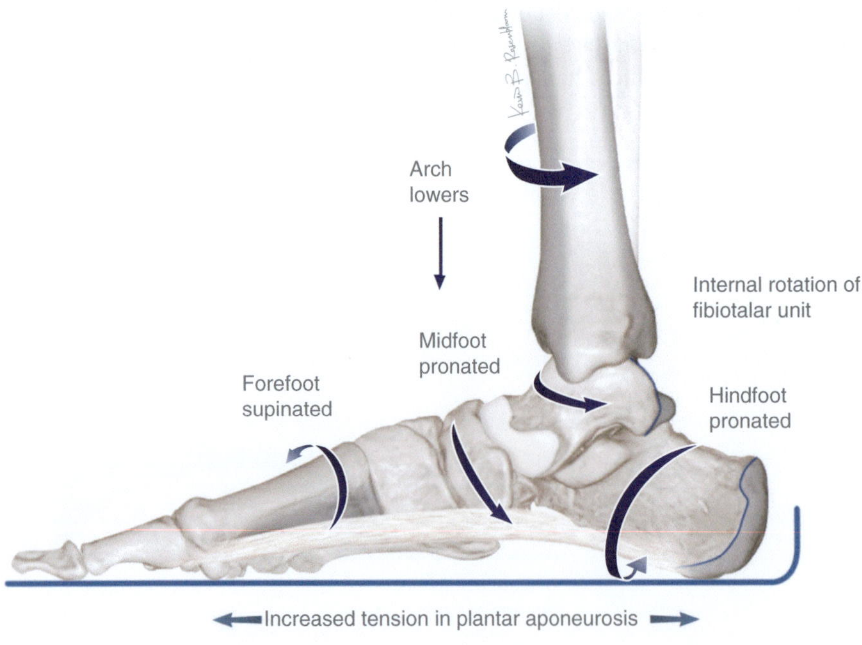

Fig. 1.10 Untwisting the plate

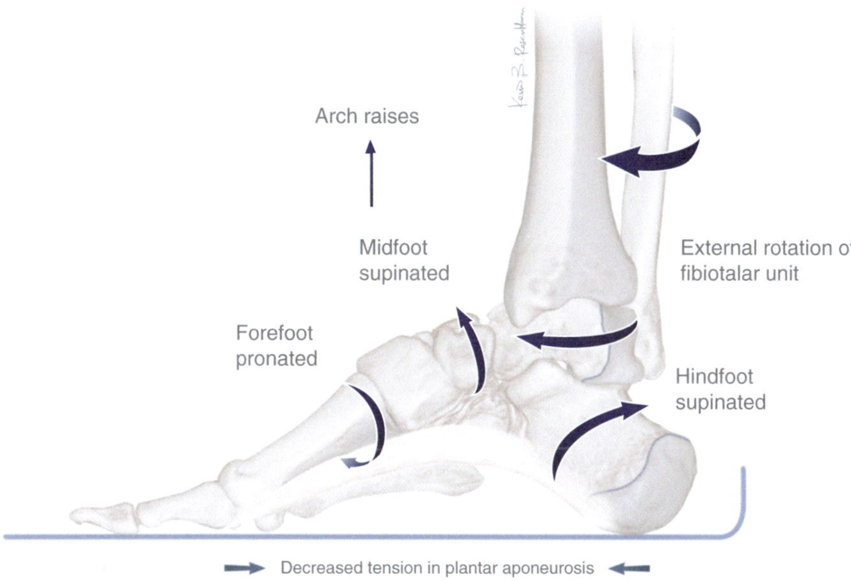

Fig. 1.11 Twisting the plate

aponeurosis, thc spring ligament, and the long plantar ligament and stiffened the entire foot for stability (Fig. 1.12). He also observed that in relaxed stance the rearfoot was pronated and the forefoot was supinated (i.e., "untwisted" footplate), enabling the joints of the rearfoot and midfoot to move into a "synarthrodial" position or close-packed position. Upon closer scrutiny, the supinated position of the forefoot on the rearfoot does tension ligaments and stabilizes the talonavicular joint which has no osseous locking. However, the calcaneocuboid joint is less stable with forefoot supination. From an anatomical standpoint, the calcaneocuboid joint is close-packed in its pronated position, not in its supinated position.

Clinical Application of the Twisted Plate Mechanism

Seringe and Wicart, both pediatric orthopedic surgeons, noted the previous work of MacConnail, Hicks, and Inman describing the response of the foot bones to internal and external rotation of the tibia [15]. They note that after MacConnail, two pediatric surgeons, Meary and Queneau, renamed the lamina pedis the "calcaneopedal unit" which includes all the bones of the foot except the talus. In agreement with MacConnail, these pediatric surgeons observe that the talus, void of any muscular attachments, is locked in the ankle mortise and moves in the transverse plane with the tibia and fibula. The combination of the tibia, fibula, and talus is termed the "talotibiofibular unit" (Fig. 1.13). These two separate functioning units provide a "horizontal segmentation" of the foot, according to Seringe and Wicart which differ from traditional segmentation of the foot into medial and lateral (sagittal) columns

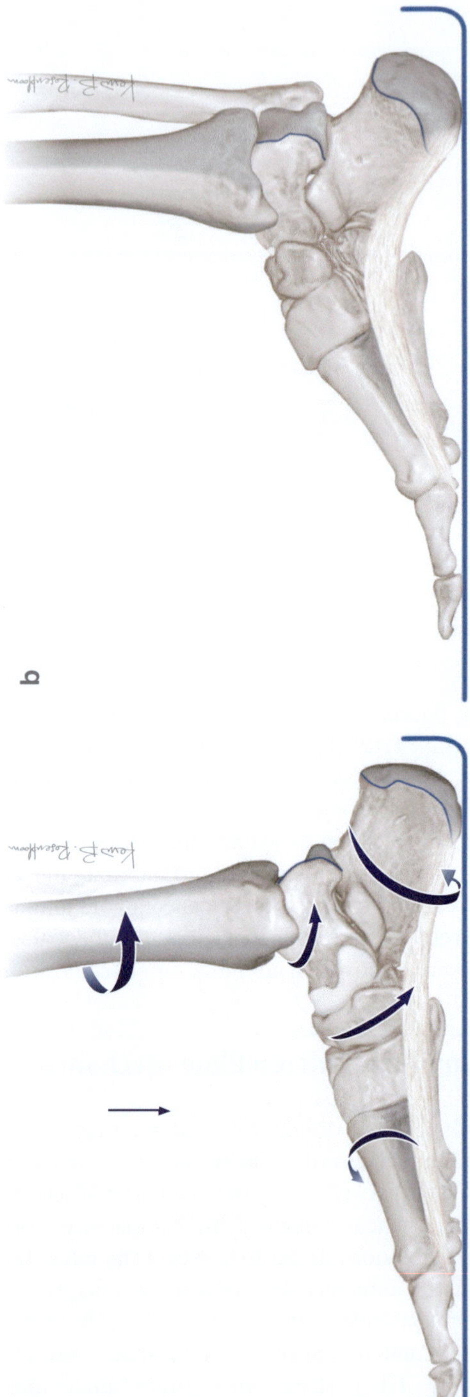

Fig. 1.12 Twisting the plate relaxes the plantar ligaments. **(a)** The untwisted plate is stiff due to maximal tension on the plantar fascia, spring ligament, and long plantar ligament, **(b)** tension in the plantar ligaments relaxes with twisting the plate

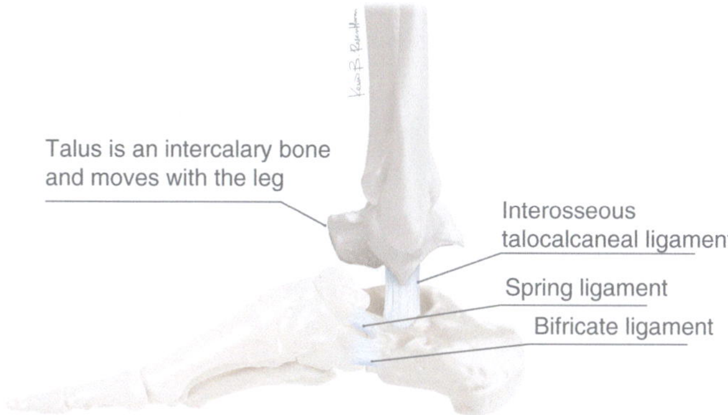

Fig. 1.13 Horizontal segmentation between the foot and the leg. The talotibiofibular unit is shaded and the calcaneopedal unit is gray. This establishes a horizontal separation of two function units, connected by the interosseous talocalcaneal ligament and the spring ligament complex. (Adapted from Seringe et al. [15])

or rearfoot-midfoot-forefoot (transverse) segments (Fig. 1.14). More importantly, the response of the calcaneopedal unit to rotation of the tibio-fibular-talar rotations is identical to what MacConnail described previously with the twisting of the lamina pedis in response to loading and unloading of the foot.

Seringe and Wicart note the misconception of segmental foot motion which occurs when only the subtalar joint is moved with inversion and eversion, in an off-weight-bearing position:

> The subtalar joints have a fairly simple function when the foot is not loaded. The calcaneus is said to pitch, turn and roll under the talus. The three basic movements of dorsiflexion/plantar flexion, abduction/adduction and pronation/supination are automatically associated in a unique eversion/inversion movement around the Henke axis (first described in 1859). Eversion combines dorsiflexion, abduction and pronation of the foot, while inversion combines plantar flexion, adduction and supination. However, this only applies to the unloaded foot (when examining a seated or lying subject). The function of the subtalar joints in the loaded foot and during walking is completely different. They have been studied relatively little and are relatively complex. Some of the distinct features of these joints have been somewhat forgotten:
>
> - previously described interaction with the TN joint;
> - role in the twisting-untwisting of the foot (flattening of the loaded foot and hollowing of the unloaded foot);
> - role in the axial rotation movements of the leg.

Seringe and Wicart describe how certain congenital deformities of the foot begin with positioning of the calcaneopedal unit relative to the talotibiofibular unit. In pes cavus, muscular imbalance leads to pronation of the forefoot (Fig. 1.15). Then, following the twisted plate mechanism, proximal rotation of skeletal segments follows. Plantarflexion of the first metatarsal along with

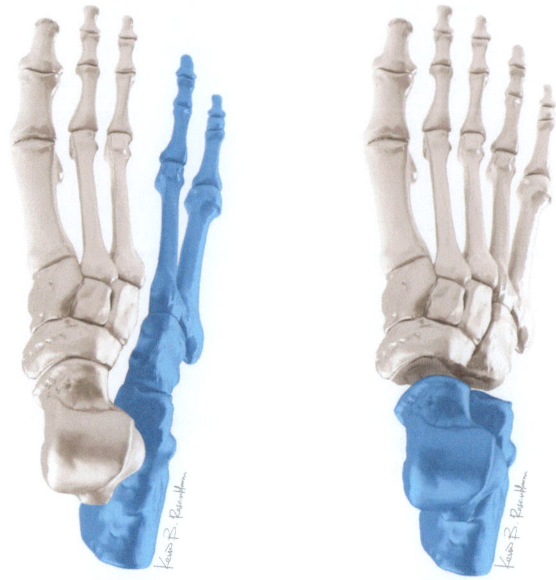

Fig. 1.14 Traditional division of functional units of the human foot

forefoot pronation creates a medial cavus deformity with the intermediate cuneiform bone being at the apex. The rearfoot responds to forefoot pronation with supination rotation which drives the talus into dorsiflexion and abduction while the leg externally rotates. The elevation of the intermediate cuneiform also corresponds to raising the transverse arch of the midfoot. The cavus deformity along with acquired claw toe deformities is accentuated in the off-weight-bearing condition when the foot is taken off the ground and the lamina pedis "untwists" as described by MacConnail.

Conversely, Seringe and Wicart observe that the flexible congenital flatfoot deformity is accentuated with weight bearing which induces "untwisting" of the calcaneopedal unit, identical to untwisting of the lamina pedis as described by MacConnail (Fig. 1.16). Furthermore, the flatfoot deformity appears to originate first from calcaneal valgus combined with internal rotation of the talotibiofibular unit. In adult acquired flatfoot, loss of stability at the talonavicular joint allows adduction and plantarflexion of the talus which is coupled to calcaneal eversion. Forefoot supination is a twisting response to rearfoot pronation. Seringe and Wicart acknowledge that in pes planus, the calcaneopedal unit will move into abduction as one single body in the child, but eventually the forefoot will move into abduction across the midtarsal and tarsometatarsal joints as the condition progresses. This gradual "untwisting" of the calcaneopedal unit described by Seringe and Wicart will be studied in detail relevant to the pathomechanics of the adult acquired flatfoot deformity in Chap. 7.

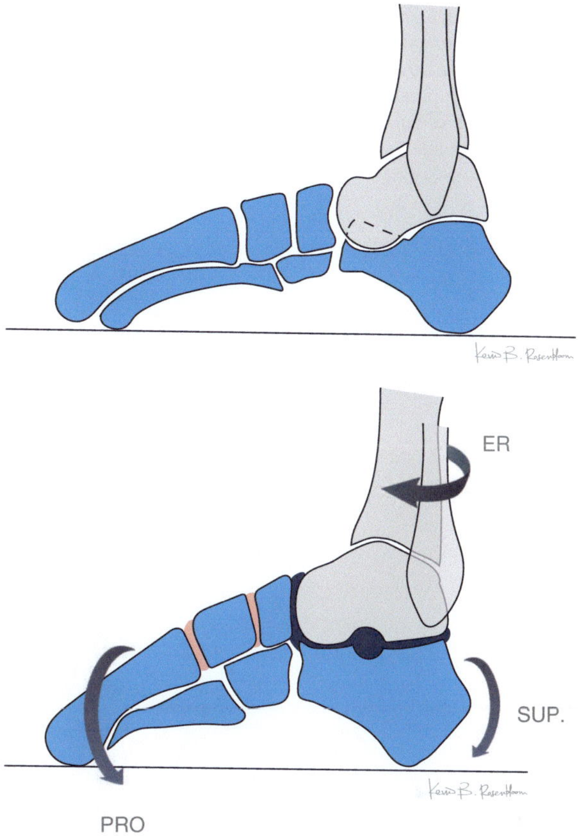

Fig. 1.15 Mechanism of pes cavus deformity. Rigid pronation of forefoot twists the rearfoot into supination and the tibiotalarfibular unit into external rotation. (Adapted from Seringe et al. [15])

Twisting the Plate and Locking for Optimal Foot Function

Bojsen-Moller proposed that the stability of the foot was almost entirely dependent upon "locking" of the calcaneocuboid joint which he observed would occur when the forefoot was pronated on the rearfoot [16]. In this case, the plate would be "twisted" to provide forefoot pronation. According to Saraffian, this twisting would relax tension on the plantar aponeurosis and the long plantar ligament. However, an osseous locking of the calcaneal-cuboid joint occurs with pronation of the forefoot on the rearfoot which is integral to optimal load transfer for a propulsive gait.

In his landmark paper, Bojsen-Moller begins with a detailed description of the calcaneocuboid joint of humans compared to the gorilla and chimpanzee foot. In the human foot, the cuboid has a "calcaneal process" located at the plantar and medial margin which extends into a recess on the calcaneus and is restrained by the plantar calcaneocuboid ligament (Fig. 1.2). The calcaneocuboid joint is a concavoconvex arrangement (saddle-shaped joint) which allows the cuboid to rotate

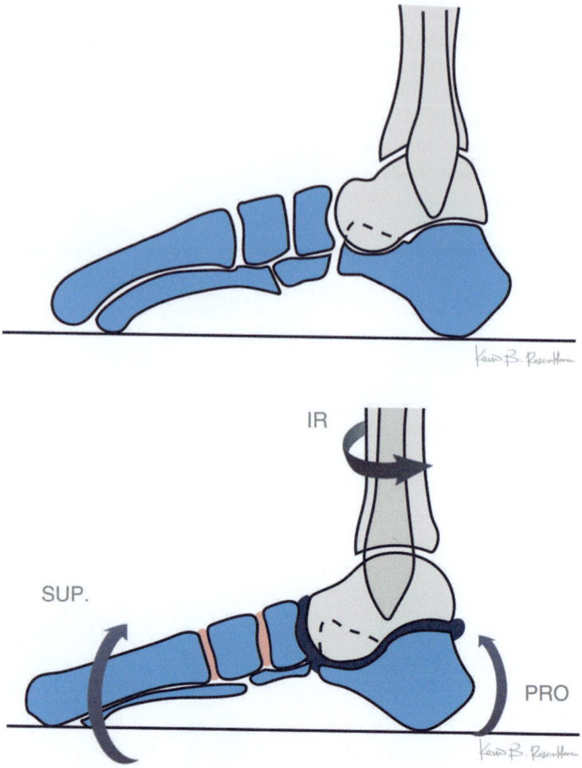

Fig. 1.16 Mechanism of pes planus. Pronation of rearfoot untwists the forefoot into supination and allows tibiofibular unit to internally rotate. (Adapted from Seringe et al. [15])

on the calcaneal process as a pivot. Pronation of the midtarsal joint moves the calcaneocuboid joint into a close-packed position, whereby the joint surfaces become congruent and stable. An anatomic "locking" of the calcaneocuboid joint occurs with pronation of the midtarsal joint, whereby dorsal excursion of the cuboid is limited by the dorsal-medial border of the calcaneus which overhangs the cuboid and by a tightening of the plantar, lateral, and dorsal calcaneocuboid ligaments (Fig. 1.17). The key soft tissue support, unique to humans, is the long plantar ligament which spans the calcaneocuboid joint plantarly. In supination of the forefoot, the calcaneocuboid ligaments are taut, but anatomic locking of the joint is not possible since the joint is not close-packed and the articular surfaces are not fully congruent.

In contrast, the calcaneocuboid joint of the ape foot as described by Bojsen-Moller surfaces appeared flat and did not have a close-packed position at pronation. In addition, the ape foot lacks a dorsal overhang of the calcaneus over the cuboid and lacks a plantar calcaneal process. Thus, motion of the forefoot on the rearfoot in the ape foot is unrestrained by any close packing or locking of the calcaneocuboid joint. Thus, a midfoot break occurs during heel rise in the ape and chimpanzee foot.

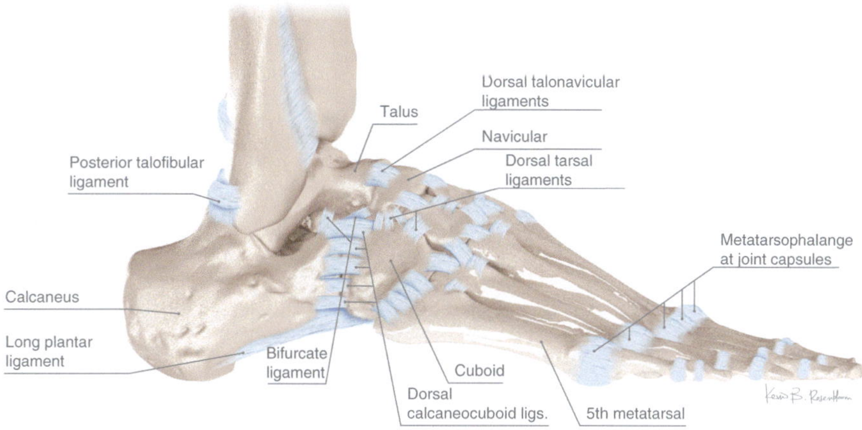

Fig. 1.17 The key ligaments of the calcaneocuboid joint

High-Gear Push Off

The second critical part of Bojsen-Moller's paper is the observation that the metatarsal parabola provides two axes of motion available across the forefoot during push off [16] (Fig. 1.18). The relative equal lengths of the first and second metatarsals establishes a transverse axis of motion across the evenly aligned first and second MTPJs with joint axes of motion oriented perpendicular to the line of progression. An oblique axis is aligned along the descending lengths of the third, fourth, and fifth metatarsals which is angulated from the line of progression. Humans have two options for push off depending upon stability of the foot prior to heel rise (Fig. 1.19).

Depending upon which axis is favored during push off, two "gears" of leverage are proposed by Bojsen-Moller which differ in ability to tension the plantar fascia and engage the windlass mechanism, previously described by Hicks [17]. Hicks observed that the plantar aponeurosis was pulled distally around the heads of the metatarsals with dorsiflexion of the digits during heel rise. The "winding" of the plantar aponeurosis around the head of the metatarsal was similar to a drum or winch on a sailing ship which maximized leverage for pulling in a line of rope. Hicks observed that when the windlass mechanism was engaged, the plantar aponeurosis shortened and the medial longitudinal arch was raised (Fig. 1.20).

According to Bojsen-Moller, the "high-gear push off," across the transverse axis of the first and second MTPJs, improves efficiency of the windlass compared to the "low-gear push off" across the oblique axis. The first metatarsal, with a larger radius compared to the lesser metatarsals, has a better mechanical advantage to increase tension and "shortening" of the plantar aponeurosis as the hallux is driven into extension during heel rise. The plantar aponeurosis invests the sesamoids, which when properly situated beneath the head of the first metatarsal, increases its radius

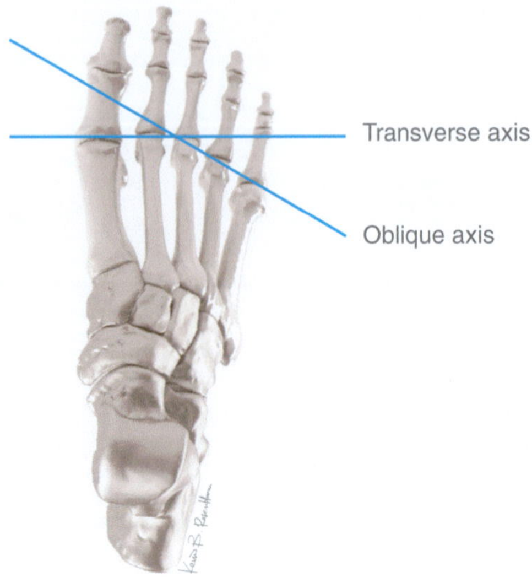

Fig. 1.18 The functional axes for push off across the forefoot

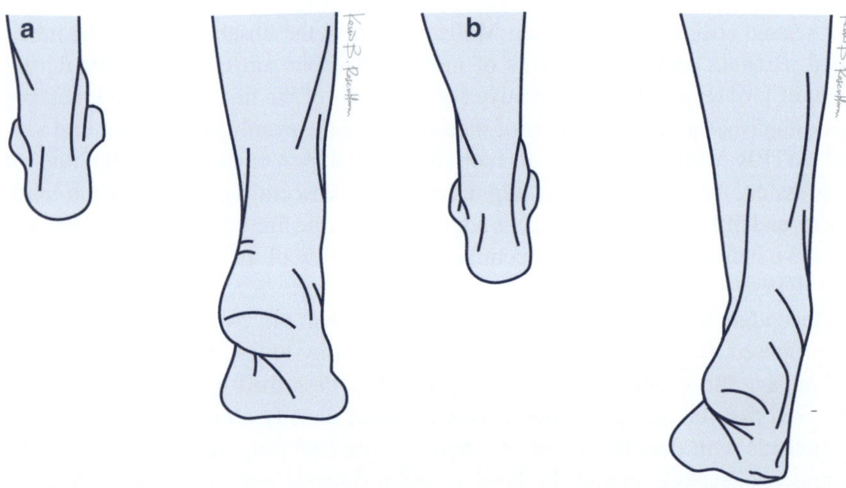

Fig. 1.19 (**a**) High gear push off across the transverse axis. (**b**) Low gear push off across the oblique axis

of curvature and greatly enhances the mechanical advantage of the hallux to engage the windlass mechanism. The sesamoids provide a radius of curvature of the first MTP which is twice that of the third metatarsal. When the sesamoids are shifted laterally in HAV deformity, this mechanical advantage of the windlass is compromised by shortening the radius of curvature of the first MTPJ.

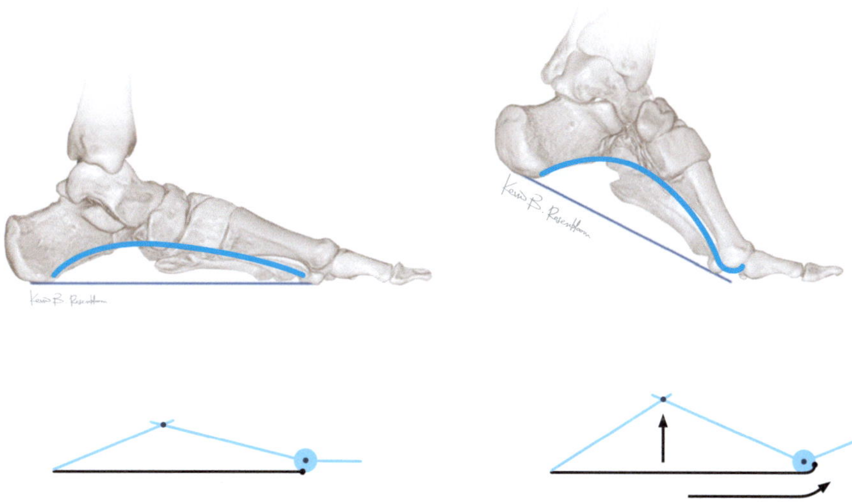

Fig. 1.20 The windlass mechanism tensions the plantar fascia with extension of the hallux during terminal stance and pre-swing

If load is directed to the first and second MTPJs during terminal stance or during the third rocker, a "high-gear" push off occurs across the transverse axis alignment of these joints relative to the line of progression. Conversely, if load is directed laterally to the MTPs 3–5 during the third rocker, a low-gear push off occurs across an oblique axis relative to the line of progression.

A final part of the study by Bojsen-Moller involved a human subject who was filmed with a high-speed camera on a walkway with a glass plate enabling view of the tissues on the plantar surface of the foot during walking gait [16].

High-gear push off across the transverse axis of metatarsals 1 and 2 was observed with the following:

- The forefoot was pronated in relation to the hindfoot.
- The lateral part of the ball was elevated very early from the ground.
- The plantar aponeurosis became tensed and could be seen through the skin.
- The tensioning of the peroneus longus tendon was visible on the lateral aspect of the foot.

At low-gear push off over the oblique axis, the following observations were made:

- Foot contact was transmitted from the heel to the lateral part of the ball of the foot.
- Push off thru occurred thru the lateral toes which were forced into dorsiflexion.
- The great toe gripped and stabilized the foot on the medial side.
- This appeared to cause a retrograde supination of the midfoot and rearfoot.

- The leg became externally rotated, the foot inverted at the subtalar joint, and the forefoot adducted in relation to the hindfoot, i.e., supination in the transverse tarsal joint.
- The medial arch became high, but neither the plantar aponeurosis nor the peroneus longus could be seen through the skin.

Bojsen-Moller compares these findings of a low-gear push off to similar observations of gait seen in nonhuman primates (Fig. 1.21). Apes and chimpanzees lack a well-developed plantar aponeurosis and cannot stabilize the medial column of their feet. Their grasping feet must move from an inverted position to a fully pronated position at the hindfoot to achieve a plantigrade position. Based upon the work of

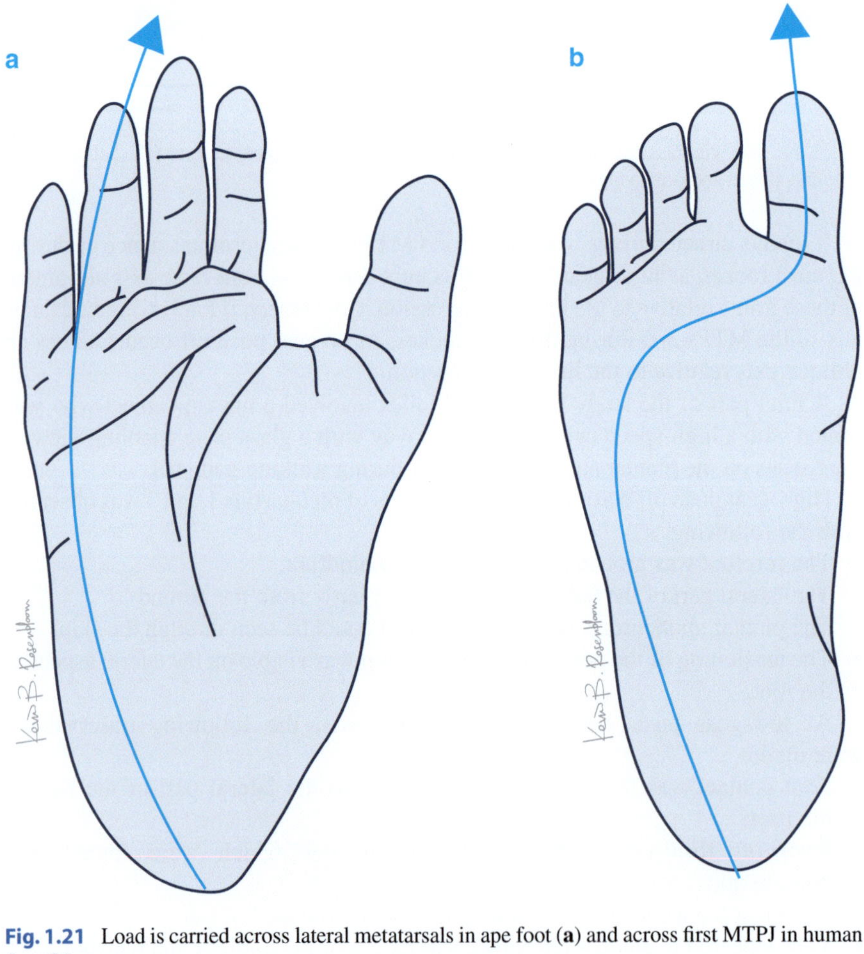

Fig. 1.21 Load is carried across lateral metatarsals in ape foot (**a**) and across first MTPJ in human foot (**b**)

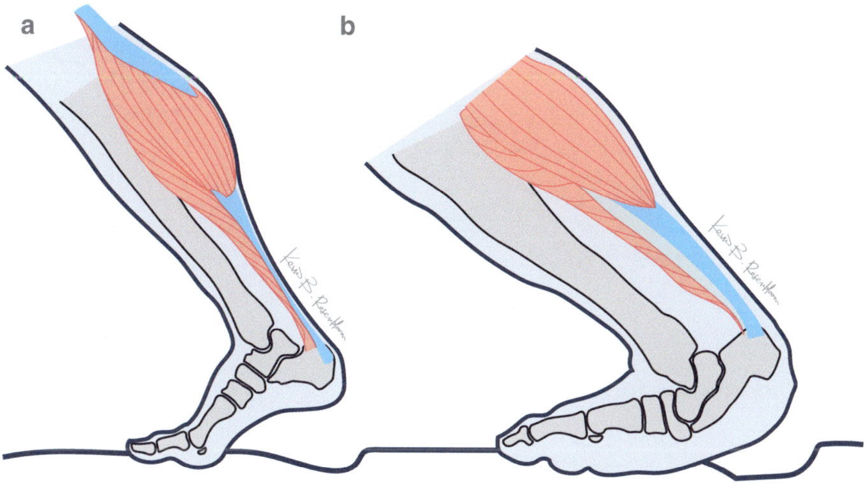

Fig. 1.22 (**a**) Pre-swing in human foot. (**b**) Midfoot break in ape foot

Elftman and Manter, Bojsen-Moller concludes that the motion of the ape foot into pronation, in order to achieve a plantigrade position, unlocks the midtarsal joint. A midfoot break occurs in the ape foot during heel rise and propulsion which directs push off thru the oblique axis of digits 2–5 (Fig. 1.22).

In terms of anatomic specialization in humans, the first metatarsal and hallux have moved into an adducted position to line up with the remaining metatarsals and thus function with the rest of the forefoot. Relative to the vertical orientation of the talus and calcaneus, the heads of the metatarsals and the entire forefoot have become "twisted" or pronated as described by MacConaill, so that the heads of the lateral metatarsals and the pulp of the lateral toes rest on the ground, while the medial and lateral longitudinal arches are established [12]. This twisting of alignment of skeletal segments is achieved partly from change of shape of certain bones and also due to rotation of these segments across certain joints within the foot itself.

What Initiates High-Gear Push Off?

The high-gear push off requires pronation of the forefoot on the rearfoot which is initiated by the peroneus longus muscle acting across the stable fulcrum of the calcaneocuboid joint. As Bojsen-Moller states:

> The functional pronation of the human forefoot for the high gear push off is the final step in the process of producing a fast and efficient lever for propulsion. Among the muscles the peroneus longus and assists in the internal rotation of the crus which forces the foot to use the transverse axes.

The peroneus longus requires a stable calcaneocuboid joint which is elevated above the substrate in order to maximize the plantarflexion lever arm of the

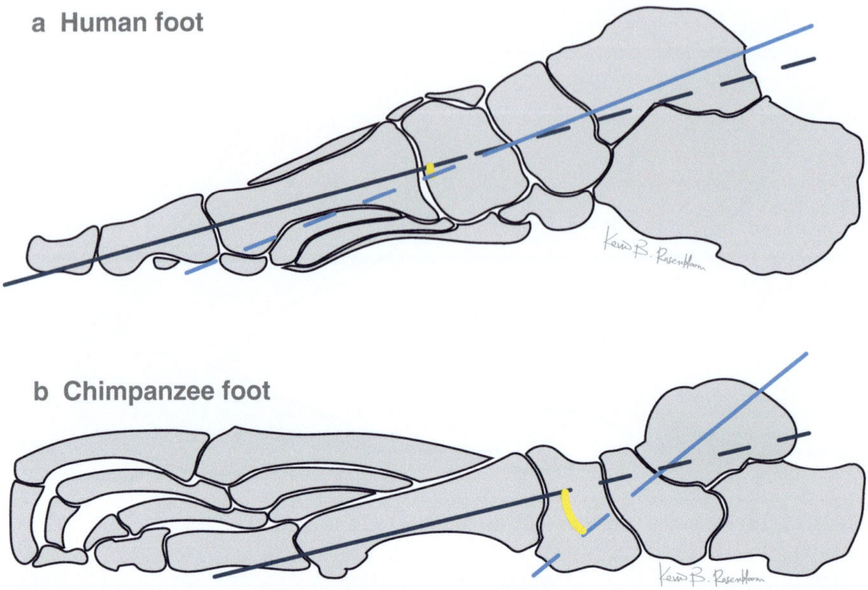

Fig. 1.23 Meary's angle: Comparison of human foot (**a**) and chimpanzee foot (**b**) in midstance. In humans, the medial column is aligned from the talus, (blue line) thru the first metatarsal (black line). The medial, lateral, and midfoot arches are established. The cuboid is elevated from the substrate. In chimpanzees, the cuboid is weight-bearing.

peroneus longus. A key specialization of the human foot is the presence of a lateral arch which elevates the cuboid from the substrate (Fig. 1.23). The ontology of the human foot twists the forefoot on the rearfoot to establish the medial and lateral longitudinal arches. The cuboid sits at the apex of the lateral arch and is supported by both osseous and ligamentous structures which are not seen in any other primate feet. In contrast, the medial arch is entirely dependent upon ligamentous support for stability.

Pronation of the forefoot on the rearfoot at the MTJ in late midstance pretightens the plantar aponeurosis before heel rise. As the rearfoot pronates, ground reaction force is increased on the medial column of the foot, and the first metatarsal will dorsiflex accordingly. Further tension is developed in the plantar aponeurosis as soon as the heel leaves the ground causing dorsiflexion moment of the entire forefoot. The larger diameter of the head of the first metatarsal combined with the additional circumference of the sesamoids favors greater mechanical advantage of the first MTPJ over the lesser MTPJs to engage the windlass mechanism which will raise the arch and stiffen the foot for propulsion (Fig. 1.20).

Testing the Theory of High-Gear Push Off

A kinematic study was carried out by Griffin and co-workers to verify Bojsen-Moller's observations [18]. Griffin et al. attempted to validate that medial transfer of weight to the forefoot and engagement of the windlass thru the first MTPJ were more efficient in humans compared to great apes (bonobos) during ambulation. 2-D digitized video capture was utilized to measure dorsiflexion of digits 1–3 from midstance thru toe-off in human and bonobo subjects.

There was significantly greater excursion of dorsiflexion of the first MTPJ in humans (range 24–53°) compared to the bonobo (range 10–38°). At the second MTPJ, dorsiflexion in humans ranged from 25 to 47° while in the bonobo, the range was 14–34°. Thus, the overall dorsiflexion of the MTPJs was greater in humans compared to the great apes and occurred in a greater magnitude at the first MTPJ.

Pressure data in this study showed a medial transfer from midstance thru toe-off in the human subjects with a peak pressure under the first metatarsal during late stance [18]. This medial shift did not occur in the bonobos, but instead pressure increased in the lateral metatarsals during toe-off. Also, a midfoot break was observed in the bonobo subjects with increased weight bearing of the lateral arch at midstance. No midfoot break was seen in the human subjects. The authors concluded that the greater excursion of dorsiflexion of all the MTPJs in humans was due to specializations of the human foot which includes:

- A stable calcaneocuboid joint
- An adducted first metatarsal (MT) that is substantially larger than the lateral metatarsals
- Distal articular regions of the MTs that are oriented in a way to allow greater metatarsophalangeal joint excursion in the sagittal plane
- Relatively short pedal phalanges

In summary, this study by Griffin et al. verified the observations which had been made by Bojsen-Moller 30 years earlier regarding the role of lateral column stability to facilitate transfer of load medially to the first MTPJ during terminal stance and propulsion [16]. Both authors verified that this was a unique human feature.

What Is the Ideal Alignment of the Human Foot?

In reality, MacConaill describes a stable position of the human foot during static bipedal stance, while Bojsen-Moller is proposing an ideal position of the foot during terminal stance in dynamic gait. MacConaill describes a natural tendency for people to instinctively spread their feet apart during prolonged, quiet standing which essentially supinates the forefoot against the rearfoot. Without muscle action during quiet stance, this positioning of the foot tenses the plantar ligaments and stiffens the foot for stability. However, this is not the ideal position of the foot for push off and engagement of the windlass, according to Bojsen-Moller. Both authors are correct in describing stiffening mechanisms which operate within the human foot during different stages of the gait cycle. During midstance with the foot in single leg support, an everted hindfoot with inverted (supinated) forefoot would maximally tension supportive ligaments while minimizing the need for active muscle tension to stiffen the foot for stability. In terminal stance and pre-swing, the foot needs to move from a stiff "untwisted" plate of bones into a more ideal position which will shift weight bearing to the medial column, engage the windlass, and facilitate high-gear push off. Stiffness is now exchanged for a springlike behavior as elastic recoil of soft tissues plantarflexes the forefoot during pre-swing.

The importance of lateral column specialization enabling the unique function of the human foot has been described by Kidd [19]. He points out that fossil evidence shows that specialization of the lateral column preceded medial column specialization in prehumans or hominid species. Kidd asserts that stabilization of the calcaneal-cuboid joint and elevation of the lateral longitudinal arch were critical precursors to the development of an adducted first ray and hallux for weight bearing of the medial column. This is in agreement with Bojsen-Moller. Indeed, when examining the sequence of chondrification and ossification of the bones of the foot in utero, this hierarchy for specialization is repeated. The central metatarsals chondrify first followed by the fifth metatarsal and the cuboid. The talus and the calcaneus are next to chondrify. The first metatarsal and medial cuneiform follow, and the navicular is the last bone of the foot to chondrify. Clearly, the medial column lags behind the lateral column in the order of maturation.

The Myth of Midfoot "Locking"

The locking of the calcaneocuboid joint described by Bojsen-Moller and specialization of the lateral column described by Kidd are part of an overall observation of a unique feature of the human foot: stiffening of the midtarsal joint. This stiffening mechanism of the midtarsal joint in humans eliminates the sagittal plane instability causing the "midfoot break" seen in apes (Fig. 1.22). Nonhuman primates show a "double heel lift" during gait where an early heel rise during midstance causes increased sagittal plane mobility as the rearfoot plantarflexes on the forefoot across the midfoot joints (Fig. 1.22). Once the midfoot eventually stiffens, a second heel rise occurs to initiate terminal stance. Elftman and Manter observed this midfoot break in chimpanzees and then proposed a mechanism explaining how the midfoot break is eliminated in human feet during ambulation [20]. They proposed that humans possess an osseous "locking mechanism" based upon the angulation of the articular facets across the midtarsal joint which allows conversion of the foot from a mobile adapter during the contact phase of gait, to a so-called rigid lever for propulsion [21]. The unique angulation of the axes of motion and the articular facets of the talonavicular joint and the calcaneal-cuboid joint was theorized by Elftman and

Manter to be made possible by the position of the calcaneus beneath the talus in the human foot, compared to the side-by-side relationship in all other primate feet.

This notion that the axes of the talonavicular and calcaneocuboid joints can be parallel or divergent, depending upon subtalar joint position, explains the locking mechanism of the midtarsal joint as proposed by Elftman and Manter. This locking mechanism of the midtarsal joint, dependent upon rearfoot alignment, has been accepted and embraced by virtually all disciplines studying or treating the foot and ankle for the past 85 years. A stiff, stable midfoot, attributed primarily to this locking mechanism, is a trait which has been proclaimed uniquely "human" compared to all other primates, enabling bipedal gait. Inexplicably, it is not just the fact that the this osseous locking mechanism based upon divergent joint axes has been refuted in more recent kinematic studies, but the fact that the uniqueness of midfoot stability in humans has also been appropriately challenged. Despite this valid research, clinicians still cling to the concept of midfoot locking. It is interesting that the challenge to this concept has not come from academic biomechanists, but rather by anthropologists who have performed elegant biomechanical research in this subject area.

Karl T. Bates, an anatomist and anthropologist, performed plantar pressure studies on healthy humans while walking and compared the findings to a bonobo and an orangutan while walking [22]. Surprisingly, there was overlap in pressure patterns where the healthy humans showed intermittent weight bearing and increased pressure under the lateral column, similar to the apes. Kinematic data from this study verified that increased lateral pressure corresponded to increased motion or compliance of the lateral arch. Clearly, the lateral midfoot is not consistently stiff in healthy humans during gait, but instead shows compliance in random fashion (Fig. 1.24). This will be further discussed in Chap. 3 with kinematic studies of foot function. Furthermore, Bates et al. showed that elevated lateral midfoot pressure was associated with decreased stiffness of the first metatarsal and hallux. However, the authors suggested that midfoot stiffening is reliant on soft tissue structures rather than osseous locking. This study verified Bojsen-Moller's concept of lateral column stiffness being requisite to shifting weight to the medial forefoot and activating the windlass during propulsion. In turn, this would lead to increased stiffness of the first ray and hallux.

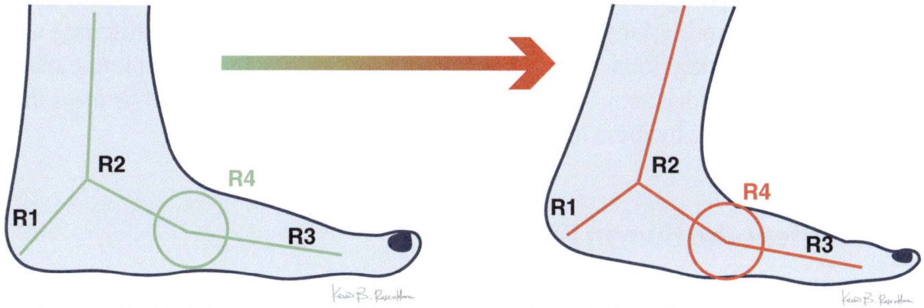

Fig. 1.24 Many humans will show a brief midfoot break with 3-D motion capture. At beginning of terminal stance, many humans will exhibit sagittal plane motion across the midtarsal joint (R4) which may be a result of gastrocnemius contracture or from loss of stability at the calcaneocuboid joint.

Jeremy DeSilva, an anthropologist, evaluated plantar pressure and midfoot kinematics in 398 healthy adults. Approximately 8% had a midfoot break as seen in apes [23]. These human subjects with a midfoot break were excessive pronators with low arch profile. Both Bates and DeSilva speculated that stiffness of the midfoot joints in humans is more dependent upon soft tissue restraints than osseous "locking" of the midfoot joints.

A direct comparison of midfoot kinematics in humans and chimpanzees was performed by Holowka, another anthropologist [24]. During midstance, the chimpanzees showed twice the sagittal plane motion across the midfoot joints, but the difference was only about 4°. Conversely, during terminal stance and pre-swing, humans actually use 10° more motion than chimpanzees across the midfoot joints. This motion across the midfoot joints of humans is in the direction of plantarflexion and adduction which is essential to generating push off power during propulsion. Adduction of the forefoot across the midfoot is coupled with external rotation of the tibia and rearfoot and directs weight bearing to the medial column to accept weight transfer and engage the windlass. Also, directing weight bearing thru the transverse axis of the longer second and first metatarsals enhances the lever arm of the triceps surae for push off and forward propulsion. This combined plantarflexion-adduction motion at the midfoot joints of humans during terminal stance and pre-swing (after heel rise) is threefold greater in magnitude than seen in the chimpanzee foot. The authors observed that this motion was of greater magnitude at the cuboid-fifth metatarsal joint than at the calcaneocuboid joint, a finding verified by later kinematic studies which will be discussed in Chap. 3.

In summary, three anthropologists have provided compelling evidence that humans can actually have a midfoot break and that motion across the midfoot joints is actually greater in humans than nonhuman primates. These anthropologists directly challenged an age-old belief that there is an osseous locking mechanism operating across the midtarsal joint of humans. Most importantly, they proposed that soft tissue structures such as the plantar fascia, the long plantar ligament, and plantar intrinsic muscles are primarily responsible for the unique ability of humans to stiffen their feet during the stance phase of gait. Stiffening, which is enabled by static and dynamic soft tissue structures, rather than from osseous locking, is the hallmark of stability of the human foot. These anthropologists also speculated that the interaction of these structures was managed by a sensorimotor system which fine-tunes and co-ordinates stiffening and acceleration across key joints of the lower extremity. As we further explore current research and knowledge about the kinematics of the human foot, it will be clear that the conclusions made by these anthropologists are entirely correct.

What Stiffens the Human Foot?

The wide acceptance of a locking mechanism of the midtarsal joint as proposed by Elftman and Manter is partly due to the fact that the average human foot does demonstrate this finding with a simple off-weight bearing examination. The midfoot joints will demonstrate either an increase or decrease in range of motion depending upon rearfoot position. Off weight bearing, the human foot will demonstrate stiffening of the midtarsal joint with loading when the subtalar joint is moved from a position of maximal pronation to maximal supination. Several studies have verified this unique change of compliance of the human midfoot joints depending upon the position of the rearfoot.

Phillips and Phillips measured increased frontal plane rotation of the forefoot on the rearfoot when the subtalar joint was moved from a neutral position to a maximally pronated position [25]. Gatt and co-workers used a 3-D multisegment foot model to determine the change in dorsiflexion of the foot as the subtalar joint was moved from a pronated position to a neutral position and finally to a supinated position [26]. Although this study was intended to measure ankle joint dorsiflexion relative to subtalar joint position, the multisegment foot model detected that more sagittal plane motion was occurring within the foot itself than at the ankle joint. Furthermore, moving the foot into a pronated position at the subtalar joint increased motion between the forefoot and the rearfoot. Gatt and co-workers disputed whether a midtarsal joint "locking" occurred since there was still significant motion between the forefoot and the rearfoot at both neutral and supinated positions of the subtalar joint. In both studies reported by Phillips et al. and Gatt et al., the major increase in forefoot to rearfoot motion occurred when the subtalar joint moved from a neutral position to a fully pronated position [25, 26]. Moving the foot from subtalar neutral to a supinated position had only a mild influence on forefoot motion. As will be demonstrated in Chaps. 2 and 3, the human foot does not move to a neutral position at the subtalar joint until the terminal stance phase of gait has been reached, well after a required stiffening of the midfoot should have taken place for efficient push off. Therefore, the human foot does not apparently rely on a change of rearfoot position to stiffen the midfoot joints during late midstance and terminal stance.

Blackwood and co-workers carried out a cadaver study which also measured greater flexibility of the forefoot in the sagittal plane when the subtalar joint was everted with less flexibility measured when the subtalar joint was inverted [27]. However, this effect was only demonstrated in the sagittal plane, not the frontal plane. Blackwood et al. concluded that the stiffening of the forefoot to the rearfoot may have involved multiple joints, not just the midtarsal joint.

If stiffening of the midfoot during gait were due to the establishment of divergent or conflicting joint axes at the midtarsal joint, kinematic studies would be able to demonstrate the asynchronous movement between the talonavicular and calcaneocuboid joints. However, this has not been the case. In a dynamic cadaver study, Nester et al. showed synchronous movement at the talonavicular joint and the calcaneal-cuboid joint during the stance phase of gait [28]. In a subsequent study of healthy human subjects, Nester was unable to measure any consistent coupling between segments of the foot but also demonstrated that the midfoot joints also showed significant motion dispelling the notion of "locking" of the midtarsal joint [29].

Okita and co-workers performed a dynamic cadaver study and also demonstrated that the midfoot joints actually become more compliant during terminal stance [30]. Interestingly, they did actually measure a divergence of joint axes in the midfoot and rearfoot, but the onset of the divergence, during the early stance phase, did not correspond with increased rigidity of the foot. Instead, the foot became rigid during heel rise when the joint axes became non-divergent. The authors concluded that soft tissue structures including tendons and ligaments would have to account for the stiffening of the human foot during the propulsive phase of gait.

The Springlike Function of the Human Foot

The human foot has been considered unique and superior to other primate feet because it carries out the dual function of being both a mobile adapter during the contact phase of gait and then converts into a rigid lever for propulsion. This notion was first proposed by Inman and Mann and has since been accepted and promoted by many authorities over the past 60 years [31–33]. However, as previously discussed in the studies of midfoot break and as will be shown in Chap. 2, the human foot is a highly compliant structure and never demonstrates true "rigidity" during any phase of the walking or running gait cycle. Rather, the human foot is unique in its ability to store and release elastic energy during ambulation. This mechanism has been termed the "arch spring" by Ker et al. [34] They showed how compression and recoil of the longitudinal arch of the human foot allow storage and return of elastic energy by key ligaments, most notably the plantar fascia. The ability to store and release elastic energy in the feet has been demonstrated in other primates but is most efficient in humans [35, 36].

The human foot contains three unique ligamentous structures which play a critical role for stiffening and elastic recoil for return of energy during walking. These three structures include the plantar aponeurosis, the long plantar ligament, and the bifurcated ligament [9]. The long plantar ligament is ideally located to provide stiffness to the lateral column (Fig. 1.25). There are two portions of the long plantar ligament: a proximal or posterior band originates from the plantar aspect of the body of the calcaneus just anterior to the tuber and inserts on the plantar aspect of the cuboid and a more superficial anterior portion continues from the posterior portion and inserts on the base of metatarsals 3, 4, and 5. The superficial and medial fibers of the long plantar ligament form the roof of the peroneus longus tendon tunnel. The short plantar ligament has a more anterior attachment to the calcaneus and fans out antero-medially to attach to the plantar surface of the cuboid, just proximal to the peroneus longus tendon tunnel (Fig. 1.26).

The lateral column of the foot lacks the longitudinal support seen in the medial column provided by the central band of the plantar aponeurosis which originates on the medial calcaneal tubercle. The primary longitudinal support of the lateral column is the long plantar ligament acting as a tie-rod across the truss of the lateral longitudinal arch. The long plantar ligament spans the calcaneocuboid joint and is critical to maintaining the elevated position of this joint from the supportive surface. Other ligaments which support the calcaneocuboid joint are the two components of

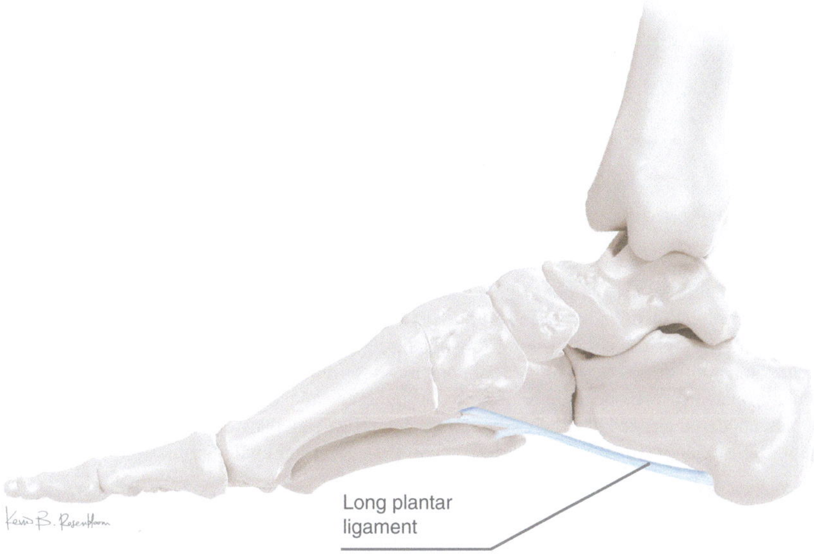

Fig. 1.25 The lateral truss structure of the human foot

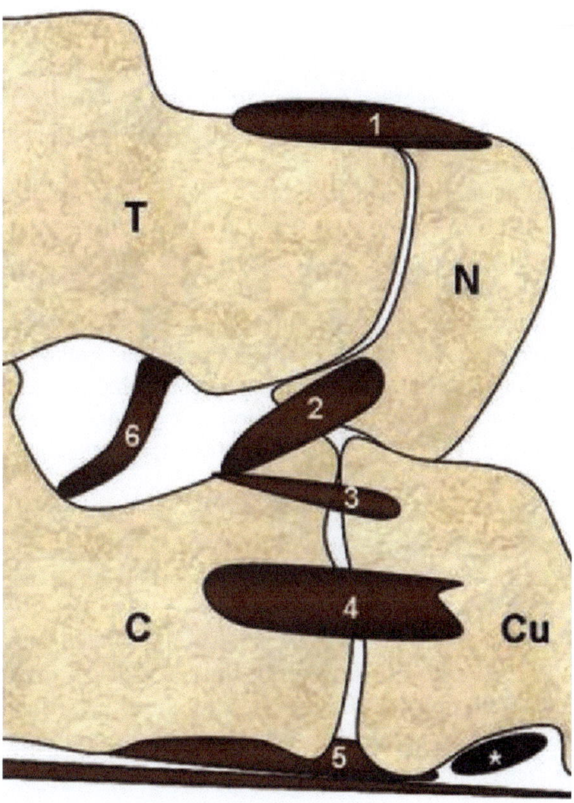

Fig. 1.26 Drawing shows lateral view of transverse tarsal joint. 1 talonavicular ligament, 2 lateral calcaneonavicular ligament, 3 medial calcaneocuboid ligament, 4 dorsolateral calcaneocuboid ligament, 5 inferior calcaneocuboid ligaments (short and long plantar ligaments), 6 cervical ligament, 2 and 3 bifurcate ligament, asterisk peroneus longus tendon, T talus, N navicular, C calcaneus, Cu cuboid. (From Melão et al. [51]. Reprinted with permission)

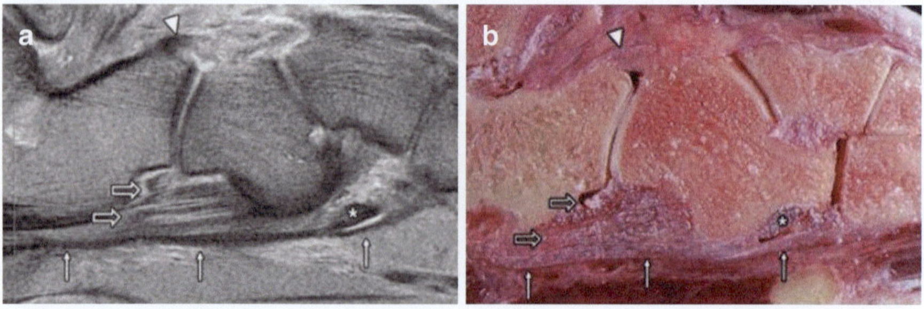

Fig. 1.27 Plantar calcaneocuboid ligaments. (**a** and **b**) Sagittal proton density MR image after intraarticular injection of gadolinium solution (**a**) and photograph of corresponding gross anatomic section (**b**) in cadaveric specimen show long plantar ligament (solid arrows), with low signal intensity, making floor of peroneus longus tendon (asterisk) tunnel. Note two bands of short plantar ligament (open arrows): one deep reinforcing articular capsule and another more superficial, both of which appeared striated. Arrowhead points to lateral calcaneonavicular ligament (medial component of bifurcate ligament). (From Melão et al. [51]. Reprinted with permission)

the bifurcate ligaments, the medial calcaneocuboid ligament and lateral calcaneonavicular ligaments, as well as the dorsolateral calcaneocuboid ligaments (Fig. 1.27).

Among other structures, the long plantar ligament is essential to resisting dorsiflexion of the forefoot on the rearfoot and is a primary structure to prevent a midfoot break [9]. The calcaneocuboid and calcaneonavicular bands of the bifurcated ligament restrain motion across the midtarsal joint. The orientation of the bifurcated ligaments suggests a function to restrict conflicting motion between the navicular and the cuboid. The plantar cuboideonavicular ligament restricts frontal plane rotation between the cuboid and navicular (Figs. 1.17 and 1.27). This ligament is not well developed in the ape foot which actually demonstrates an articular facet for motion between the calcaneus and navicular. The human foot demonstrates an interosseous ligament between the cuboid and navicular to further restrict frontal plane motion.

The apparent locking of the midtarsal joint via supination of the subtalar joint as demonstrated by Phillips and Phillips [25] as well as Gatt et al. [26] might be due to tensioning of all of these ligaments surrounding the calcaneocuboid joint. As the rearfoot is moved from inversion to eversion, alignment and tension on the long plantar and bifurcate ligaments will change which allow increased motion across the midfoot joints.

The critical differentiating feature of the human foot is the medial longitudinal arch spanned by the plantar aponeurosis (i.e. plantar fascia), providing twice the contribution to arch stability as the spring ligament and plantar ligaments combined [37]. The plantar fascia has been shown to undergo some stretch during midstance but must remain stiff for activation of the windlass mechanism [38].

There is also some recent evidence showing the importance of the transverse arch of the midfoot for providing overall stiffness of the human foot [39]. The transverse arch begins medially at the medial cuneiform and then extends laterally across the cuneiforms ending at the cuboid. The second cuneiform is shaped in a fashion to provide a keystone function to the transverse arch. Venkadesan and co-workers make a convincing argument with experimental evidence that a transverse arch curvature will impart more stiffness on the human foot than will a longitudinal arch curvature. They show that the presence of a transverse arch in the human foot accounts for the significant increased stiffness compared to other primate feet which lack a transverse arch. They verified these assumptions by performing a cadaver study measuring vertical deformation of the foot with loading. When the plantar intertarsal ligaments supporting the transverse arch were cut, a significant 40% loss of stiffness in the foot occurred. Venkadesan compared this finding to the work of Ker et al. who showed that stiffness of the foot decreased by just 23% when the plantar fascia was transected [34].

Besides passive ligamentous structures, certain dynamic mechanisms contribute to energy storage and return during human gait. Compared to other primates, the human Achilles is well developed and elongated for improved storage and release of elastic energy. Studies have shown that the stretch and recoil of the Achilles tendon provides a 35% energy recovery. This energy return increases to 52% when the plantar aponeurosis and plantar ligaments recoil after loading [35, 40].

Kelly and co-workers conducted a series of studies showing the ability of the plantar intrinsic muscles to affect stiffness of the longitudinal arch of the foot [41, 42]. These studies showed activation of the plantar intrinsics which occurs in response to loading of the foot and elongation of the medial arch. However, an in vivo study by these researchers showed that the plantar intrinsic muscles do not affect arch structure during midstance but do stiffen the foot during terminal stance and push off [43]. Kelly et al. suggest that the plantar intrinsic muscles are critical for providing plantarflexion moment across the MTPJs during terminal stance and pre-swing which enables the ankle to perform positive work for push off. This link between forefoot stiffness and efficient energy release at the ankle has been demonstrated in previous investigations [44, 45].

Hick's model of the windlass mechanism assumes that the plantar fascia is stiff and resistant to elongation. Indeed, studies show that increased dorsiflexion of the first MTPJ increases stiffness of the plantar aponeurosis [46, 47] (Fig. 1.20). However, other studies have demonstrated that the plantar aponeurosis can undergo elongation with loading which could compromise function of the windlass [48, 49]. Welte and co-workers developed an apparatus to measure arch deformation and stiffness in nine human subjects [50]. As the hallux was dorsiflexed to engage the windlass, the length of the arch shortened as expected. However, when the foot was loaded while the first MTPJ was dorsiflexed and the windlass was engaged, stiffness of the arch was actually reduced compared to the plantar-flexed position of the hallux. Activating the windlass mechanism actually increased flexibility of the longitudinal arch of the foot when vertical load was applied in this study. A higher arch has the capacity to flatten and elongate, allowing absorption and storage of energy. Welte et al. propose that the shortening and elongation of the arch which occurs with activation of the windlass mechanism may optimize the length of key ligaments which can then undergo stretch/shortening for energy storage and return [50] (Fig. 1.28).

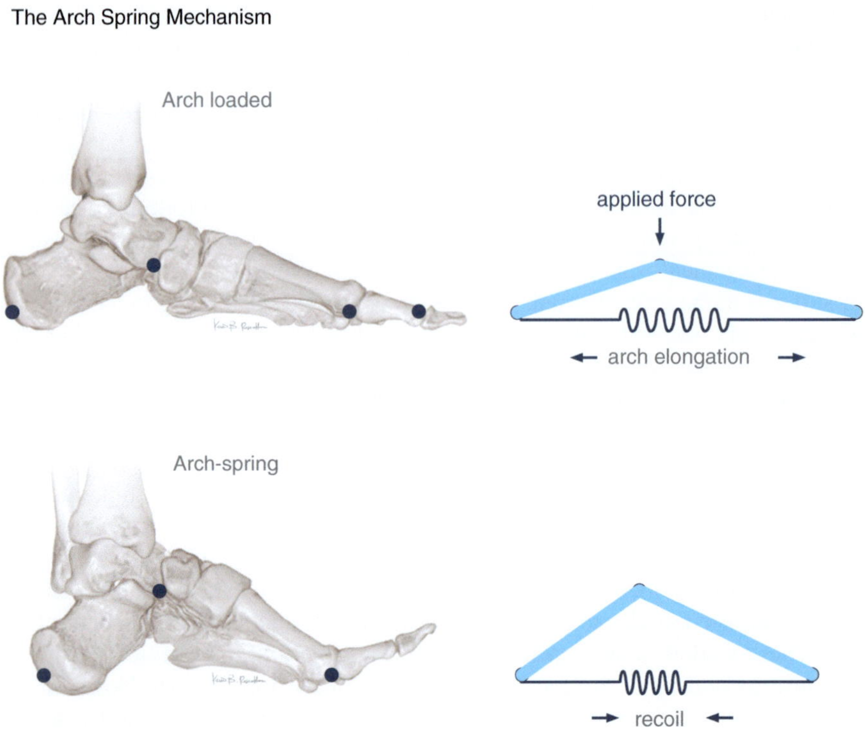

Fig. 1.28 The arch spring mechanism

The work of Kelly and others has shown that, contrary to popular belief, the arch of the human foot is not totally stiff or rigid during terminal stance. The original model of a "locking mechanism" between two divergent axes of the midtarsal joint has been challenged. The windlass mechanism by itself does not explain the raising of the arch and triplane motion of the foot during push off.

Summary

In summary, the human foot demonstrates unique anatomic structures which optimize upright bipedal ambulation. Some of these features, such as anatomic locking of the calcaneal-cuboid joint, impart stiffness on the midfoot and allow transfer of load from lateral to medial for push off across the transverse axis of the metatarsophalangeal joints. Other key ligamentous structures unique to the human foot, particularly the plantar aponeurosis, function as an engine for energy storage and return, thus enabling a propulsive gait. Further springlike action is provided by the tendo Achilles after storage of elastic energy from eccentric loading during midstance. For the Achilles to provide positive work across the ankle during push off, the forefoot must be stiff to resist ground reaction forces.

The plantar intrinsic muscles appear to play a key role in providing forefoot stiffness via plantar stabilization of the metatarsophalangeal joints against the supportive surface.

References

1. Jones FW. Structure and function as seen in the foot. 2nd ed. London: Baillière, Tindall and Cox, 1949; 1944. p. 2.
2. McNutt EJ, Zipfel B, DeSilva JM. The evolution of the human foot. Evol Anthropol. 2018:1–21.
3. Bramble DM, Lieberman DE. Endurance running and the evolution of Homo. Nature. 2004;432:345–52.
4. Latimer B, Lovejoy CO. The calcaneus of Australopithecus afarensis and it's implications for the evolutionary development of bipedality. Am J Phys Anthropol. 1989;78:369–86.
5. Raichlen DA, Armstrong H, Lieberman DE. Calcaneus length determines running economy: implications for endurance running performance in modern humans and Neanderthals. J Hum Evol. 2011;60:299e308.
6. Latimer B. The perils of being bipedal. Ann Biomed Eng. 2005;33(1):3–6.
7. Kunos CA, Latimer B. Material and structural properties of human and African ape cortical and cancellous bone: implications for the evolution of bipedality. Am J Phys Anthropol. 2003;36(Suppl):133.
8. Inman VT. The joints of the ankle. Baltimore: Williams & Wilkins; 1976.
9. Gomberg DN. Functional differences of three ligaments of the transverse tarsal joint in hominoids. J Hum Evol. 1985;14:553–62.
10. Crelin ES. The development of the human foot as a resume of its evolution. Foot Ankle. 1983;3(6):305–21.
11. Sarrafian SK, Kelikian AS. Development of the foot and ankle. In: Kelikian AS, editor. Sarrafian's anatomy of the foot and ankle. 3rd ed. Philadelphia: Wolters Kluwer; 2011. p. 3–39.
12. MacConaill MA. The postural mechanism of the human foot. Proc R Ir Acad. 1944-45;50B:265–78.
13. Steindler A. The supinatory compensatory torsion of the forefoot in pes valgus. J Bone Joint Surg Am. 1929;2(11):272–6.
14. Sarrafian SK. Functional characteristics of the foot and plantar aponeurosis under tibiotalar loading. Foot Ankle. 1987;8:4–18.
15. Seringe R, Wicart P. The talonavicular and subtalar joints: the "calcaneopedal unit" concept. Orthop Traumatol Surg Res. 2013;99(6):S345–55.
16. Bojsen-Møller F. Calcaneocuboid joint and stability of the longitudinal arch of the foot at high and low gear push off. J Anat. 1979;129:165.
17. Hicks JH. The mechanics of the foot: II. The plantar aponeurosis and the arch. J Anat. 1954;88:25–30.
18. Griffin NL, D'Aout K, Richmond B, Gordon F, Aerts P. Comparative in vivo forefoot kinematics of Homo sapiens and Pan paniscus. J Hum Evol. 2010;59:608e619.
19. Kidd R. The past is the key to the present: thoughts on the origins of human foot structure, function and dysfunction as seen from the fossil record. Foot. 1998;8:75–84.
20. Elftman H, Manter J. Chimpanzee and human feet in bipedal walking. Am J Phys Anthropol. 1935;20:69–79.
21. Elftman H, Manter J. The evolution of the human foot with special reference to the joints. J Anat. 1935;70:56–67.
22. Bates KT, Collins D, Savage R, McClymont J, Webster E, Pataky TC, et al. The evolution of compliance in the human lateral mid-foot. Proc R Soc B. 280:20131818.
23. DeSilva JM, Bonne-Annee R, Swanson Z, Gill CM, Sobel M, Uy J, Gill SV. Midtarsal break variation in modern humans: functional causes, skeletal correlates, and paleontological implications. Am J Phys Anthropol. 2015;156:543e552.

24. Holowka NB, O'Neill MC, Thompson NE, Demes B. Chimpanzee and human midfoot motion during bipedal walking and the evolution of the longitudinal arch of the foot. J Hum Evol. 2017;104:23–31.
25. Phillips RD, Phillips RL. Quantitative analysis of the locking position of the midtarsal joint. J Am Podiatry Assoc. 1983;73(10):518–22.
26. Gatt A, Chockalingam N, Chevalier TL. Sagittal plane kinematics of the foot during passive ankle dorsiflexion. Prosthetics Orthot Int. 2011;35:425–31.
27. Blackwood CB, Yuen TJ, Sangeorzan BJ, et al. The midtarsal joint locking mechanism. Foot Ankle Int. 2005;26:1074–80.
28. Nester CJ, Liu AM, Ward E, Howard D, Cocheba J, Derrick T, Patterson P. Invitro study of foot kinematics using a dynamic walking cadaver model. J Biomech. 2007;40(9):1927–37.
29. Nester CJ, Jarvis HL, Jones RK, Bowden PD, Anmin L. Movement of the human foot in 100 pain free individuals aged 18–45: implications for understanding normal foot function. J Foot Ankle Res. 2014;7:51.
30. Okita N, Meyers SA, Challis JH, Sharkey NA. Midtarsal joint locking: new perspectives on an old paradigm. J Orthop Res. 2014;32(1):110–5.
31. Inman VT. Human locomotion 1966. Clin Orthop Relat Res. 1993;288:3–9.
32. Mann R, Inman VT. Phasic activity of intrinsic muscles of the foot. J Bone Joint Surg Am. 1964;46:469–81.
33. Root ML, Orien WP, Weed JH, Hughes RJ. Normal and abnormal function of the foot. Los Angeles: Clinical Biomechanics Corp; 1977.
34. Ker RF, Bennett MB, Bibby SR, Kester RC, Alexander RM. The spring in the arch of the human foot. Nature. 1987;325:147–9.
35. Alexander RM. Elastic energy stores in running vertebrates. Am Zool. 1984;24:85–94.
36. Vereecke EE, Aerts P. The mechanics of the gibbon foot and its potential for elastic energy storage during bipedalism. J Exp Biol. 2008;211:3661–70.
37. Huang CK, Kitaoka HB, An KN, Chao YS. Biomechanical evaluation of longitudinal arch stability. Foot Ankle Int. 1993;14(6):353–7.
38. Caravaggi P, Pataky T, Goulermas JY, Savage R, Crompton R. A dynamic model of the windlass mechanism of the foot: evidence for early stance phase preloading of the plantar aponeurosis. J Exp Biol. 2009;212:2491–9.
39. Venkadesan M, et al. Nature. 2020; https://doi.org/10.1038/s41586-020-2053-y.
40. Ishikawa M, Komi PV, Grey MJ, Lepola V, Bruggermann G-P. Muscle-tendon interaction and elastic energy usage in human walking. J Appl Physiol. 2005;99(2):603–8.
41. Kelly LA, Cresswell AG, Racinais S, Whiteley R, Lichtwark G. Intrinsic foot muscles have the capacity to control deformation of the longitudinal arch. J R Soc Interface. 2014;11:20131188.
42. Kelly LA, Lichtwark G, Cresswell AG. Active regulation of longitudinal arch compression and recoil during walking and running. J R Soc Interface. 2014;12:20141076.
43. Farris DJ, Kelly LA, Cresswell AG, Lichtwark GA. The functional importance of human foot muscles for bipedal locomotion. PNAS. 2019;116(5):1645–50.
44. Takahashi KZ, Worster K, Bruening DA. Energy neutral: the human foot and ankle subsections combine to produce near zero net mechanical work during walking. Sci Rep. 2017;7:15404.
45. Zelik KE, Honert EC. Ankle and foot power in gait analysis: implications for science, technology and clinical assessment. J Biomech. 2018;75:1–12.
46. Carlson RE, Fleming LL, Hutton WC. The biomechanical relationship between the tendoachilles, plantar fascia and metatarsophalangeal joint dorsiflexion angle. Foot Ankle Int. 2000;21:18–25.
47. Cheng H-YK. Finite element analysis of plantar fascia under stretch—the relative contribution of windlass mechanism and Achilles tendon force. J Biomech. 2008;41:1937–44.
48. Fessel G, Jacob HAC, Wyss C, Mittlmeier T, Müller-gerbl M, Büttner A. Changes in length of the plantar aponeurosis during the stance phase of gait—an in vivo dynamic fluoroscopic study. Ann Anat. 2014;196:471–8.
49. Gefen A. The in vivo elastic properties of the plantar fascia during the contact phase of walking. Foot Ankle Int. 2003;24:238–44.

50. Welte L, Kelly LA, Lichtwark GA, Rainbow MJ. Influence of the windlass mechanism on arch-spring mechanics during dynamic foot arch deformation. J R Soc Interface. 2018;15:20180270.
51. Melão L, Canella C, Weber M, Negrão P, Trudell D, Resnick D. Am J Roentgenol. 2009;193(3):662–71.

Human Walking: The Gait Cycle

2

Walking is convenient means for traveling short distances. In the absence of pathology, gait appears coordinated, efficient, and effortless.

– Jacqueline Perry

Key Points

- There are eight phases of the walking gait cycle.
- Understanding the terminology of the phases of gait is important when reviewing studies of the pathomechanics of foot pathologies.
- External moments created by ground reaction force drive the motion of skeletal segments, while muscles are activated to restrain that motion.
- Evaluation of walking gait in the clinical setting will reveal many of the underlying mechanisms contributing to lower extremity pathology.

Introduction

The primary function of the human foot is to support the body for upright, bipedal ambulation. Walking and running gait has been extensively studied over the past 50 years, and many of the pathologies discussed in this book are related to abnormal kinematics or kinetics during ambulation. For this reason, a chapter is devoted to the walking gait cycle to review the critical events which are commonly referred to in medical publications which discuss foot and ankle pathologies.

Critical to understanding normal and pathologic gait is recognizing the role of ground reaction forces (GRF) as well as the role of muscles and ligaments acting on specific joints of the lower extremity. Terminology has evolved over the years to

D. H. Richie Jr, *Pathomechanics of Common Foot Disorders*,
https://doi.org/10.1007/978-3-030-54201-6_2

describe the various phases of the walking gait cycle. Understanding this terminology and the important events of the gait cycle will allow better understanding of the mechanisms which cause various foot pathologies.

Kinetics

A force acting at a distance from the rotational axis of a joint is known as a *moment of force*. The terms *moment* and *torque* essentially describe moment of force, which is calculated by M = F x D, where M is the moment of force in newton-meters (N-m), F is the force in Newtons (N), and D is the distance in meters (m).

Two types of moments are described in the scientific literature. *Internal moments* are generated by structures within the body. Active internal moments are produced by muscular contraction, while passive internal moments are produced by bone and cartilage within the joint as well as ligaments surrounding a joint. *External moments* originate outside the body and are produced by gravitational forces or by ground reaction force (GRF).

A simple example of joint moments is provided by a description of the motion of the knee joint during walking. Shortly after heel strike during the walking gait cycle, the knee flexes due to an external flexion moment created by ground reaction forces passing posterior to the knee joint axis. This external flexion moment is resisted by an internal extension moment produced by contraction of the quadriceps muscle.

Internal joint moments can also be produced by ligaments which develop tension during the translation of certain skeletal segments. A patient with a genu valgum deformity will develop abnormal *external* knee abduction moment as GRF passes lateral to the knee joint. This external knee abduction moment is resisted by the medial collateral ligaments of the knee which develop tension and thus exert an *internal* knee adduction moment. For clinicians, appreciating internal moment of force produced by bones, cartilage, ligaments, and tendons, in response to external joint moments, is critical to understanding the pathomechanics of injury.

Not only understanding how soft tissue structures produce internal joint moments, but also the timing and type of muscle contraction acting on joints have a critical role in pathomechanics of injury. Muscles can contract and produce active tension while shortening, which is termed a *concentric contraction*. Conversely, a muscle can produce active tension while it is being lengthened by joint movement which is termed an *eccentric contraction*. Muscles create more power and are more vulnerable to injury during an eccentric contraction than during a concentric contraction.

Muscle contraction can impart "stiffness" to a joint or skeletal segment. However, muscles themselves cannot store energy or have elastic recoil, while tendons can. A good example of muscular stiffening and tendon recoil is the action of the gastrocnemius and soleus during the midstance and terminal stance phases of the gait cycle. These muscles contract eccentrically to restrain ankle joint dorsiflexion during the stance phase of gait. Elongation of the Achilles tendon during midstance stores elastic energy which then recoils rapidly in pre-swing, providing push off or propulsion. Push off is not due to active concentric contraction of the calf muscles, but rather due to elastic recoil of the Achilles tendon.

As we study the phases of the gait cycle, it will become apparent that the primary stabilization of the joints of the lower extremity is provided by eccentric contraction of muscles. The essence of human walking is characterized by muscular restraint of joints during a forward fall. Specifically, the forward fall is controlled by eccentric muscular contraction rather than by active push off created by concentric contraction. Instability of joints results more often from failure of eccentric muscular stabilization rather than concentric muscle contraction.

The gait cycle was originally broken down into three phases of the stance phase, contact, midstance, and propulsion, while the swing phase had no subdivisions [1, 2]. Now, with the development of gait labs using high-speed cinematography, it became clear that distinct events occur during both stance and swing requiring differentiation into eight subdivisions or gait phases [3] (Fig. 2.1). These phases accomplish three tasks: weight acceptance, single limb support, and swing limb advancement. An understanding of the gait phases is important when interpreting research relevant to the pathomechanics of foot and ankle problems.

Many clinicians and researchers refer to a specific "rocker" phase of the gait cycle. The rockers describe anatomic motions during a specific phase of the stance period of the gait cycle (Fig. 2.2). The rockers allow progression of the stance phase limb over the supportive foot. Referring to a rocker has some value when discussing pathomechanics

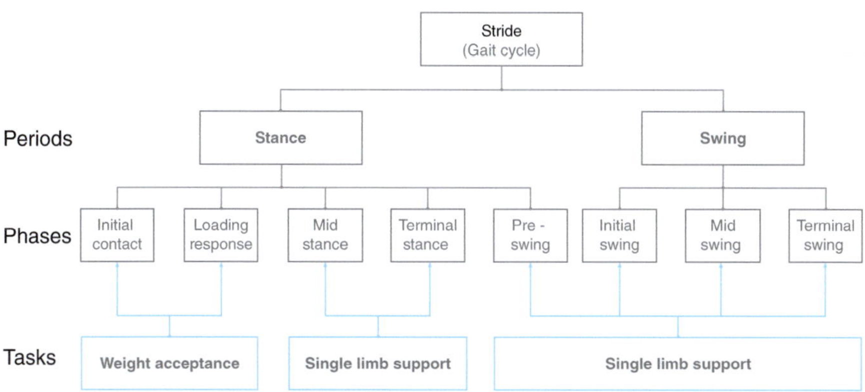

Fig. 2.1 Divisions of the gait cycle

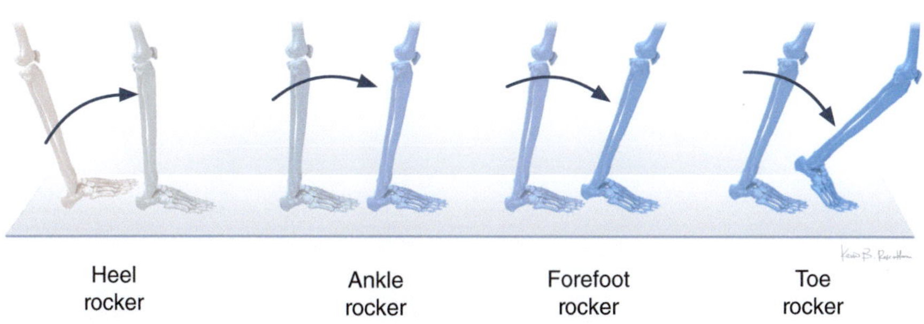

Fig. 2.2 The rockers: progression over the supporting foot

because it emphasizes a segment of the foot where motion is primarily occurring. Conversely, the concept of a rocker infers that the foot is primarily rotating as a rigid body and ignores other substantial movement which is occurring at multiple joints of the foot during the entire stance phase of gait. This will be discussed in detail in Chap. 3.

Key Events in the Walking Gait Cycle

A description of each phase of walking will focus on the location of the ground reaction force (GRF) vector which generates external moment at three key joints of the lower extremity: the hip, knee, and ankle. Within the foot, a more detailed description of kinematics will be provided in Chap. 3. The external moments created by GRF cause rotational moments at the major joints of the lower extremity which are counteracted by muscular contraction. Muscle contractile force provides stability via controlled deceleration and acceleration. Neuromuscular control is provided by the sensorimotor system which will also be further studied in Chap. 3.

Phase 1 Initial Contact 0– 2% of the Gait Cycle (Fig. 2.3)

Initial contact is an important descriptive term because it includes those people who strike the ground with their heel, i.e., heel striker, versus those who might strike

Fig. 2.3 Initial contact

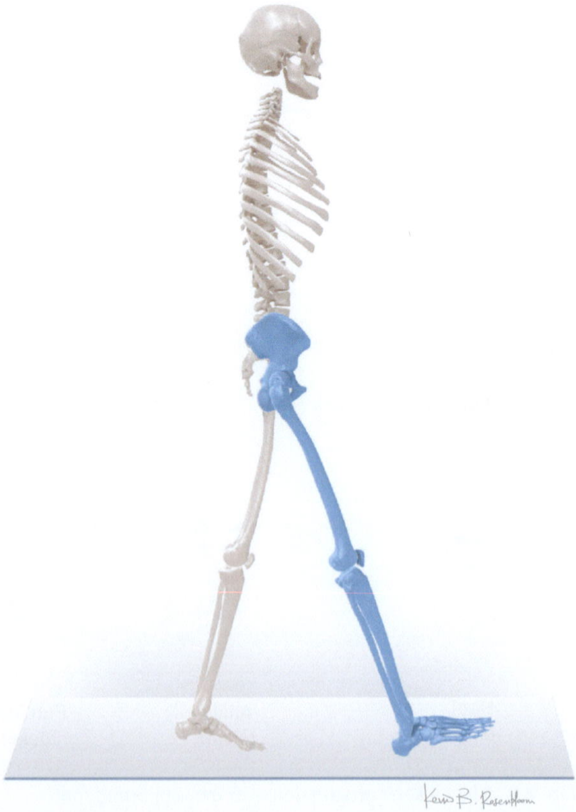

with their forefoot or midfoot. In order to heel strike, there must be adequate extension of the hip, and there must be adequate range of motion of the ankle to allow dorsiflexion of the foot to a position which is at least 90° to the leg.

At initial contact, the knee is extended and the ankle is dorsiflexed. The gastrocnemius is at maximal length. A tight gastrocnemius would limit extension of the both the hip and knee, and the foot may be unable to dorsiflex at the ankle. Thus, a shortened stride along with a midfoot or forefoot strike may be observed as compensation for a tight gastrocnemius at initial contact. A shortened stride can also be the result of restricted range of motion at the hip.

At initial contact, GRF is directed anterior to the hip, creating flexion moment. Also, GRF is directed anterior to the knee, creating external extension moment. Muscle activity prior to foot strike is primarily located in the quadriceps, providing knee extension, as well as with the tibialis anterior which is dorsiflexing the foot.

Phase 2 Loading Response 2–12% of the Gait Cycle (Fig. 2.4)

Loading response is characterized by controlled plantarflexion of the ankle as the foot rolls forward over the heel bringing the forefoot to the ground. This event is

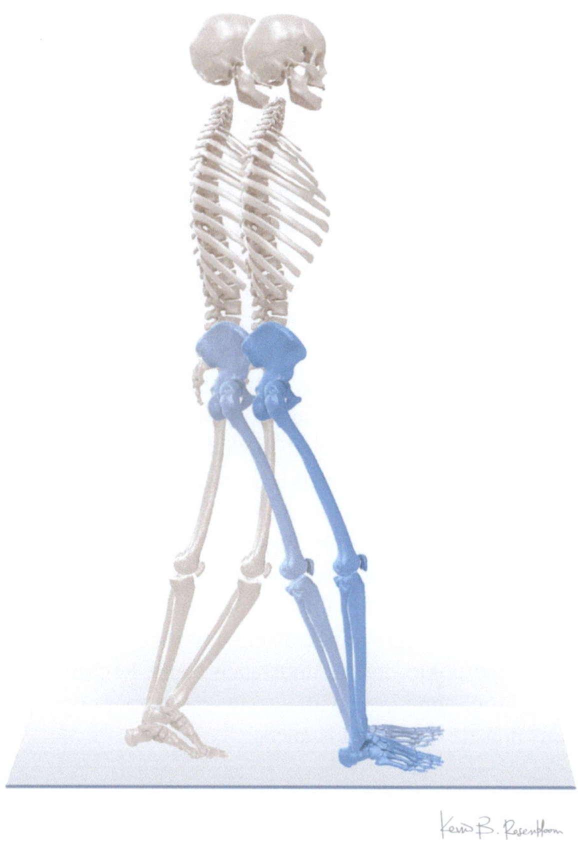

Fig. 2.4 Loading response

known as the heel rocker or the "first rocker." The body is falling forward over the rounded surface of the calcaneus, and GRF is suddenly shifted posterior to the knee but remains anterior to the hip. Thus, GRF creates external flexion moment at the hip. Yet, the hip begins moving into extension during loading response and throughout the remainder of the stance phase. Hip extension occurs because concentric contraction of the hamstrings and gluteal muscle groups overpowers the external hip flexion moment from GRF vector. As a result, the hip begins extending and continues to do so until just before toe-off.

During loading response, the GRF moves posterior to the knee, moving the knee into flexion. The subtalar joint is moving in the direction of pronation which is coupled with internal tibial rotation and flexion of the knee joint. Internal rotation of the tibia is required to "unlock" the knee from full extension. Loss of coupling between the foot and leg has been observed in some pathologies such as adult acquired flatfoot deformity as discussed in Chap. 7.

Plantarflexion at the ankle is resisted or decelerated by the leg extensors: the tibialis anterior (TA), the extensor hallucis longus (EHL), and the extensor digitorum longus (EDL) which all contract eccentrically as they are lengthened.

Phase 3 Midstance 12– 31% of the Gait Cycle (Fig. 2.5)

Loading response ends when the forefoot contacts the ground. This event also corresponds with toe-off of the opposite limb. Thus, loading response ends at the same time that double support ends. Midstance marks the beginning of single support and is one of the most critical phases of the walking gait cycle.

During midstance, the body is rotating like an "inverted pendulum" over a single supporting foot which is fixed on the ground. The sensorimotor system is critical to maintaining balance and postural control during this precarious support phase. The motion of the tibia over the foot at the ankle joint during midstance is known as the "ankle rocker" or the second rocker of the stance phase of gait. Clinicians also describe this ankle rocker motion as "tibial progression" over the fixed foot which can be restrained or inhibited by contracture of the gastrocnemius muscle. Tibial progression requires a stiff, stable midfoot.

With loss of weight bearing of the contralateral limb, GRF affecting the single supporting limb is directed medial to the hip joint, creating external adduction moment. The trunk and pelvis are stabilized over the supportive limb by contraction of the ipsilateral (stance limb) hip abductors. Weakness of these muscles will shorten the single support phase.

The knee changes direction during midstance, moving from flexion into extension. This is driven by GRF which moves anterior to the knee joint. GRF is still anterior to the ankle joint causing dorsiflexion. Extension of the knee and dorsiflexion of the ankle during midstance cause lengthening of the gastrocnemius. This elongation, combined with eccentric contraction increases strain in the gastrocnemius muscle during midstance.

Fig. 2.5 Mid stance

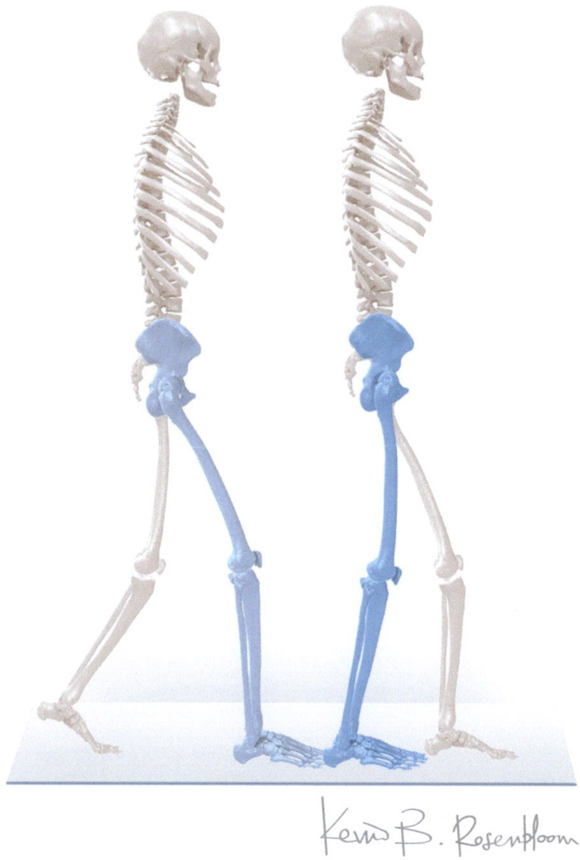

Forward momentum causes passive dorsiflexion at the ankle joint which is restrained by eccentric contraction of the triceps surae. Tension develops in the posterior calf musculature which creates internal plantarflexion moment at the ankle. GRF at the forefoot and plantarflexion moment at the ankle increase plantar pressures in the forefoot during late midstance.

The soleus is primarily responsible for restraining tibial advancement or ankle joint dorsiflexion during midstance. As tibial advancement is decelerated, the forward fall of the body moves the femur over the restrained tibia, and the knee moves into extension during the second half of midstance. Therefore, a stable tibia is essential for knee extension during midstance. Weakness of the soleus will cause the knee to remain in flexion, resulting in a "drop down" or crouched compensation of the entire body during midstance.

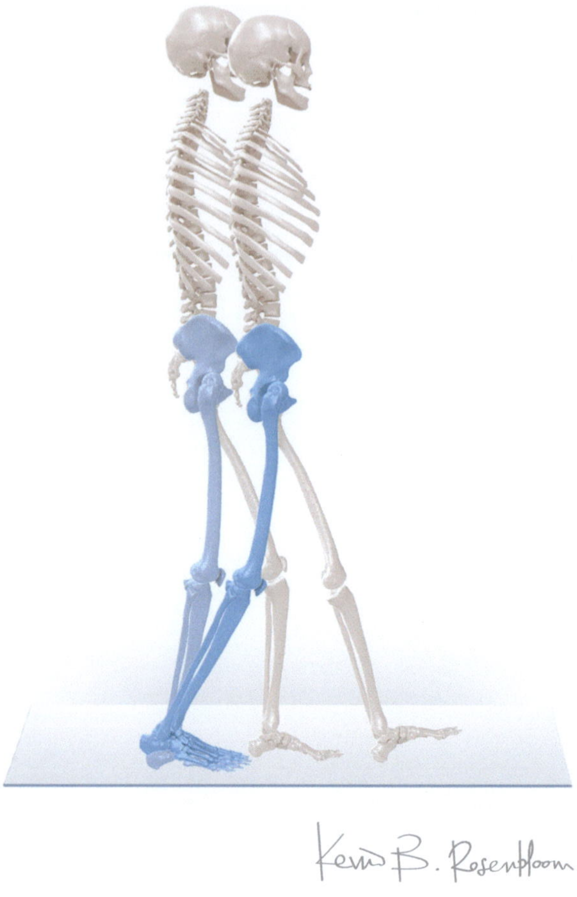

Fig. 2.6 Terminal stance

Phase 4 Terminal Stance 31–50% of the Gait Cycle (Fig. 2.6)
Terminal stance is commonly referred to as the *heel rise* phase of the walking gait cycle. Up until this phase, the rockers have relied on a rounded anatomic surface upon which the foot pivots over, i.e., the calcaneus and the dome of the talus. The round contour of the metatarsal heads provides the third rocker or the "forefoot rocker" during terminal stance.

The timing of heel rise varies among individuals depending on the stability of the midfoot joints and available range of motion of the ankle joint. Timing of heel rise also varies according to walking speed. Terminal stance begins when the swing phase limb passes by the stance phase limb and ends when the swing phase limb strikes the ground. *Thus, heel rise should not occur until the swing phase limb has reached or passed forward of the stance phase limb.*

Heel rise also marks the initiation of subtalar joint supination which occurs at the beginning of terminal stance. In terms of timing, this important event occurs at 31% of the total gait cycle and 60% of the stance phase of gait (see Fig. 3.1, Chap. 3). Clinically, inversion of the rearfoot will not be observed until the swing phase limb has passed beyond the stance phase limb. Pelvic rotation to advance the swing phase limb will externally rotate the tibio-fibular-talar unit of the stance phase limb to supinate the hindfoot.

The hip reaches full extension at the end of terminal stance, when the contralateral swing phase foot strikes the ground. The hip abductors remain active during terminal stance to support the pelvis and trunk.

Heel rise during early terminal stance is not accompanied by ankle joint plantarflexion. In fact, the ankle joint is at maximum dorsiflexion during initial heel rise. While this may appear confounding, the heel will rise from the ground due to motion at joints proximal to the ankle. This is explained by the following mechanism:

Knee flexion tilts the tibia anterior, causing the heel to rise from the ground during terminal stance. Knee flexion results from the release of powerful eccentric contraction of the gastrocnemius which has been building during midstance. This contraction overpowers external extension moment which has been directed at the knee joint during midstance. Both the gastrocnemius and soleus reverse to a concentric contraction as the ankle joint begins plantarflexing-halfway thru terminal stance, or 45% of the gait cycle. Therefore, the initial heel rise which initiates terminal stance is brought about by knee flexion, not by ankle plantarflexion. Ankle plantarflexion does not occur until later in terminal stance.

In summary, the calf musculature undergoes an eccentric contraction to decelerate ankle joint dorsiflexion during midstance and early terminal stance. Halfway into terminal stance, the calf muscles contract concentrically which generates push off power. This also corresponds to the timing of peak plantar pressure under the forefoot during this phase of gait (Fig. 2.7).

Phase 5 Pre-swing 50–62% of the Gait Cycle

Pre-swing begins another double support period when the opposite foot has made contact with the ground, and this phase ends when the previous single supportive foot has completely left the ground.

The final fourth rocker or "toe rocker" occurs during pre-swing which is actually an advancement of the foot over the first metatarsophalangeal joint. Thus, it should be termed the "great toe rocker." The great toe rocker has also been described as "high-gear push off" by Bojsen-Moeller as opposed to "low-gear push off" over the lateral lesser metatarsophalangeal joints.

The concept of "push off" during walking gait has been debated for many years by researchers. The terminal stance and pre-swing phases of gait were previously combined and called the "propulsive" period of gait or the "push off" phase [4]. This description was later challenged by Jacqueline Perry who asserted that the peak of GRF at the forefoot during this period was not due to active tension from the triceps providing a downward thrust, but rather due to body position exerting increased center of pressure on the forefoot [3]. This is verified by the lack of EMG activity in the calf musculature during pre-swing. Perry proposes that the gastroc-soleus contractile force "locks" the ankle in terminal stance so that the limb and foot rotate on the forefoot rocker.

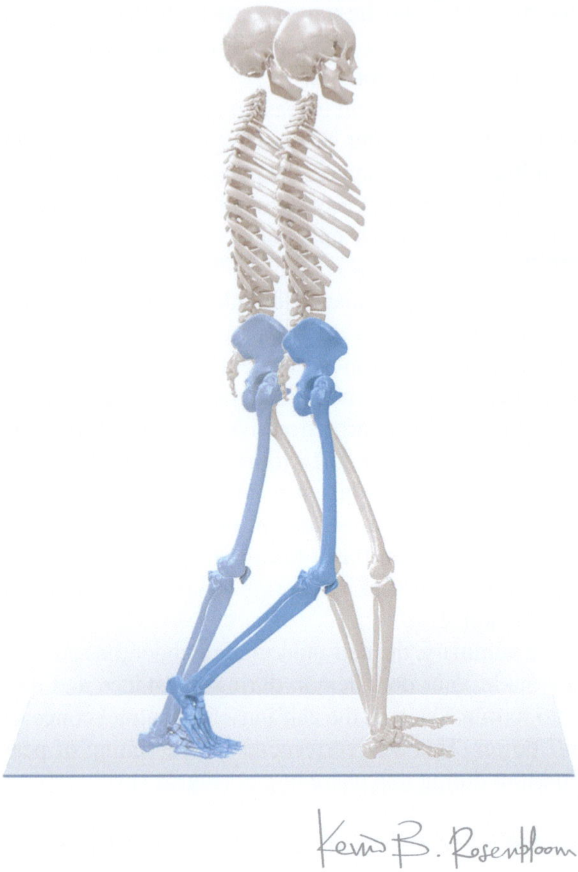

Fig. 2.7 Pre swing

Later, other researchers verified Perry's assertion that push off power during terminal stance and pre-swing results from elastic recoil of soft tissue structures rather than active contractile muscle activity [5, 6]. The spring mechanism of the human foot is discussed in detail in Chap. 1.

Roll off occurs during the forefoot rocker, or terminal stance. The ankle is not plantarflexing in early terminal stance, but the heel rises due to knee flexion. The ankle is stiff due to contractile activity of the triceps surae resisting ground reaction forces against the forefoot during heel rise.

Push off occurs in pre-swing. The description of a "push off" phase of gait will be used synonymously with "pre-swing" in this book. In this phase, abrupt cessation of contraction of the gastroc-soleus causes elastic recoil of the Achilles, generating significant plantarflexion power across the ankle [3]. Thus, an actual propulsive event occurs. The ankle will be observed to rapidly plantarflex during pre-swing, at 50% of the gait cycle.

The segment of the foot between the ankle and the metatarsal heads must be a rigid lever to support the body mass which is now forward positioned on the foot and providing a significant dorsiflexion force across the midfoot and ankle joints. The plantarflexion force of the calf during heel rise cannot be effective if the midfoot joints are not stable during this critical phase of gait. Excessive motion across the midfoot joints will inhibit heel rise.

Stability of the midfoot joints during terminal stance and pre-swing is provided by several dynamic and passive mechanisms. These mechanisms are specialized in the human foot to provide resistance to the "midfoot break" seen in other primates. This topic is discussed in detail later in Chap. 1, but the primary contribution to midfoot stability is the combination of dynamic muscle activity and elastic recoil of specific ligamentous structures.

Subtalar joint motion accelerates in the direction of supination during terminal stance and pre-swing [2]. Maximal supination of the rearfoot averages 7° and is achieved at the end of pre-swing, or at 62% of the gait cycle [1]. Supination of the rearfoot has a stiffening effect across the midfoot joints. Research has shown that stiffening of the foot during terminal stance and pre-swing is not due to an osseous "locking mechanism" but rather due to tensioning of key ligaments. Some of the stiffening of ligaments across the midfoot joints is facilitated by subtalar joint supination. This topic is discussed in Chap. 1 and in Chap. 3.

Efficient gait requires stability or stiffness of the metatarsophalangeal joints. As the center of pressure moves forward during terminal stance, dorsiflexion moment increases across the MTP joints. Controlled mobility of these joints is critical to progression thru the forefoot rocker. Proper functioning of the toe flexors, both extrinsic and intrinsic, as well as proper tension from the plantar aponeurosis controls or restrains dorsiflexion of the digits on the metatarsal heads. Excessive transverse plane and frontal plane motion of the first metatarsal compromises the plantarflexors of the hallux in HAV deformity. This loss of purchase by the hallux leads to overload of the lesser metatarsals during terminal stance and pre-swing and is a common cause of metatarsalgia. This will be discussed in Chaps. 5 and 7.

During pre-swing, the hip begins to flex, the knee is already flexing, and the ankle joint is already plantarflexing which started halfway thru terminal stance. Flexion of the knee and hip literally "pulls" the foot off the ground during pre-swing. Arthritic conditions of the knee and hip can give the false impression of reduced ankle joint motion as the foot appears to lack push off or propulsion.

Powerful concentric contraction of the triceps ceases at the end of terminal stance, allowing the release of stored elastic energy which causes the ankle to plantarflex during pre-swing. The foot rotates on the metatarsal phalangeal joints which are moving into extension which tightens the plantar aponeurosis. This engages a "windlass mechanism," whereby the plantar aponeurosis winds around the metatarsal heads like a winch, effectively shortening and stiffening the fascia to resist bending dorsiflexion moments across the midfoot joints and metatarsals themselves (see Fig. 1.19, Chap. 1). There is also an elastic recoil of the plantar aponeurosis as the heel is lifted from the substrate (see Fig. 1.26, Chap. 1).

Fig. 2.8 Initial swing

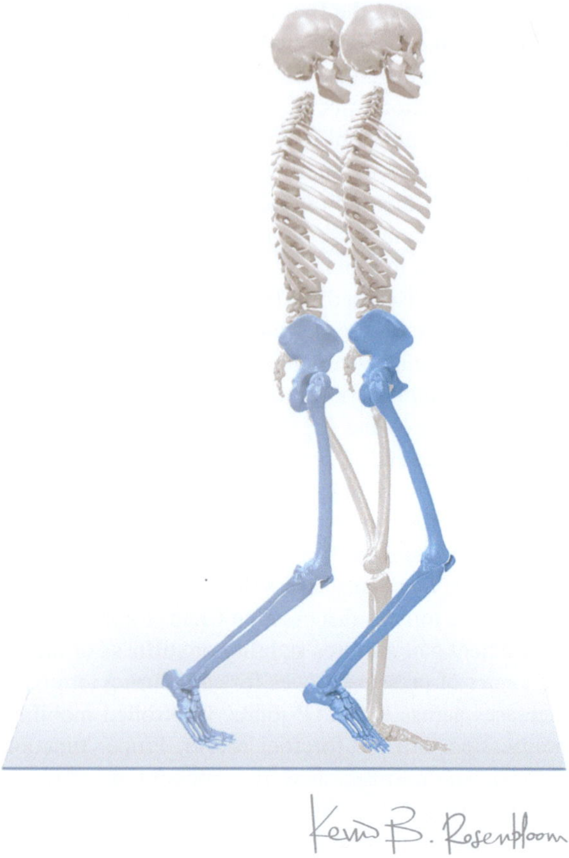

The subtalar joint continues to move in the direction of supination which is coupled with external rotation of the tibia. The stability of the foot during this critical phase of the gait cycle is dependent upon several mechanisms including stiffening from elastic recoil of the plantar aponeurosis, contraction of the plantar intrinsics, elastic recoil of the tibialis posterior, and favorable tensioning of the plantar ligaments from a supinated position of the subtalar joint.

Phase 6 Initial Swing 62–75% of the Gait Cycle Events (Fig. 2.8)
Also known as toe-off, this marks the end of the stance phase of gait which occurs at 60% of the gait cycle. It also marks the beginning of the swing phase of gait.

The hip continues to flex due to contraction of the rectus femoris and the adductor magnus which has been active during pre-swing. Ground reaction force has reduced to almost zero at toe-off. Hip flexion causes further knee flexion.

The "ankle" reaches maximum plantarflexion, approximately 30° at toe-off. In reality, the ankle and multiple midfoot joints plantarflex during pre-swing and initial

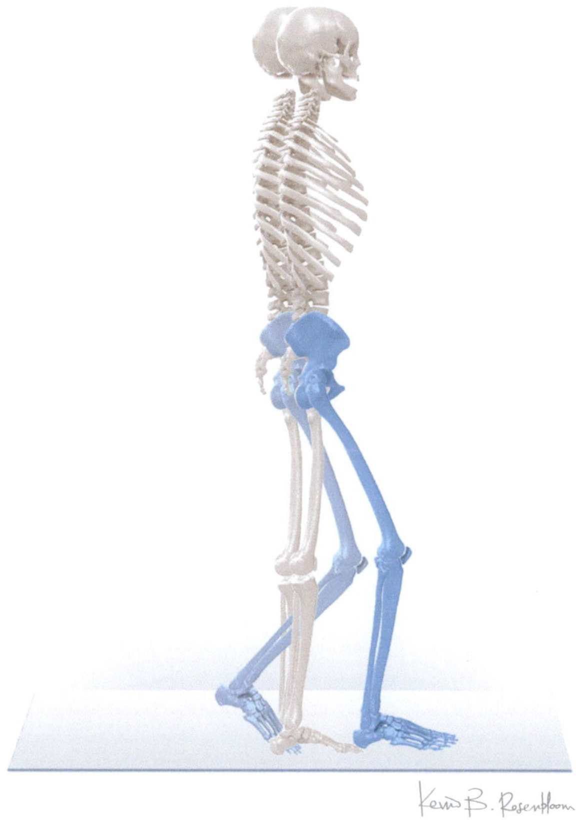

Fig. 2.9 Mid swing

swing. Plantarflexion of the midfoot joints in humans is in the range of 10° during pre-swing and initial swing, providing a high level of mobility which is not seen in other primates. This is discussed in Chap. 1.

The triceps surae has ceased contracting before initial swing, and the tibialis anterior becomes the major lower leg muscle reversing ankle plantarflexion and initiating dorsiflexion which will continue thru swing.

Phase 7 Mid-swing 75–87% of the Gait Cycle (Fig. 2.9)
This phase begins when the swinging foot becomes even with, or adjacent to the opposite stance foot and ends when the swing limb moves forward and the tibia becomes vertical to the ground. The toes have their closest clearance to the ground when the feet are adjacent during swing.

The hip reaches 20° of flexion aided by the iliopsoas contraction. There is very little muscle contraction around the knee during this phase. The knee is carried passively into flexion by hip flexion with the foot lagging behind. The tibia assumes a vertical alignment because the hip flexion and knee flexion magnitudes

Fig. 2.10 Terminal swing

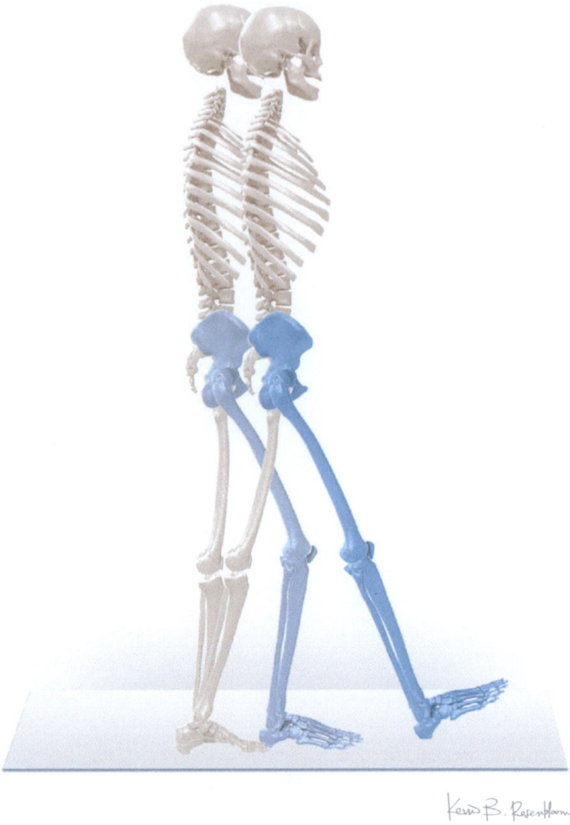

are equal. The ankle has been moving into dorsiflexion after toe-off due to contraction of the tibialis anterior. The subtalar joint remains slightly supinated during swing, primarily due to the tibialis anterior.

Phase 8 Terminal Swing 87–100% of the Gait Cycle (Fig. 2.10)

The tibia moves from a vertical position to the ground to a forward position as the knee extends and the foot "reaches" forward. Thus, the foot is ahead of the knee joint at foot strike during walking. In running, the foot remains directly beneath the knee.

The hip achieves maximum flexion as the swing phase limb reaches forward during terminal swing. The knee moves into extension due to inertia of the swinging shank of the lower leg. The hamstrings contract eccentrically to decelerate this motion. Contracture of the hamstrings or the gastrocnemius will inhibit full knee extension during terminal swing. The tibialis anterior continues to contract, holding the ankle somewhere around neutral or 90-degree foot-to-leg alignment at foot strike.

Common Compensatory Changes Observed in Gait

Gait evaluation will give insight into the contributing mechanisms which cause most of the pathologies discussed in this book. In each chapter, the pathomechanics of the condition will be studied relevant to specific events during the walking gait cycle. Clinicians should be able to conduct a brief gait evaluation to look for certain compensation or movement of a skeletal segment which will help with the diagnosis and with the formulation of a treatment plan. Many of the studies cited in this book giving insight into the pathomechanics of common foot problems were performed in sophisticated gait labs with 3-D motion capture, EMG, and plantar pressure measurement. While these resources are not readily available to the average clinician, much information can be gained from a cursory gait evaluation of the patient walking up and down a hallway or across the floor of the clinic.

Besides detecting overall impairment, an experienced clinician can observe motion and alignment of key skeletal segments of the lower extremity during gait evaluation. Watching a patient walk can verify whether pain is causing compensation in gait and whether deformity might be either a cause or the result of another underlying pathology.

Upper body posturing can suggest a neurologic deficit or a balance disorder. A "hip strategy" will be used by a balance - impaired patient during single support as the upper body sways to maintain postural control. Limb length discrepancy can be detected by uneven pelvic alignment or heel off which might appear asymmetrical. Congenital or acquired frontal plane deformity of the knee can have profound effects on foot posture and must be recognized before planning reconstructive foot and ankle surgery or when prescribing custom orthotic devices.

Contracture of the gastrocnemius is believed to cause many foot and ankle pathologies and can often be detected with a cursory gait evaluation. Traditionally, clinicians will look for an "early heel off" which would actually be an observation of heel rise occurring before the opposite limb reaches a side-by-side relationship with the stance limb. Sometimes a gastrocnemius contracture will cause the midfoot to compensate with excessive sagittal plane motion, such that an early heel off does not occur (Fig. 2.11). However, these patients may demonstrate a midfoot strike rather than heel strike during touchdown which can be a sign of equinus deformity.

These observations are best made when viewing the patient from both anterior and posterior perspectives as they walk away from and back to the observer. Further information can be obtained from a medial/lateral or side view of the patient walking:

1. At initial contact, a tight gastrocnemius will cause diminished ankle joint dorsiflexion and restriction of full knee extension. In the "normal" subject, the ankle should be dorsiflexed to a neutral or 90-degree position on the leg, and the knee should be at full extension at heel strike.

2. Contracture of the gastrocnemius will limit tibial progression during the second rocker. This will cause an early heel off seen before the swing phase foot passes by the stance phase foot. This heel rise is initiated by knee flexion, not ankle plantarflexion. Early knee flexion is often caused by contracture of the gastrocnemius crossing posterior to the axis of the knee joint.
3. However, tibial progression can also be misleading as it can occur without a "rocker" occurring at the ankle joint. In adult acquired flatfoot deformity, there is often an unstable midfoot which may undergo excessive sagittal plane motion. This condition will reveal a "midfoot break" observed from a side view of the patient. There will be a delay in heel rise as the entire tibia and rearfoot plantarflex across the midtarsal joint. Until the midfoot stiffens and sagittal plane motion at the midtarsal joint stops, the heel will not rise off the ground. Thus, a delay in heel rise occurs in most flatfoot conditions even though there may be gastrocnemius contracture.
4. Heel rise marks the beginning of the terminal stance phase, and this event should occur after the swing phase foot has passed by the stance phase foot. Heel rise also begins the "third rocker" as the tibia stops moving over the foot, and the entire foot plantarflexes across the metatarsophalangeal joints (MTPJs). At this point, the foot should have achieved maximal stiffness due to tensioning of extrinsic and intrinsic muscles. This enables the tibia and foot to move together as one rigid body over the metatarsal heads and toes. Ideally, the third rocker demonstrates motion across the transverse axis of the first and second MTPJs. There are some pathologies which cause lack of push off thru this transverse axis. Patients will appear to roll off laterally across the forefoot (see Fig. 1.19, Chap. 1).
5. Pre-swing is the most important phase of the gait cycle in terms of foot stability and generation of power. Push off during pre-swing should be directed thru the hallux. Dorsiflexion of the first metatarsophalangeal joint should achieve 40° while engaging the windlass mechanism. Proper functioning of this mechanism should be evidenced by rear foot supination coupled with external rotation of the tibia.
6. Loss of motion across the first MTPJ with full engagement of the windlass may be difficult to measure with direct observation. However, looking at the hip and opposite foot from a side view can help detect lack of efficient push off. Incomplete or premature ending of the third rocker will stop full extension of the hip at the end of terminal stance. Normally, the hip continues full extension thru heel rise until heel strike of the opposite foot. Hallux rigidus will stop hip extension over the stance phase foot prematurely. Lack of push off across the transverse axis may be a result of midfoot instability. When viewed from the side, the patient may demonstrate excessive midfoot motion or "midfoot break" during late midstance and terminal stance (Fig. 2.11). This finding is also seen as a compensation for gastrocnemius contracture with the midfoot joints providing sagittal plane motion for tibial progression when ankle joint dorsiflexion is blocked.

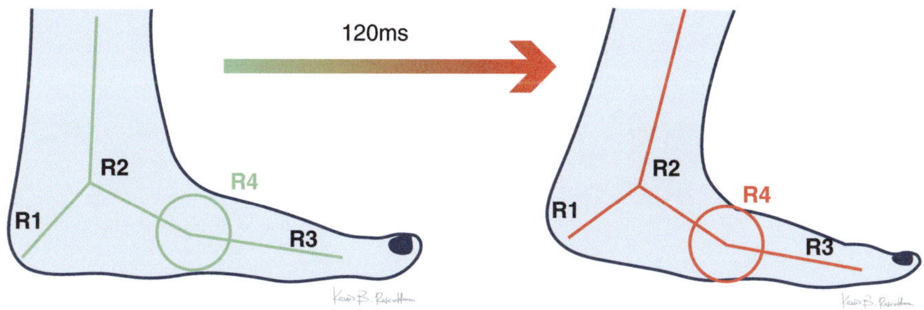

Fig. 2.11 The midfoot break during late midstance
At beginning of terminal stance, many humans will exhibit sagittal plane motion across the midtarsal joint (R4). (From Amis [7])

References

1. Root, M.L., Orien, W.P., and Weed, J.H.: Clinical biomechanics: Normal and abnormal function of the foot. Vd. 2. Los Angeles, Clinical Biomechanics Corp., 1977, pp. 138–139.
2. Inman VT, Ralston HJ, Todd FN. Human walking. Baltimore: Williams and Wilkins Company; 1981.
3. Perry J. Gait analysis. Normal and pathological function. 2nd ed. Thorofare: SLACK; 2010.
4. Cavagna GA, Margaria R. Mechanics of walking. J Appl Phys. 1966;21:271–8.
5. Winter DA. Biomechanical motor patterns in normal walking. J Mot Behav. 1983;15:303–30.
6. Winter DA. Energy generation and absorption at the ankle and knee during fast, natural, and slow cadences. Clin Orthop. 1983;(175):147–54.
7. Amis J. The split second effect: the mechanism of how equinus can damage the human foot and ankle. Front Surg. 2016;3:38. https://doi.org/10.3389/fsurg.2016.00038.

Motion of the Foot: Joints, Muscles, and Sensorimotor Control

3

The human foot is a masterpiece of engineering and a work of art.

–Leonardo da Vinci

Key Points

1. Kinematic studies show that many joints in the foot move during the stance phase of gait in multiple planes. Some joints move more than previously recognized:
   - The talonavicular joint and the calcaneocuboid joints move more than the subtalar joint.
   - The midfoot joints contribute more sagittal plane motion than the ankle.
   - The medial column moves more than the ankle.
   - The lateral metatarsals move more than the medial metatarsals.
   - The navicular-cuneiform joints move more than the midtarsal joint.
   - The ankle is not a pure mitered hinge joint. It moves in the transverse and frontal planes.
   - There is more frontal plane motion in the ankle than the subtalar joint.
2. The torque producing capacity of a muscle is dependent upon cross-sectional area and moment arm.
3. Moment arms for the lower extremity muscles change as foot position changes.
4. The tibialis posterior has a longer moment arm for transverse plane adduction of the foot than for inversion of the foot.

D. H. Richie Jr, *Pathomechanics of Common Foot Disorders*,
https://doi.org/10.1007/978-3-030-54201-6_3

5. The plantar intrinsic muscles along with the plantar aponeurosis and the plantar ligaments store and release elastic energy in response to deformation of the longitudinal arch of the foot.
6. Healthy feet display considerable compliance and variability of motion across multiple joints. There is no distinct pattern of motion which defines a "normal" foot.
7. The sensorimotor system has ultimate control over motion of the skeletal segments of the lower extremity during ambulation. The timing of muscular contraction, the force of contraction, the activation of antagonist muscles, and the pre-activation of muscles, all of which contribute to stiffness across joints creates multiple options for wide variation among the general population.

Introduction

The human foot has often been described as an engineering marvel converting from a "mobile adapter" during the first half of stance to a "rigid lever" during the second half. This dichotomous behavior has been considered to be the singular and most differentiating feature of the human foot compared to all other primates and living creatures. However, recent research studying the kinematics of the human foot has dispelled the notion that the human foot actually becomes "rigid" during the stance phase of gait. The notion of "rigidity" of the human foot and the myth of midtarsal joint "locking" has been explored in Chap. 1 and will be further studied in this chapter. Thanks to modern technology, we now have the ability to measure motion of all the bones of the human foot as well as the physical forces acting upon the joints of the foot to enable bipedal ambulation.

General Motion of the Foot Segments

The concept of a dichotomous behavior of the human foot during walking gait was proposed by Root et al. in 1977 [1]. They based their assumptions on their original cadaver studies as well as clinical observations of patients [2]. Root et al. also relied on previous published work of many authors including Close, Inman, and Wright who were the original pioneers studying lower extremity kinematics when they developed a gait lab at the University of California in the 1960s [3–5]. Although Root et al. did perform cadaver studies, their own gait studies were limited to estimation of segmental motion based upon gross visualization of bone landmarks in real time. Root et al. basically observed that the rearfoot, i.e., the subtalar joint, moved in the direction of eversion from heel strike to 30% of the stance phase of gait. This rearfoot motion was theorized to "unlock" the midtarsal joint according the previous observations of Elftman and Manter [6]. Unlocking of the midtarsal joint would enable adaptation to irregular surfaces while also assisting in shock absorption. Also, subtalar joint pronation is coupled with internal rotation of the tibia which is necessary to "unlock" the knee from full extension and allowing flexion after foot touchdown.

At 30% of the stance phase of gait, Root et al. proposed that the subtalar joint begins inverting and would achieve a "neutral position" by heel rise or terminal stance which occurs at 60% of the stance phase [1]. Continued supination of the rearfoot would "lock" the midtarsal joint converting the foot into a "rigid lever" for propulsion during push off. Critical to this model of normal foot function was the notion that a pronated or everted hindfoot position carried beyond 30% of the stance phase of gait would be considered abnormal.

Root et al. relied on simple visual observation of the feet of many subjects and drew conclusions about how and when the joints of the foot moved during gait. Much of the focus was on the rearfoot where alignment of the posterior aspect of the calcaneus relative to the lower leg could be marked with a pen and then visualized, while a patient walked away from the observer. Clearly lacking in the 1970s was technology which could more accurately measure position and motion of bone segments of the foot.

In the 1980s, new technology was developed in gait labs around the United States which would allow accurate tracking of skeletal segments using external markers and high-speed cameras. The development of this new technology and surge of biomechanics research was largely funded by running shoe companies, and the insights gained from this research changed much of what we had previously learned about function of the foot during walking and running.

In the early 1990s, several studies were published which refuted some of the observations of Root et al. regarding motion of the foot during walking gait. Specifically, these studies using 2-D motion tracking of the rearfoot angle to the lower leg disputed whether the rearfoot begins re-supination as early in the stance phase of gait as proposed by Root et al. [7–9].

Specifically, McPoil and Cornwall showed that the rearfoot pronated after heel strike and then remained in a pronated position until the end of midstance [9]. In this study, the rearfoot did not begin inverting or supinating until 60% of the stance phase which is the beginning of terminal stance and did not reach a "neutral" position until the start of swing phase. The pronated position of the rearfoot during the first 60% of the stance phase was in the same alignment as seen in the relaxed stance position of the subjects. This observation of minimal rearfoot motion during the stance phase of gait was verified by several other studies using the same 2-D motion tracking methodology [10–12]. However, these early kinematic studies used technology which was limited to evaluating frontal plane motion of the rearfoot only, during the first 60% of the gait cycle.

In 1999, multisegment foot models using three-dimensional stereophotogrammetric analysis were developed which enabled measurement of motion between three segments: the leg, the rearfoot, and the forefoot [13, 14]. Early studies using 3-D multisegment foot models demonstrated significant movement of the midfoot joints which exceeded the total motion of the subtalar joint in human subjects [14–16]. Also, studies of human subjects using (3-D) multisegment foot models demonstrated that within the foot, more motion occurs in the sagittal plane than the frontal or transverse plane in gait studies of human subjects [17, 18].

3-D multisegment foot model studies enable tracking of skeletal segments throughout all phases of the walking gait cycle. Previous insight into frontal plane rearfoot motion from 2-D studies was verified with 3-D studies showing approximately 5 degrees of calcaneal eversion at 60% of the stance phase. The rearfoot is then observed to move to a 6° inverted position at toe-off, although there is wide variation among individuals (Figs. 3.1 and 3.2). Overall frontal plane motion of the rearfoot during the walking gait cycle with skin-mounted markers is approximately 11°. This amount of motion is hardly detectable with the human eye.

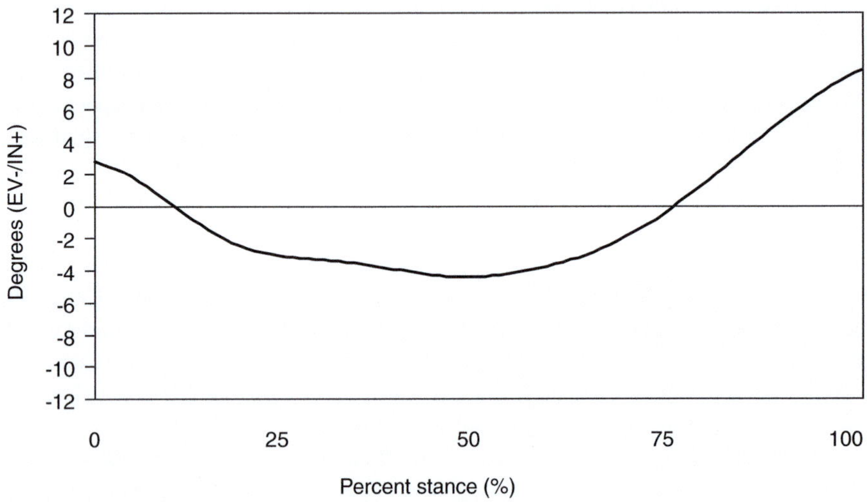

Fig. 3.1 Studies of healthy individuals show that the rearfoot pronates until 60% of the stance phase of gait and then supinates thru pre-swing

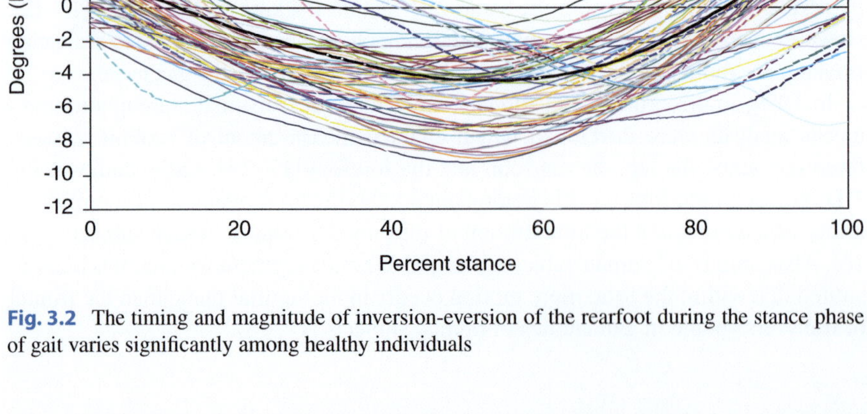

Fig. 3.2 The timing and magnitude of inversion-eversion of the rearfoot during the stance phase of gait varies significantly among healthy individuals

Multisegment foot models demonstrated that significant motion occurs between the forefoot and the rearfoot during the stance phase of gait. In particular, studies of patients with rheumatoid arthritis and posterior tibial tendon dysfunction (PTTD) showed that the forefoot supinated as the rearfoot pronated, indicating an "untwisting motion" as previously described by MacConnail [19, 20].

The significant motion of joints distal to the subtalar joint during walking disputed the notion that the foot behaves as a rigid body during the stance phase of gait and that evaluation of rearfoot alignment alone would overlook substantial motion occurring in other joints.

The early multisegment foot model studies had shortcomings. There is a lack of consistency of methodology and selection of segments within the foot. Many studies modeled the forefoot as one rigid body, which proved to be misleading as future studies showed significant motion within the joints of the forefoot. The plane of reference for motion in each body plane varies among studies, and the motions described can be misleading. In some cases movement of bone segments are reported relative to the supportive surface and in other cases relative to other bones within the foot or the tibia. For the foot and ankle surgeon, some bone realignment procedures are important relative to the supportive surface such as a medial displacement calcaneal osteotomy. In other cases such as hallux abductovalgus surgery, realignment of the first metatarsal is performed relative to the lesser metatarsals.

Kinematic studies of multisegmented foot models mostly rely on surface markers placed on the skin which have movement artifact. Also, certain bones such as the talus, navicular, and the cuboid do not have natural osseous axes and are difficult to mount surface markers.

The Bone Pin Studies

The solution to movement artifact from surface skin markers is the use of pins mounted into the bones themselves to more accurately track motion. This practice of placing bone pins in human subjects is prohibited in most countries. An alternative is implanting pins in cadaver specimens, but the challenge would be replicating dynamic gait to simulate true life motion of the skeletal segments.

Ward, Cocheba, and Patterson developed one of the first dynamic cadaveric gait simulators and published a series of studies using bone pins and strain gauges to better understand the kinematics and kinetics of human walking [21, 22]. Ward teamed with Chris Nester, a PhD biomechanist, to conduct a landmark study of joint movement within the human foot during gait using a dynamic cadaver walking simulator. [22] Subsequently, Nester was a subject and co-author of a similar bone pin study of the kinematics of walking using human subjects in the country of Sweden [23].

The findings of the two bone pin studies are remarkably similar and consistent, yet one study was performed using a dynamic cadaver walking simulator, while the other was performed on a small group of human subjects in Sweden. The findings of the two studies were eye-opening as they revealed considerable movement across joints of the foot which were previously thought to be relatively rigid and immobile (Table 3.1).

Table 3.1 Contribution to motion from the individual joints

Summary of the Bone Pin Kinematic Studies		
Summary from the studies: Nester et al. [24] and Lundgren et al. [23]		
	Joint	Range in degrees * = Primary contributor
Sagittal		
	Tib-talar	18.1*
	Cub-5th met	12.9
	Nav-med cun	11.5
	Talo-nav	10.3
	Calc-cuboid	9.75
	Tal-calc	7.3
	Met 1-med cun	5.4
Frontal		
	Talo-nav	13.6*
	Tib-tal	11.7
	Cub-5th met	11.6
	Talo-calc	9.8
	Calc-cub	9.45
	Nav-med cun	9.3
	Met 1-med cun	6.1
Transverse		
	Talo-nav	16.5*
	Tib-tal	8.9
	Calc-cub	8.1
	Tal-calc	7.8
	Cub-5th met	7.45
	Met 1-med cun	5.6
	Nav-med cun	5.3

* = Primary joint contributor

Here are the key findings of the bone pin studies [22, 23]:

The Lateral Metatarsals Move More than the Medial

The dynamic cadaver study showed that there is significant motion at the calcaneal-cuboid joint (9.8 Sagittal, 7.6 Frontal, 8.0 Transverse), but overall motion was half that of the talonavicular joint [22]. Both joints moved in tandem, verifying the concept of a midtarsal joint. As shown in Chap. 1, strong ligament attachments between the cuboid and navicular are unique features of the human foot. The more profound finding from this study was the amount of significant sagittal plane motion detected at the fourth and fifth metatarsal-cuboid articulation which measured 11.5°, exceeding the calcaneal-cuboid joint by 2°. This compares with sagittal plane motion at metatarsals 1–3 at their respective cuneiforms which averaged 6.1°. Clearly, the metatarsals do not all move together as one rigid body segment, and the lateral metatarsals move in a greater range than the medial.

The Navicular-Cuneiform Joints Move More than the Midtarsal Joint

At the medial column, both studies showed that there was considerable sagittal plane motion at the navicular-cuneiform joint in the sagittal plane, measuring 11.4°, while sagittal plane motion at the first metatarsal-cuneiform joint was less than half, measuring only 5.6°. Overall, the motion between the cuneiforms and navicular in the sagittal plane was equivalent to motion between the talus and navicular or between the calcaneus and the cuboid. In other words, the navicular-cuneiform joints contribute as much sagittal plane motion as the midtarsal joint. This finding has surgical implications when deciding which joint fusion on the medial column of the foot might have the most profound stabilizing influence.

The Midfoot Joints Contribute More Sagittal Plane Motion than the Ankle

Sagittal plane motion across the talonavicular, navicular-cuneiform, first metatarsal-cuneiform joints, as well as in the lateral column at the calcaneal-cuboid and cuboid fourth and fifth metatarsal joints exceeds the sagittal plane motion contributed by the subtalar joint and ankle joint combined. This places serious doubt about the validity of using the age-old technique of measuring "ankle joint dorsiflexion" by using the entire foot as the distal reference. When measuring foot-to-leg dorsiflexion, at least half the motion is actually occurring in the foot, not the ankle. Furthermore, this study casts doubt on other multisegment fool model investigations which assumed that the forefoot functions as one rigid body.

The Medial Column Moves More than the Ankle

The second bone pin study was conducted by Lundgren, Nester, and co-workers who utilized human subjects with bone pins implanted in their feet, and tracking of movement was accomplished with 3-D (three-dimensional) stereophotogrammetric analysis. Verifying the previous observations of the cadaver bone pin study from Nester et al. [23] this in vivo study showed that the overall contributions of the joints of the medial column of the foot (talonavicular, navicular-medial cuneiform, and first metatarsal medial cuneiform) were equivalent to the sagittal plane motion of the ankle joint. In other words, the human foot moves more during the stance phase of gait at the medial arch than at the ankle joint.

Pure Ankle Joint Motion Can Now Be Measured

Bone pin studies allow the placement of markers on the talus which can allow isolated measurement of the talocrural joint. Previous 3-D multisegment foot model studies using skin-mounted markers could not evaluate pure frontal and transverse plane motion at the ankle, separate from the subtalar joint. Sagittal plane motion at the ankle joint in the bone pin studies verified findings from previous investigations showing approximately 8° of dorsiflexion and 8° of plantarflexion.

The Ankle Moves in the Transverse and Frontal Planes

The two bone pin studies show motion which contradicts the "mitered hinge" perception of ankle joint motion in humans. Specifically, there is more than just pure sagittal plane motion available at the talocrural joint. Previous kinematic studies have shown surprising high degree of transverse plane motion at the ankle joint [25, 26]. This observation was verified in the two bone pin studies showing and average of 8.9° of transverse motion of the talus relative to the tibia.

There Is more Frontal Plane Motion in the Ankle than the Subtalar Joint

There is also a surprising amount of frontal plane motion in the ankle joint during walking which averaged 8.1° in the Lundgren study. This is a similar amount of frontal plane motion measured in the subtalar joint which averaged 9.8° in the same study. The bone pin study from Nester actually measured more frontal plane motion in the ankle during walking compared to the subtalar joint (15.3° vs 9.7°) [23]. It should now be recognized that frontal plane motion of the "rearfoot" observed clinically with gait analysis is actually contributed equally by both the ankle joint and the subtalar joint.

The Talonavicular Joint and the Calcaneocuboid Joints Move More than the Subtalar Joint

The bone pin studies measured greater overall motion at the talonavicular joint than motion at the talocalcaneal (subtalar) joint. In fact, the talonavicular joint has more motion in all three body planes than any of the rearfoot or midfoot joints. In comparison with the subtalar joint, the talonavicular joint moves over 40° in all three body planes, while the subtalar joint moves 24°. In terms of frontal plane motion, the talonavicular joint moved an average of 13.6°, while the subtalar joint moved 9.8°. Surprisingly, the calcaneal-cuboid joint moves 9.4° in the frontal plane which is equivalent to frontal plane motion at the subtalar joint. In fact the calcaneal-cuboid joint moves in a larger range of motion in all three planes compared to the subtalar joint.

Majority of First Ray Motion Is as the Naviculocuneiform Joint

Motion between the navicular and the medial cuneiform is twice that of the motion occurring at the first metatarsal medial cuneiform joint. This dispels the myth that the first metatarsal medial cuneiform joint is the primary contributor to "hypermobility" of the first ray in many foot pathologies. In fact, the bone pin studies verify that the first metatarsal medial cuneiform joint moves less than any other joint of the foot in all three body planes. Therefore, fusion of this joint would not be expected to have any significant effect on overall motion or function of the foot.

The Lateral Metatarsals Move More than the Medial Metatarsals

The cuboid-fifth metatarsal joint has twice the sagittal plane motion than the first metatarsal medial cuneiform joint. This observation, combined with the measured motion across the calcaneal-cuboid joint, dispels the notion that the lateral column of the foot is relatively rigid during the stance phase of gait. In fact, the higher rate of nonunion with fusion of the calcaneal-cuboid joint versus the first metatarsal medial cuneiform joint might be explained by the significant difference between these joints in overall range of motion measured by the bone pin studies.

The Lateral and Medial Columns Have the Same Sagittal Plane Motion

In the sagittal plane, the lateral column moves 23° using combined motion at fifth met cuboid and calcaneal-cuboid joints. Along the medial column, the combined motion of the first metatarsomedial cuneiform joint, the navicular-cuneiform joint, and the talonavicular joint contribute 25°. Therefore, motion in the sagittal plane is relatively equivalent when comparing the medial columns with the lateral. As with previous studies, the overall sagittal plane motion contributed by the joints of the foot exceed the motion contributed by the ankle by almost 50%. Measuring reduced foot-to-leg dorsiflexion range of motion clinically and diagnosing an ankle equinus condition are an erroneous assumption since the joints of the foot are providing most of the motion during this standard measurement technique.

The two bone pin studies conducted by Nester et al. and Lundgren et al. provide the most precise and reliable information about the movement of the bones of the human foot during walking gait [23, 27]. Besides reaffirming previous observations about rearfoot motion during walking, new insights into the motion of the lateral column as well as the metatarsal cuneiform joints were provided by these two studies which dispelled previous myths and misconceptions. Furthermore, a new understanding of the complexity of ankle joint motion provides essential information for the foot and ankle surgeon. A summary of both bone pin studies showing range of motion in three body planes is provided in Figs. 3.3, 3.4, and 3.5.

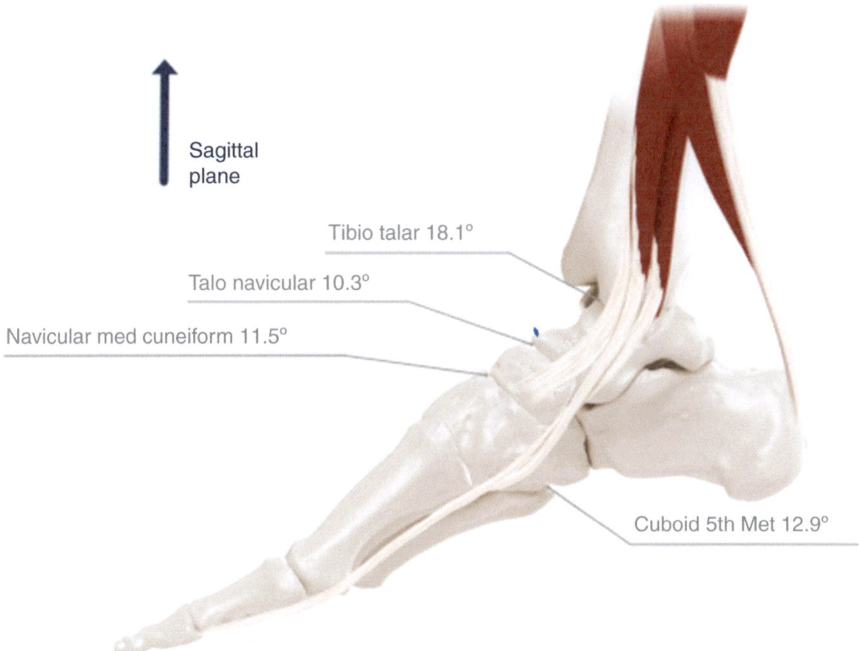

Fig. 3.3 Sagittal plane motion of the foot: top 4 contributing joints

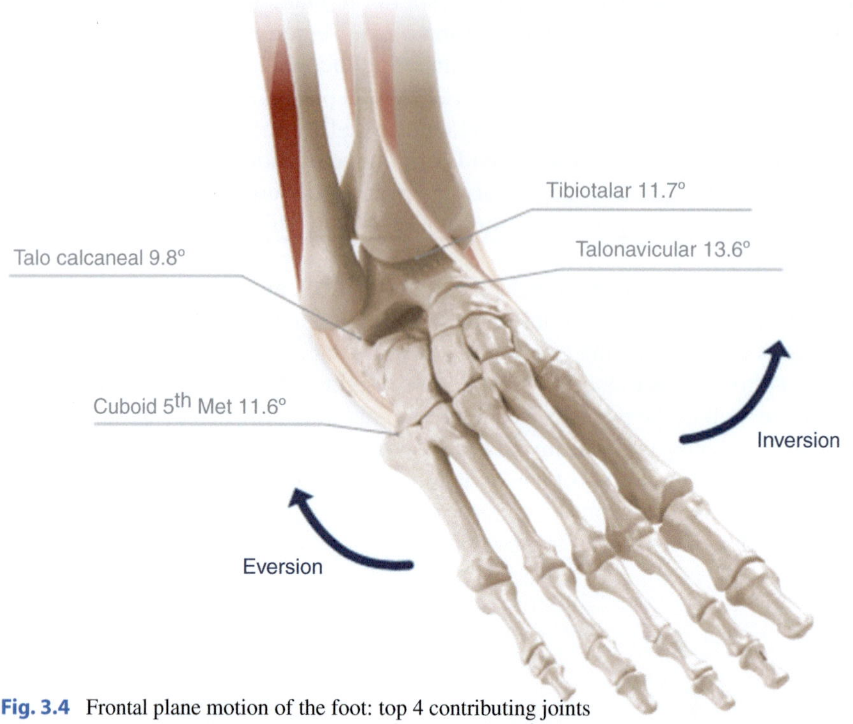

Fig. 3.4 Frontal plane motion of the foot: top 4 contributing joints

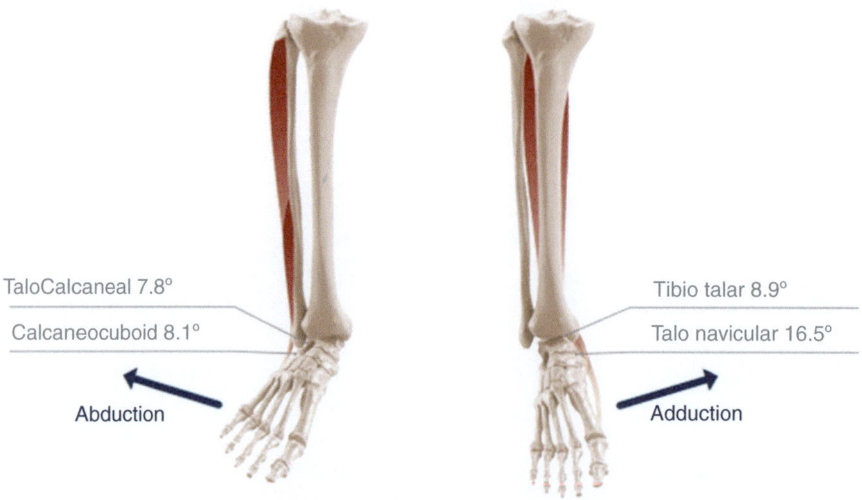

Fig. 3.5 Transverse plane motion of the foot: top 4 contributing joints

The bone pin studies reaffirmed previous investigations about the contribution of various joints of the foot to overall range of motion. Clearly, the subtalar joint is not the major contributor to overall motion of the foot. This contradicts a long-held notion that the subtalar joint is the primary contributor to overall alignment and performance of the human foot in dynamic gait.

Lundberg et al. conducted a kinematic study of human subjects by implanting metal beads in certain bones of the feet and then tracking motion with radiographs as the foot was moved passively on a platform [27]. This investigation measured almost twice the range of motion in the talonavicular joint as the subtalar joint. Studies of arthrodesis of the midfoot and hindfoot joints in cadaver specimens have verified the minimal role of the subtalar joint in providing overall motion to the human foot. When the subtalar joint is fused, at least 50% residual motion is still available in the talonavicular and calcaneal-cuboid joints [25, 26]. Conversely, fusion of the talonavicular joint essentially eliminates all motion in the hindfoot including the subtalar joint [25, 26]. Clearly, the contribution of the subtalar joint to overall foot motion is significantly less than the talonavicular joint.

How Does the "Normal" Foot Function in Gait?

Multisegment foot model studies have primarily focused on subjects with pathologies such as posterior tibial tendon dysfunction and rheumatoid arthritis. Most of these studies have used relatively small numbers of subjects. There has been a void in understanding "typical" or normal kinematics of the human foot. At the same time, understanding how the bones and joints move in healthy feet allows the foot and ankle surgeon to identify abnormal function and perhaps choose surgical procedures which help restore the foot to a more healthy or "normal" function.

A study performed by Nester and co-workers of 100 healthy, asymptomatic individuals provided insight into defining the kinematics of the "normal foot." [28] Unique to this study was not only the large number of healthy subjects but also the 3-D multisegment model which divided the foot into six segments using external skin-mounted markers: leg, calcaneus, midfoot (navicular and cuboid), the medial forefoot (first metatarsal), the lateral forefoot (fourth and fifth metatarsals), and the hallux (Fig. 3.6).

Despite the potential shortcomings of using skin-mounted markers, the findings of this study were quite similar to the bone pin studies showing overall sagittal plane motion dominates over frontal and transverse plane motion among all the joints of the foot. An important and critical finding to this study is the observation that the medial forefoot (first ray) and the lateral forefoot (fourth and fifth metatarsals at the cuboid) have significant motion independent of each other, and there is more frontal and transverse plane motion across the forefoot-midfoot junction than across the midtarsal joint (Fig. 3.7).

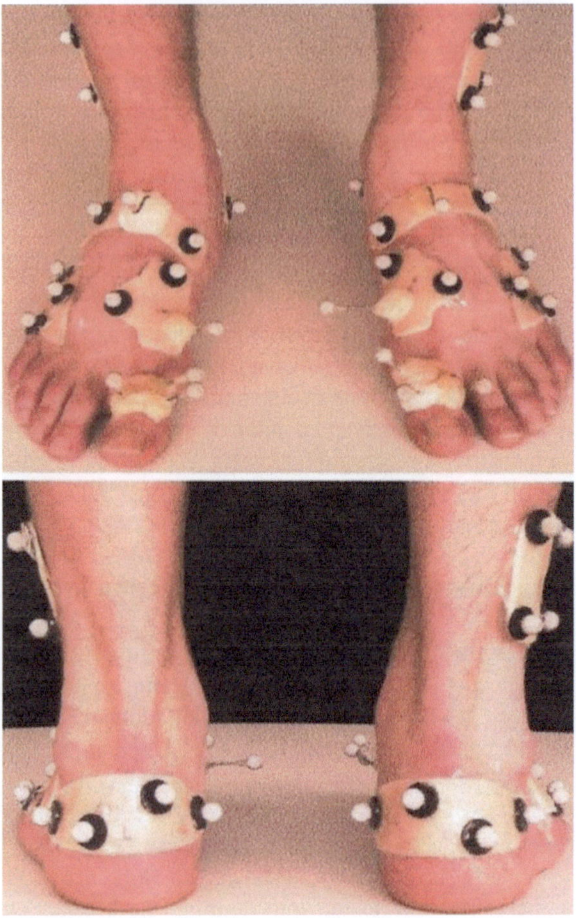

Fig. 3.6 Nester et al. studied 100 healthy adults using a model which divided the foot into six segments. (From: Nester et al. [28])

Furthermore there was no evidence that the forefoot acted as a rigid body. And, as demonstrated in the bone pin studies, there was considerable motion between the forefoot and the rearfoot during terminal stance, refuting the concept of a "locking mechanism" of the midtarsal joint.

The goal of this study was to try to identify if there was a single pattern of movement which predominated across all 100 healthy subjects .In terms of identifying predictable magnitude of "normal" motion at various joints of the human foot, this study of a large number of healthy subjects could not find any consistency. There is no pattern of motion to define a normal healthy foot. The notion of coupling between segments of the foot could not be verified. Rearfoot eversion averaged less than 5° during the stance phase, but there was wide variation (Fig. 3.7). In fact there is wide variation of magnitude of motion among all the joints of the foot in healthy asymptomatic individuals.

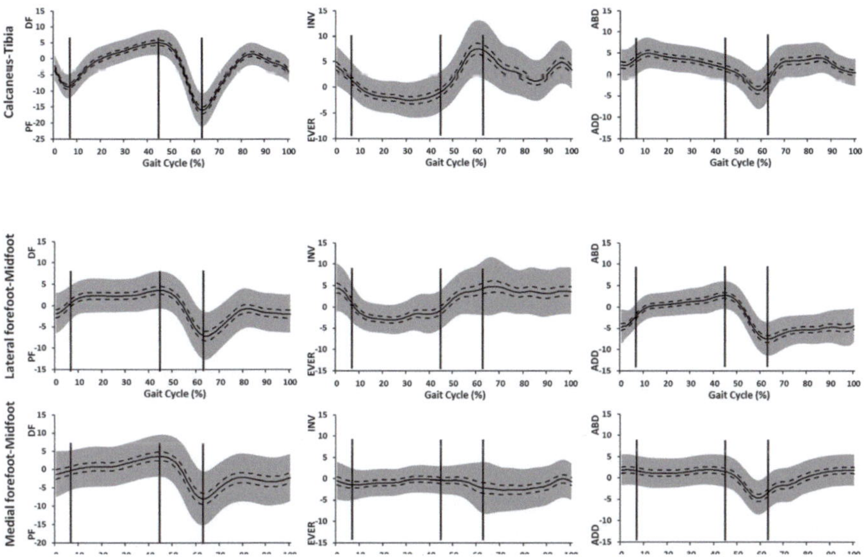

Fig. 3.7 Healthy individuals show significant motion between the forefoot and midfoot during walking gait. (From: Nester et al. [28])

That is not to say that certain foot types do display more predictable motion in certain planes. Kinematic studies have verified certain patterns of motions seen in planus versus cavus foot types. These studies have measured increased frontal plane eversion of the rearfoot in pes planus foot types as well as increased transverse plane motion of the forefoot on the rearfoot in the direction of abduction [29]. Cavus feet on the other hand demonstrate a more inverted position of the rearfoot during gait [30]. However, beyond these simple and well-accepted observations of flat-footed and high-arched individuals, movements of bone segments within the foot are not predictable based upon foot type.

Nester et al. suggested that neuromuscular mechanisms can affect direction and magnitude of joint motion within the human foot to explain the wide variation seen among healthy human subjects [28]. Variability of movement within a certain range is theorized to allow the body to adapt to environmental challenges and reduce overuse of specific structures [31, 32]. Variability in foot strike and loading of the plantar surface of the foot reduces focal repetitive loading on specific structures and thus reduces the likelihood of skin ulceration [33]. In other words, variability is a healthy finding in human locomotion.

Conversely, reduced variability of motion of skeletal segments can suggest a pathologic condition. Patients with chronic ankle instability (CAI) demonstrate reduced stride-to-stride variability suggesting a more constrained neuromuscular system [34]. This reduced kinematic variability may be due to less variability in activation of lower extremity muscles, particularly the peroneus longus and tibialis anterior [35]. A rigid, almost spastic contracture of a muscle group is not healthy and seen in many pathologic conditions. Reduced variability of kinematic and muscular activation during gait is thought to render patients with CAI more vulnerable to unexpected perturbation and risk of injury [36].

The mechanisms which contribute to neuromuscular control of the trunk and extremities during standing and walking are complicated and not yet fully understood. However, this subject is worthy of study for any clinician treating the foot and ankle as it is integral to the pathomechanics and treatment of most pathologies which will be encountered.

Neuromuscular Control

Ultimately the alignment and function of the skeletal segments of the lower extremity during upright standing, walking, and running is under the control of the sensorimotor system. This system utilizes components of the peripheral and central nervous systems and then coordinates muscle activity. To understand this intricate system, we will start with a review of the postural control mechanism which is just one part of the sensorimotor system.

Postural control is a neuromotor mechanism which functions to keep the body's center of mass within its base of support [37]. In other words, postural control is a mechanism to keep the body upright during standing and walking. Postural control relies upon sensory input from multiple levels including the visual, the vestibular, and the somatosensory system. In the lower extremity, the sensorimotor system provides afferent input to the central nervous system from multiple sources including the skin, tendons, muscles, and ligaments. The components of the sensorimotor system are depicted in Fig. 3.8.

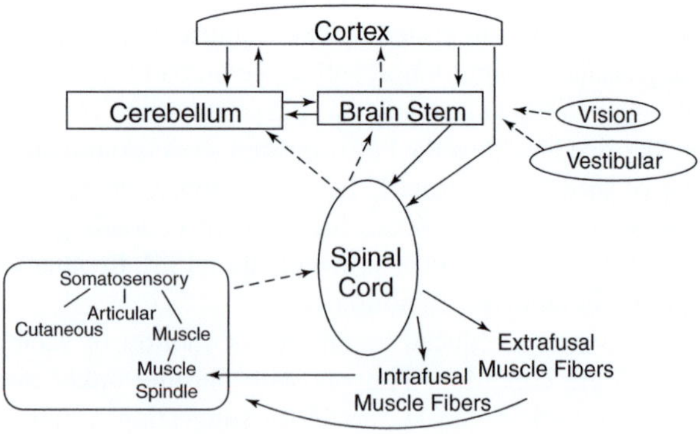

Fig. 3.8 The sensorimotor system: dotted lines are afferent pathways; solid lines are efferent pathways. Reflexes occur at the spinal level, while more advanced motor control involves higher levels at the cerebellum, brain stem, and cerebral cortex

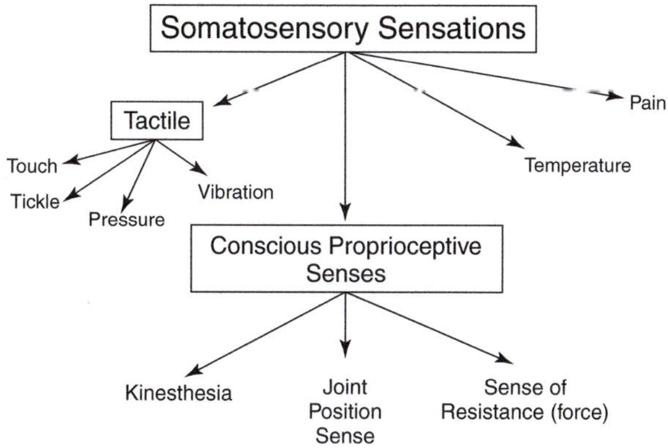

Fig. 3.9 Afferent input from the somatosensory system in the lower extremity

The somatosensory system provides afferent input from tactile, pain, temperature, and proprioceptive sensors located in the lower extremities (Fig. 3.9).The term proprioception refers to three sensory perceptions: joint position sense, joint movement, and sense of resistance. The somatosensory system receives proprioceptive input from the ligaments surrounding the joints as well as input from other sensors including muscles, tendons, and retinacula [32]. Pain, temperature, and tactile sensations are all part of the somatosensory system. Tactile sensation alone has three components: touch, pressure, and vibration. The plantar surface of the foot is rich with tactile sensors which provide essential feedback for balance and postural control during standing and walking.

An essential component of the somatosensory system is the interaction of the Golgi tendon organs and the muscle spindles which connect to the gamma motor neuron loop [38]. These structures regulate muscle stiffness and respond to changes in length of muscle which occur in joint movement (Figs. 3.10 and 3.11). The muscle spindles and Golgi tendon organs are critical proprioceptors for joint position. Joint movement places stretch on tendons which initiates a protective reflex to activate skeletal muscles to control movement. A sudden perturbation during walking will cause excessive joint motion which also activates compensatory muscular response to restore postural control and/or prevent potential joint injury.

Joint stiffness provides functional stability by limiting abnormal motion [39]. Joint stiffness is provided in part by muscle "stiffness" contributed by co-contraction of antagonistic muscles. This results in compression between articular surfaces which are the second part of the mechanism of joint stiffness [40]. Stiffer muscles thru pre-activation resist sudden joint displacement more effectively [41]. This is accomplished thru the gamma motor neuron loop.

Active muscles recruit the muscle spindle reflex loop to shorten the reflex time needed to prevent joint subluxation and injury. The muscle spindles keep the muscles in a state of activation. In other words, the *electromechanical delay* in muscle

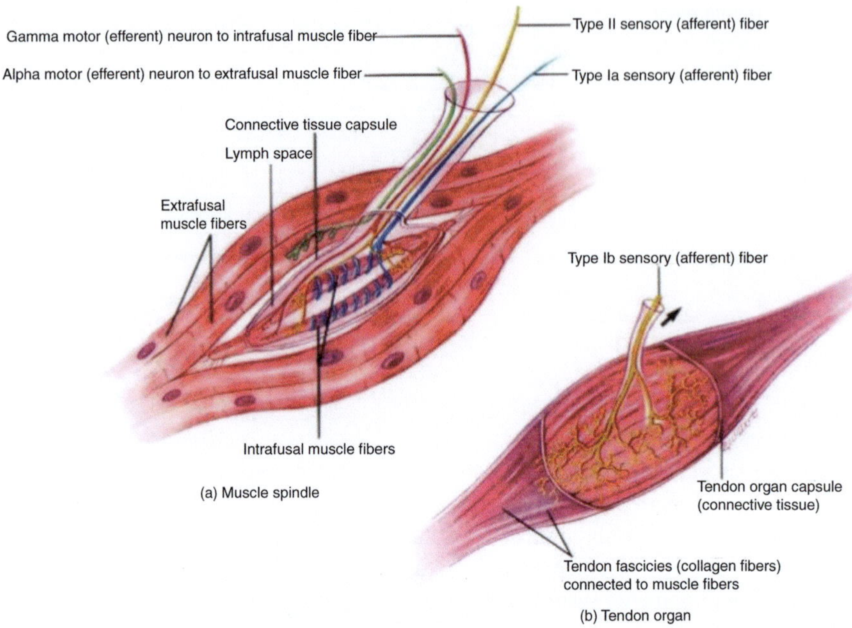

Fig. 3.10 The muscle spindles and Golgi tendon organ

activity is shortened when muscles are properly activated before a perturbation occurs, thus reducing the risk of joint injury [42, 43]. This pre-activation of muscles is controlled at supraspinal levels, often part of a "feed-forward" mechanism.

Motor control, or activation of motor neurons, occurs at three primary levels of the nervous system: the spinal cord, the brainstem, and the cerebral cortex. Immediate activation occurs at the spinal cord level via reflexes which control many of the muscular activities during gait. The brainstem modulates inhibitory muscle tone and is an essential area for postural control. The cerebral cortex regulates more fine-tuned muscular activity for complex and discrete voluntary movements.

The complexity of the sensorimotor mechanism explains why gait studies cannot define "normal" motion of the lower extremity skeletal segments during walking and running. The timing of muscular contraction, the force of contraction, the activation of antagonist muscles, and the pre-activation of muscles, all of which contribute to stiffness across joints creates multiple options for wide variation. Even when all of the elements of the sensorimotor system are functioning optimally, the ultimate effects on walking or running gait will vary significantly among healthy individuals.

When trauma disrupts key elements of the sensorimotor system, compensation or dysfunction mediated by higher (supraspinal) centers will occur. Injury to joint

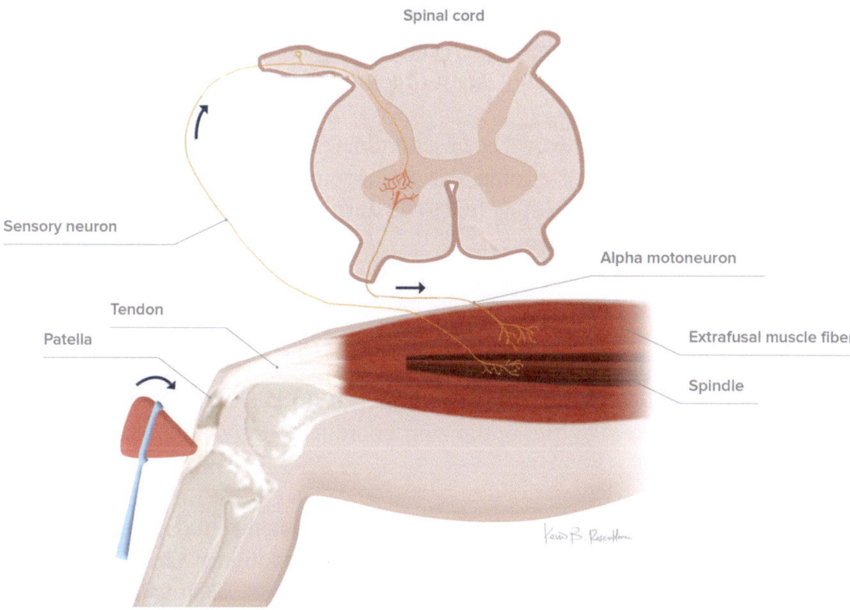

Fig. 3.11 The patellar tendon-stretch reflex, via the muscle spindles, activates the alpha motor neuron innervating the quadriceps muscle Joint movement

mechanoreceptors will disrupt neuromuscular control mechanisms. Injury can also occur to other levels of the somatosensory system, altering afferent input. Pain from injury can affect multiple levels of the sensorimotor system including the gamma motor neuron loop.

Two common orthopedic injuries have been extensively studied in terms of their effects on the sensorimotor system: the ankle sprain and anterior cruciate ligament (ACL) rupture. Subjects studied immediately after ligament disruption at the ankle show impaired proprioception and delayed lower extremity muscular activation [44]. Subjects who fail to recover from an ankle sprain develop chronic ankle instability which is a complex disorder affecting all levels of neuromuscular control of the lower extremity [45]. Most notably, these patients demonstrate alteration in muscle activation around the ankle and in proximal joints indicating changes at higher motor centers [46]. After ACL injury, changes in muscle firing patterns are seen distally in the soleus and tibialis anterior as well as proximally in the hip musculature [47]. These studies show that after orthopedic injury, alterations in motor control occur which must be identified and treated by the treating clinician. These changes may be the result of interruption of somatosensory input or loss of integration for activation of motor response. There is significant evidence that balance training is effective in restoring sensorimotor control after orthopedic injury [48–51].

Ultimately, the sensorimotor system coordinates the activation of agonist and antagonist muscles to stabilize joints during locomotion. According to Perry [52], each weight-bearing limb has four functions during locomotion which are modulated by the sensorimotor system:

1. *Upright stability* is maintained despite an ever-changing posture.
2. *Progression* is generated by the interaction of selected postures, muscle force, and tendon elasticity.
3. *The shock of ground impact* at the onset of each stride is attenuated.
4. *Energy is conserved* by the optimizing coordination and minimizing muscular effort.

Not only is the timing of activation critical to this system, but the eventual power and velocity of contraction of a muscle will dictate the motion at various joints. The muscles of the human body vary significantly in their physical structure which dictates the amount of tension developed during active contraction. When clinicians consider off-loading or relieving "tissue stress" with treatment interventions, they must have an in-depth knowledge of the phasic muscle activity, strength and excursion of the muscles, and the moment arm of the muscle/tendon unit acting across various joints.

Do Joint Axes Determine Direction and Range of Motion?

A popular theory in podiatric biomechanics is that the orientation of a joint axis, specifically in the subtalar joint, will predict behavior and loads experienced by the entire foot during gait [53]. This theory has some validity but also has shortcomings which should be considered. Firstly, no joint anywhere in the human body has an axis of motion. Rather, motion of a joint will describe the alignment of the theoretical axis around which the skeletal segments are rotating. In this regard, motion defines the axis, and not vice versa. Secondly, passively moving the foot at only one joint, off weight bearing, cannot accurately predict the direction, range, and ultimate axis of rotation of that joint during dynamic gait. All of the joints of the foot affect motion of the other joints, and moment arm of the muscles crossing those joints changes as joint position changes. Finally, the subtalar joint does not contribute the majority of motion of the human foot, and motion of the subtalar joint does not predict overall motion of the entire foot.

Motion occurring at any joint in the lower extremity is the result of the magnitude and direction external forces acting across the joint. These forces or moments are counteracted upon by both dynamic muscle-tendon structures as well as passive joint capsule and ligamentous structures. Even the shape and quality of the articular surface of the joint will dictate the overall joint motion. Finally, the strength and timing of contraction of agonist and antagonist muscle groups coordinated by the sensorimotor system will have the final influence on the ultimate direction and range of motion of a joint during gait. It is the summary of all of these influences and not a simple theoretical axis which will ultimately determine how a skeletal segment will perform in dynamic gait.

Unfortunately, how the sensorimotor system will regulate upright stability and progression during ambulation is not predictable in each and every patient.

However, a general understanding of muscle function in the lower extremity will allow identification of impairment and will improve implementation of treatment interventions which could restore function to a more optimal level. Muscle function is dependent upon several key factors including phasic activity, muscle strength, and moment arm across various joints.

Muscle Function in the Lower Extremity

Phasic Activity

The timing and intensity of muscle activity in the lower extremities are depicted in Figs. 3.7 and 3.8 below and taken from the work of Perry [52]. In general, the extensors work in concert with each other; however the tibialis anterior (TA) is most active at initial contact, while the extensor hallucis longus (EHL) is most active at mid swing (Fig. 3.12).

The ankle plantar flexors show more specialization of phasic activity (Fig. 3.13). The tibialis posterior (TP) contracts immediately at initial contact to decelerate or restrain internal rotation of the tibia and pronation of the subtalar and midtarsal joints. The flexor digitorum longus (FDL) does not contract until the start of midstance and fails to provide any restraint during initial contact. This raises one of several questions about the suitability of the FDL for tendon transfer in adult acquired flatfoot (AAF) deformity.

The gastrocnemius and soleus peak during terminal stance or heel rise and then drop off sharply before pre-swing. This complete absence of muscular activity during pre-swing or toe-off verifies that push off is not due to active contraction of the calf musculature.

Muscle Strength

In a landmark study, Silver et al. evaluated strength and excursion of the muscles of the lower extremity [54]. Excursion was determined by measuring the length of the muscle fibers. Strength was a measurement of muscle volume divided by muscle fiber length. Thus, muscle mass by itself does not determine strength. The length of the fibers of the muscle counteracts the strength or tension the muscle can develop. Essentially, *longer* fiber length equates to *decreased* muscle strength.

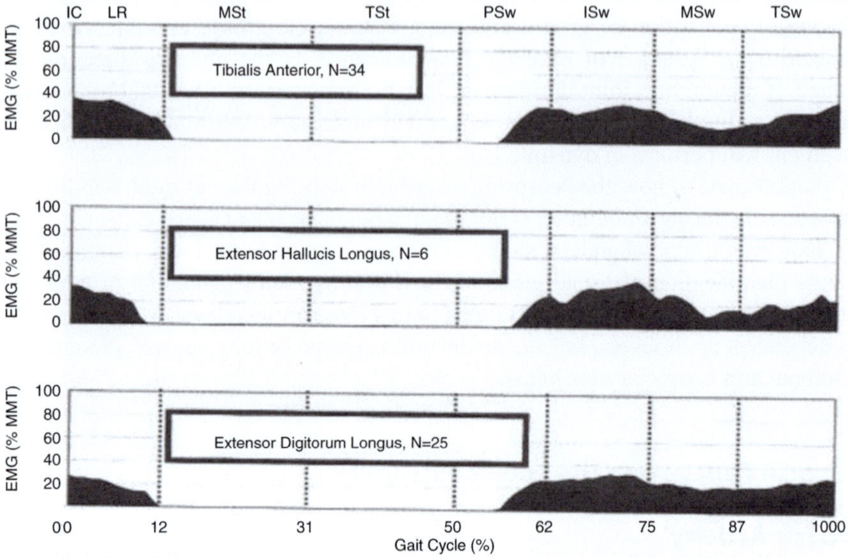

Fig. 3.12 Ankle dorsiflexor muscles. Normal mean intensity and timing during free walking (quantified EMG). Intensity as a percent of maximum manual muscle test value (% MMT) indicated by height of shaded area. The dark shading indicates the activity pattern of the majority of subjects. Vertical bars designate the gait phase divisions. N = samples included in data. (From Perry and Burnfield [52] Figs. 4–5 Page 57 and 59. Reprinted with permission)

Interesting insight is gained by the Silver study regarding the importance of certain muscle groups in the lower extremity. The ankle plantarflexors account for 54.5% of overall lower extremity muscle strength, while the dorsiflexors account for only 9.4%. This underscores the critical function of the posterior leg musculature to control forward fall during gait. The soleus is the strongest muscle crossing the ankle joint accounting for 30% of the strength of the lower leg musculature, while the gastrocnemius provides 19.2% of overall strength. The medial head of the gastrocnemius provides 72% of the entire strength of the gastrocnemius.

Although overall percent of strength contributed by the ankle dorsiflexors is relatively minor, there is a primary reliance upon the tibialis anterior (5.6%) which exceeds the combined power of the extensor hallucis longus (1.2%) and extensor digitorum longus (1.7%). The strength of the peroneus longus (5.5%) is more than twice that of the peroneus brevis (2.6%). The mean fiber lengths of the peroneus brevis and the tibialis posterior are identical. Despite an obvious strength difference, the flexor digitorum longus (1.8%) is commonly transferred as a replacement for the tibialis posterior (6.4%) in adult acquired flatfoot surgery.

Wickiewicz expanded on Silver's study to provide more insight into the performance of muscles of the lower extremity [55]. Instead of looking at gross length of the muscle fibers, Wickiewicz performed microscopic measurements to count the number of sarcomeres. Velocity of muscle contraction is dependent upon fiber

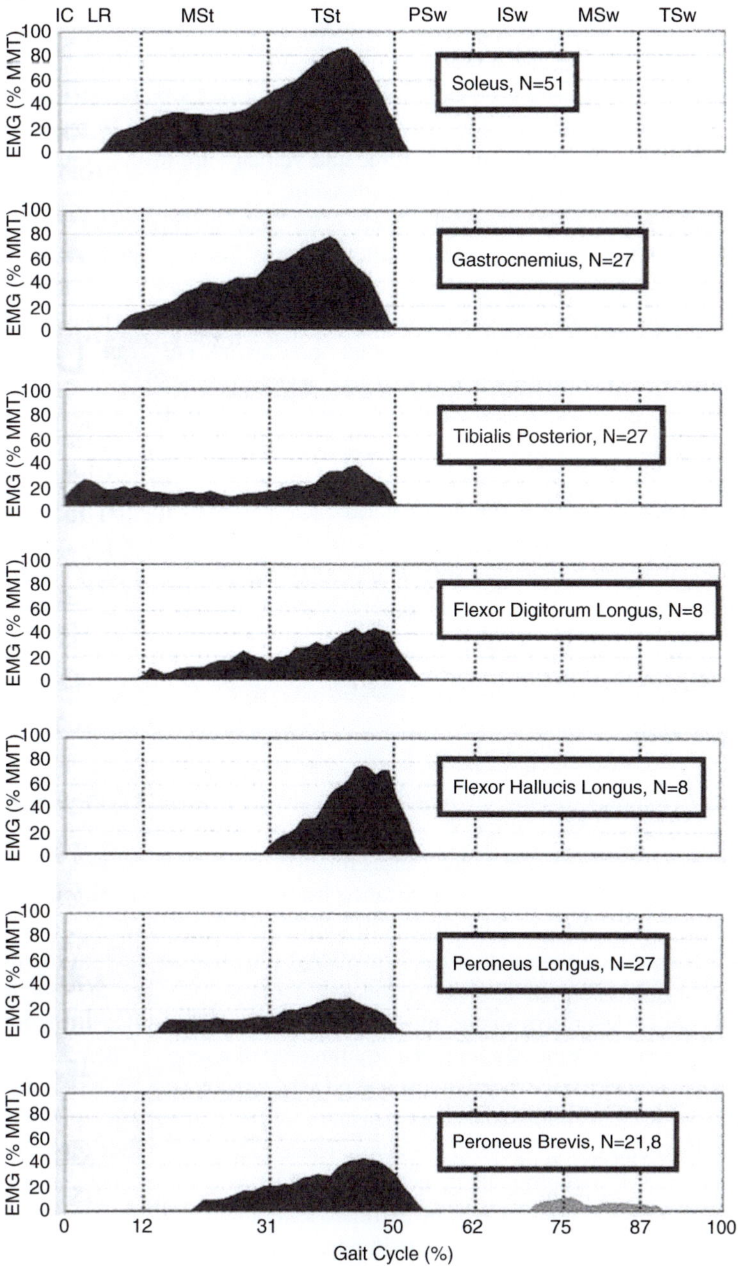

Fig. 3.13 Ankle plantar flexor muscles including the triceps surae (soleus, gastrocnemius) and peri-malleolar (TP, FDL, FHL, pernoneus longus, pernoneus brevis). Normal mean intensity and timing during free walking (quantified EMG). Intensity as a percent of maximum manual muscle test value (% MMT) indicated by height of shaded area. The dark shading indicates the activity pattern of the majority of subjects. The light gray area indicates less frequent activity patterns. Vertical bars designate the gait phase divisions. N = sample included in the data (dominant pattern, less frequent pattern if present). (From Perry and Burnfield [52] Figs.4–6 Page 57 and 59. Reprinted with permission)

length with longer fibers providing higher velocity. Longer fibers are actually made up of sarcomeres arranged end to end, or as Wickiewicz describes, they are sarcomeres *in series* providing higher velocity. Force or tension is determined by the number of sarcomeres arranged side to side, or *in parallel*, which then increases the muscle cross-sectional area. Wickiewicz evaluated lower extremity muscle performance based upon the principle that force and velocity are inversely proportional: high-velocity muscle contraction has less tension.

The Wickiewicz study showed that there is no "balance" between muscle groups in the lower extremity. The ankle dorsiflexors have fast-shortening velocities but develop only 15% the tension as the ankle plantarflexors. The EHL stands out above all the other dorsiflexors in velocity as it has three times the fiber length compared to cross-sectional area compared to the other dorsiflexors. Therefore, the EHL is the weakest of the ankle dorsiflexors. The soleus and gastrocnemius have the shortest muscle fiber length followed by the tibialis posterior and the peroneus longus. These are the four muscles which are the top tension producers of the lower leg. Based upon muscle fiber length, the flexor digitorum longus is designed for higher velocity and lower tension than the tibialis posterior. The peroneus longus is designed for better tension than the peroneus brevis. The lateral head of the gastrocnemius is composed of fibers which are 46% longer than the medial head. Therefore, the medial head of the gastrocnemius generates twice the tension compared to the lateral.

Moment Arm

Another critical factor in muscle function of the lower extremity is the alignment of a tendon which enables a muscle to produce moment or torque at a joint. Torque is a force applied over a distance which causes rotation of a joint about an axis. Moment arm is defined as the perpendicular distance from the joint's axis of rotation to the line of action of the muscle. The amount of torque generated by a muscular contraction is dependent on three variables: The strength of the contraction, the angle of application of force by the muscle-tendon unit across a joint, and the length of the moment arm or lever arm of the muscle-tendon unit. The optimal situation for a muscle to generate torque at a joint is a combination of a long moment arm which is oriented at a 90 degree (perpendicular) angle to the axis of joint rotation [56].

There are various methods used to measure muscle moment arms in both cadaveric (in vitro) and living subjects (in vivo). These methods include direct measure with radiographic imaging, load measuring, and tendon displacement measures [57–60]. Studies of moment arm of extrinsic tendons of the foot have challenges and shortcomings. Some studies have attempted to isolate movement at a specific joint by fusing other joints of the foot. Other studies manually moved the foot in one specific body plane, while other studies allowed the foot to move freely in all three body planes.

Notwithstanding, consistent observations have been made from these muscle moment arm studies which are helpful to consider when implementing offloading strategies for treatment of injury. It is important to recognize that foot position and deformity will have significant effect on the moment arms of key muscles in the lower extremity. Reconstructive foot and ankle surgery, by changing position of skeletal segments, will affect moment arms and performance of certain muscles.

A study by Hinterman et al. showed that positioning of the ankle in dorsiflexion or plantarflexion has profound effects on changing moment arms of specific lower leg muscles [59]. As the ankle plantar flexes past neutral or 90°, eversion moment arm of both peroneal muscles (PL and PB) significantly *decreases.* This verifies that the plantar-flexed ankle is at risk for inversion injury not only due to change of mechanical alignment within the talocrural joint but also due to compromise in peroneal muscular efficiency for stability.

Hinterman also showed that as the ankle moves into plantarflexion, there is greater *increase* in plantarflexion moment arm for the gastrocnemius and soleus [59]. As the ankle moves into dorsiflexion, the moment arm of all the invertors of the ankle *decreases*. This study showed that as the foot moves into eversion, moment arm of the invertors, particularly the tibialis posterior, *decreases.* In summary, the triceps surae and the tibialis posterior function optimally when the ankle is plantar-flexed and the foot is in an inverted position.

A study by Lee et al. looked at the changes in moment arm of the tibialis anterior and the two heads of the gastrocnemius when the foot is inverted or everted [61]. The gastrocnemius has a small (4 mm) inversion moment arm at the rearfoot when the ankle is at 90° and the subtalar joint is neutral. As the foot moved into inversion, the inversion moment arm of the gastrocnemius increases particularly at the medial head. Eversion of the foot causes an eversion moment arm of the lateral head only, while the medial head always maintains a small inversion moment arm [61]. Eversion of the foot creates an eversion moment arm of the TA, while inversion of the foot changes this action and creates a significant inversion moment arm. This is an important consideration in flatfoot surgery. The TA can become a significant invertor of the foot provided that foot position, relative to the leg, is corrected from an everted to an inverted position.

Wang et al. performed a dynamic gait study of human subjects to measure the changes in muscle force in response to subtalar joint inversion and eversion [62]. This was not a moment arm study, but instead measured joint accelerations during gait. Therefore, this study measured the effects of change of muscle force in response to foot position which then changes the muscle moment arm. Inversion of the foot decreases the efficiency of the tibialis anterior for dorsiflexion but improves the efficiency for inversion. This was verified when the researchers measured a decreased dorsiflexion acceleration in the tibialis anterior when the foot was inverted. Eversion of the foot increased efficiency of the tibialis anterior to dorsiflex the ankle and also increased the ability of the gastrocnemius and soleus to plantarflex the ankle and extend the knee. Therefore, inversion of the foot actually decreases efficiency of the sagittal plane stabilizers of the foot, while eversion improves this efficiency. Finally this study verified a long-held belief that the triceps will further evert the subtalar joint with contractile activity if that joint is already positioned into eversion.

A comprehensive study of moment arms of the lower leg muscles was conducted by McCullough and co-workers [63]. Using a simulator which allows continuous motion in three planes, moment arms were measured in all nine extrinsic muscles of the foot in five cadaveric specimens. For the first time, moment arm was measured relative to transverse plane displacement of the foot. In this plane, the largest moment arm for abduction (external rotation) of the foot was contributed by the peroneus brevis (23 mm). For adduction (internal rotation of the foot), the flexor digitorum longus and the tibialis posterior had similar moment arms, 22.1 mm and 21.4 mm, respectively. For eversion of the foot, the peroneus longus had the largest average moment arm of 31 mm. For inversion of the foot, the largest moment arm was contributed by the tibialis anterior (16 mm). Surprisingly, the FDL and tibialis posterior had identical inversion moment arms of 10 mm which were 30% lower than the tibialis anterior. In the sagittal plane, the Achilles tendon had the largest average plantarflexion moment arm of 53 mm, while the extensor hallucis longus had the largest dorsiflexion moment arm of 43 mm. The peroneal tendons had surprising large moment arms for dorsiflexion with the peroneus brevis contributing 16.7 mm and the peroneus longus 21.1.

In the transverse plane, there is similarity of moment arms for adduction compared to abduction. The tibialis posterior and flexor digitorum longus have an adduction moment arm of 22 mm. The peroneus longus and brevis have an abduction moment arm of 16 mm and 23 mm, respectively. The transverse plane moment arms of all these muscles is appreciated when looking at the distal course of the tendons which cross 45° to the adduction-abduction axis of the foot (Fig. 3.14).

Applying more recent knowledge about muscle moment arms as well as muscle strength measures, a hierarchy of contribution from each muscle to foot stability can be developed. While moment arms may appear similar within a group of muscles of the lower leg, factoring in the cross-sectional area or strength of the muscle with the moment arm will allow a calculation of torque capacity of the muscle [65]. The first step is to estimate the isometric force output of the muscle. Zajac calculated the peak muscle stress of skeletal muscle, assuming that all motor units would be activated simultaneously, and proposed a baseline value of 35Ncm2 [56]. This value can then be multiplied by the cross-sectional area of any muscle to yield an estimate of maximal isometric force expressed in Newtons [62]. By multiplying this force by the moment arm, the total moment or torque can calculated, expressed in Newton-meters. Table 3.2 lists the overall torque or moment capacity of each of the lower extremity muscles for each of six planes of motion. From this table and from studies previously cited regarding change of position of moment arms, we can summarize the major contributors of muscular control of the foot in all six body planes (Fig. 3.15).

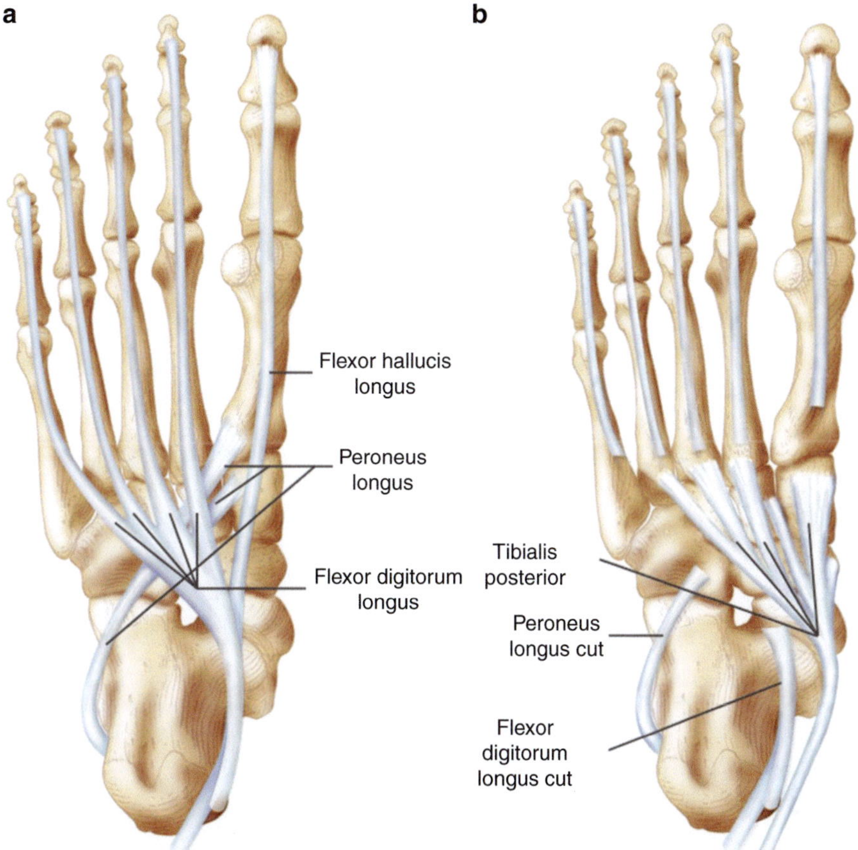

Fig. 3.14 The distal course of the FDL and TP tendons cross obliquely across the long axis of the foot, allowing production of transverse plane moment. (Taken from: McKeon et al. [64])

Major Muscle Contributors in Six Planes of Motion (Summary of Table 3.2)

Sagittal Plane

The plantar flexion muscles provide fivefold greater torque than the dorsiflexors (Fig. 3.16). Contrary to a popular assumption, there is little evidence for "balance" of power between agonist and antagonist muscles of the lower leg. Furthermore, there is no balance in phasic muscle activity comparing agonists and antagonists as they fire at different times. Other factors dictate timing of muscle activity during gait including forward momentum and ground reaction forces.

The significant strength advantage of the plantar flexion muscles underscores the critical requirement for the soleus and the gastrocnemius to restrain dorsiflexion and

Table 3.2 Capacity for moment production in three planes

	Muscle moment arm (mm)	Cross sect. area (cm^2)	Isometric force (N)	Moment (Nm)
Dorsiflexion				
TA	36.5	10	350	12.7[a]
EHL	43	2	70	3.0
EDL	32	5.3	175	5.6
PB	16.7	6	210	3.5
PL	21.1	12	420	8.8
Plantarflexion				
SOL	53	58	2030	107.5[a]
GAST	53	32	1120	59.3
FHL	25	5	175	4.3
FDL	12	5	175	2.1
TP	9.4	21	735	6.9
Inversion				
TP	10.2	21	735	7.5[a]
TA	16.6	10	350	5.8
Eversion				
PL	31.3	12	420	13.14[a]
PB	20.5	6	210	4.3
EDL	5	5.3	175	0.8
Internal rotation (adduction)				
TP	21.4	21	735	15.7[a]
FDL	22.1	5	175	3.86
External rotation (abduction)				
PL	16	12	420	6.72[a]
PB	23	6	210	4.83

Based upon: McCullough et al. [63] and Wickiewicz et al. [55]
[a]Largest moment capacity

control forward fall of the body which is the driving force of the walking gait cycle. To further understand the importance of the plantarflexors vs. the dorsiflexors, one only has to compare the impairment of an Achilles tendon rupture compared to the dropfoot condition resulting from loss of the ankle extensor muscle group. The dropfoot patient compensates quite effectively with a steppage gait, while patients with an Achilles tendon rupture are apropulsive and relatively incapacitated.

The soleus contributes two-fold the plantarflexion moment as the medial head of the gastrocnemius. This moment increases when the ankle moves into plantarflexion and the moment arm of the Achilles moves to a perpendicular alignment to the ankle joint axis. The medial head of the gastrocnemius is a different muscle than the lateral head. The medial head has shorter muscle fibers, develops twice the ankle plantarflexion moment than the lateral, and has more inversion moment arm in all positions at the ankle compared to the lateral.

All other ankle flexors have far less torque capacity in comparison to the soleus and gastrocnemius: the flexor hallucis longus (FHL) provides only 5% the ankle flexion moment of the soleus, yet it is the preferred tendon to transfer to the Achilles in neglected rupture. Despite a shorter moment arm, the tibialis posterior is actually a stronger ankle plantar flexor than the FHL.

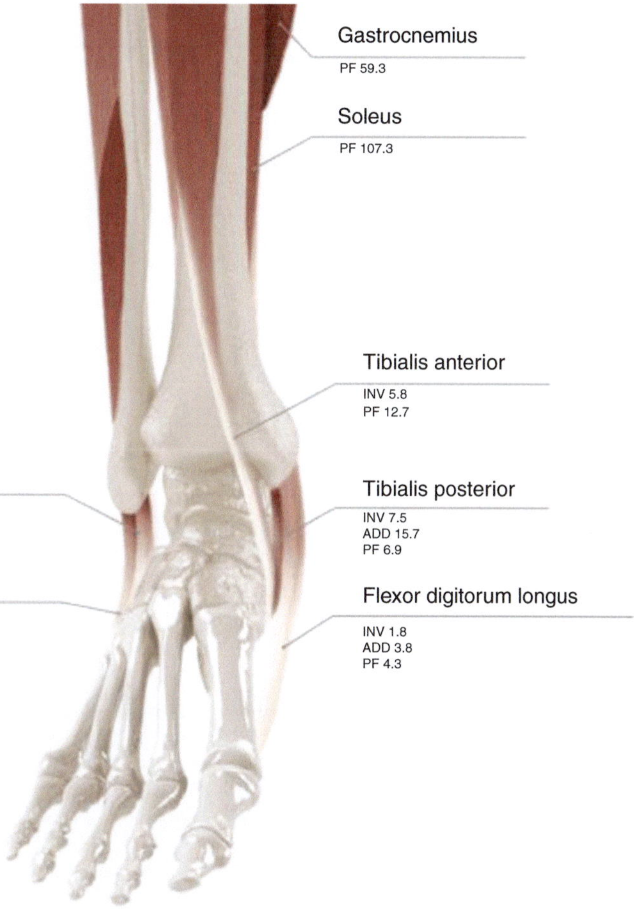

Fig. 3.15 Summary of muscle moment capacity (Newton-meters) in 3 planes

The tibialis anterior (TA) is the primary ankle dorsiflexor. However, dorsiflexion moment of the TA weakens with foot inversion. This may contribute to foot drop seen in patients with cavus deformity. Despite the longest moment arm, the minimal cross-sectional area of the extensor hallucis longus (EHL) makes this muscle a negligible ankle dorsiflexor. The peroneus longus (PL) has surprising ability to dorsiflex the ankle, better than either the EHL or the EDL. However, this dorsiflexion motion provided by the PL is accompanied by significant eversion of the foot.

Frontal Plane

The invertors and evertors are evenly balanced in muscle power indicating that walking places equal burden on the lower leg musculature in the frontal plane (Fig. 3.17). Contrary to popular belief, the tibialis posterior and the tibialis

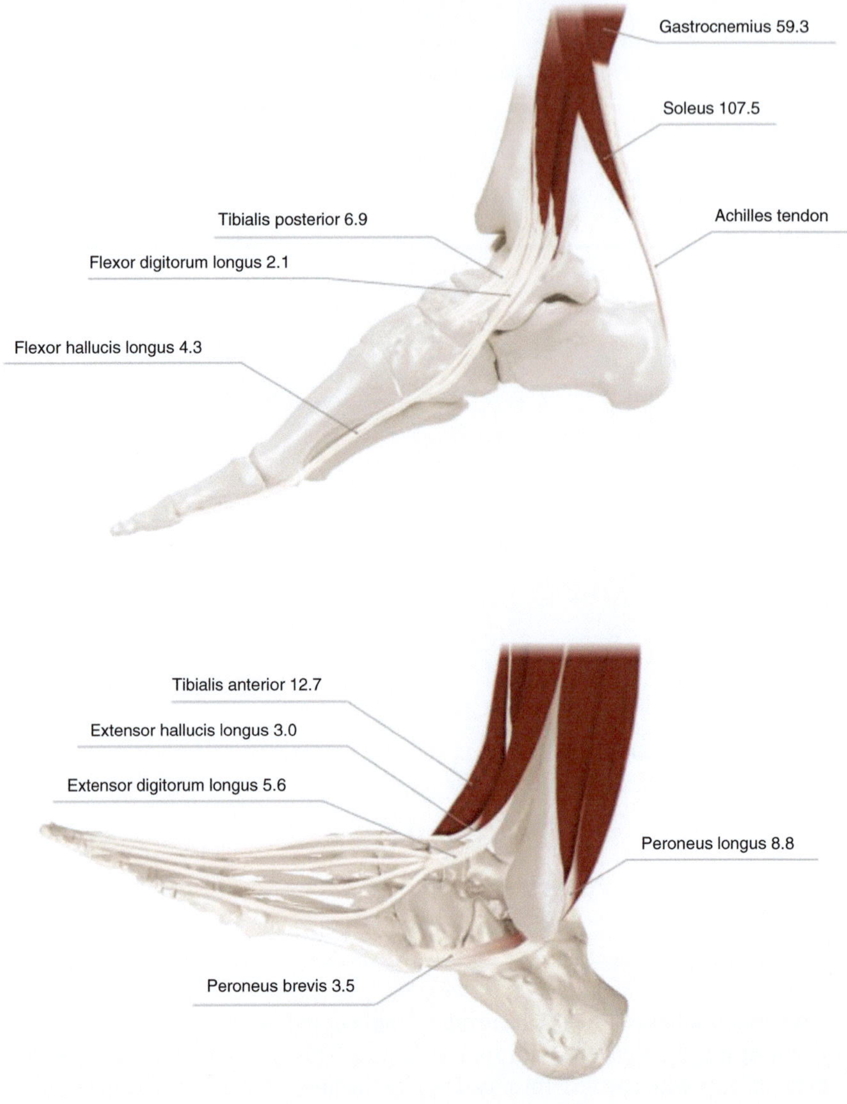

Fig. 3.16 Muscle moment capacity (Newton-meters): sagittal plane

anterior have similar inversion moment production capacity. However, the TA becomes an evertor when the foot is everted. The tibialis posterior continues to be an invertor with eversion of the foot, but moment arm for this action becomes significantly reduced with further eversion [59]. Due to favorable moment arm and larger muscle cross section, the peroneus longus (PL) is the primary everter of the foot exceeding the peroneus brevis (PB) by over threefold in overall torque production capacity. This eversion power of the peroneals is significantly reduced with ankle joint plantarflexion.

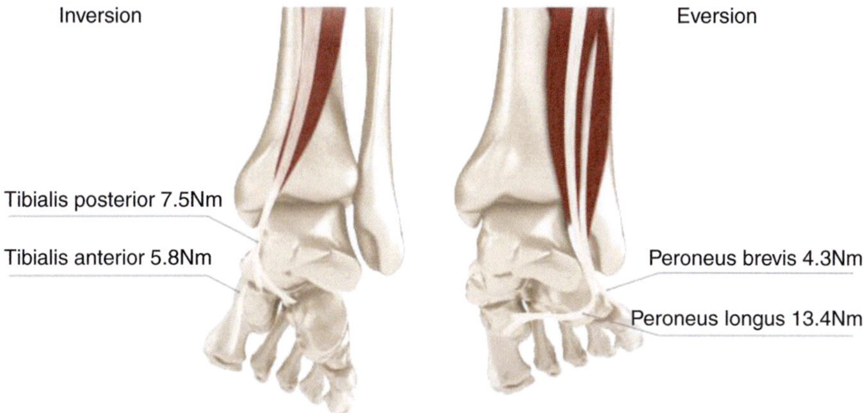

Fig. 3.17 Muscle moment capacity (Newton-meters): frontal plane

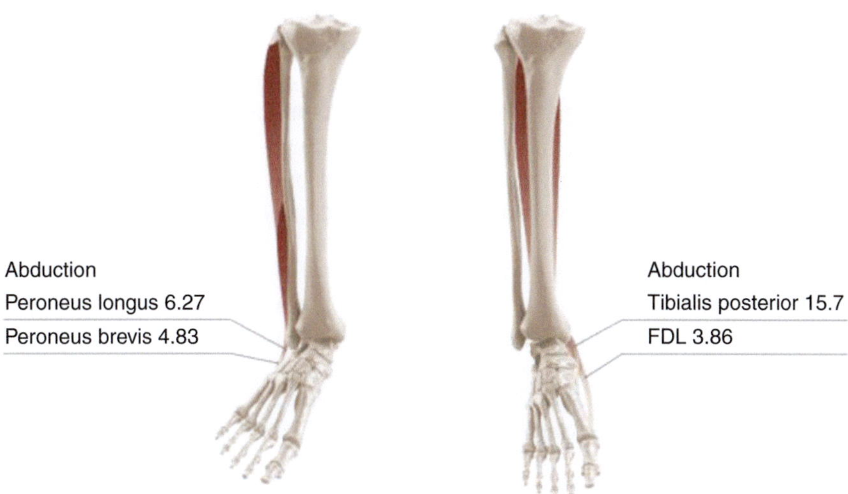

Fig. 3.18 Muscle moment capacity (Newton-meters): transverse plane

Transverse Plane

The internal rotators have almost twice the moment or torque on the foot than the external rotators (Fig. 3.18). Thus, walking demands significantly greater restraint of external rotation than internal rotation. The alignment of the foot relative to the leg during walking causes the foot to naturally abduct which places greater demand on the tibialis posterior, which is the primary restraint to external rotation. This explains why the dominant plane of deformity assumed by the foot after posterior tibial tendon rupture is external rotation or abduction rather than frontal plane valgus deformity [20]. Furthermore, overall deformity resulting from rupture of the posterior tibial tendon is significantly more severe than what occurs in the foot after rupture of the abductors or external rotators, i.e., the peroneal tendons. This may explain why tenodesis procedures using the peroneal tendons to treat lateral ankle instability do not sacrifice stability of the foot in the transverse plane.

The tibialis posterior produces twofold greater adduction (internal rotation) moment than it does inversion moment. This muscle should be considered first and foremost as an adductor in the transverse plane rather than its secondary role as an invertor. Despite the largest moment arm for abduction, the PB does not produce the greatest moment in this direction because the PL has twice the muscle cross section and still has significant moment arm to become the primary external rotator of the foot.

Muscle Activity/Demand and Foot Type

Hypertrophy of the FDL muscle has been observed with MR imaging of patients with posterior tibial tendon insufficiency [66]. Ultrasound study revealed increased cross-sectional area of the FDL and FHL muscles in patients with pes planus compared to patients with normal arch feet [67]. This verifies the role of these muscles as arch supporters which are recruited for increased activity in pes planus.

Decreased activity of the peroneus longus and brevis has been measured in patients with pes planus compared to normal arched individuals [68, 69]. Similarly, decreased cross-sectional area of the peroneal musculature has been demonstrated with ultrasound study of patients with pes planus [67]. This suggests a decrease load on the foot "pronators" in pes planus. Conversely, increased activity of the ankle inverter muscles has been shown to increase in pes planus [68].

The Plantar Intrinsic Muscles

The plantar intrinsic muscles include the: abductor hallucis, quadratus plantae, flexor digitorum brevis, abductor digiti minimi, flexor digiti minimi, opponens digiti minimi, adductor hallucis, dorsal and plantar interossei, and lumbricales (Fig. 3.19).

The role of these muscles is just now being elucidated as key stabilizers of the foot. Understanding of the function of these muscles has been challenging due to their reduced size and depth within the foot-making surface EMG studies difficult. Early studies demonstrated that the plantar intrinsics function later in the stance phase of gait to stabilize the digits during heel rise (Fig. 3.20) [70].

There is also evidence that these muscles control pronation of the subtalar joint [71]. Later studies verified the role of pronation control of the plantar intrinsics as well as an emerging concept that these muscles were important stabilizers of the medial longitudinal arch of the foot [72, 73].

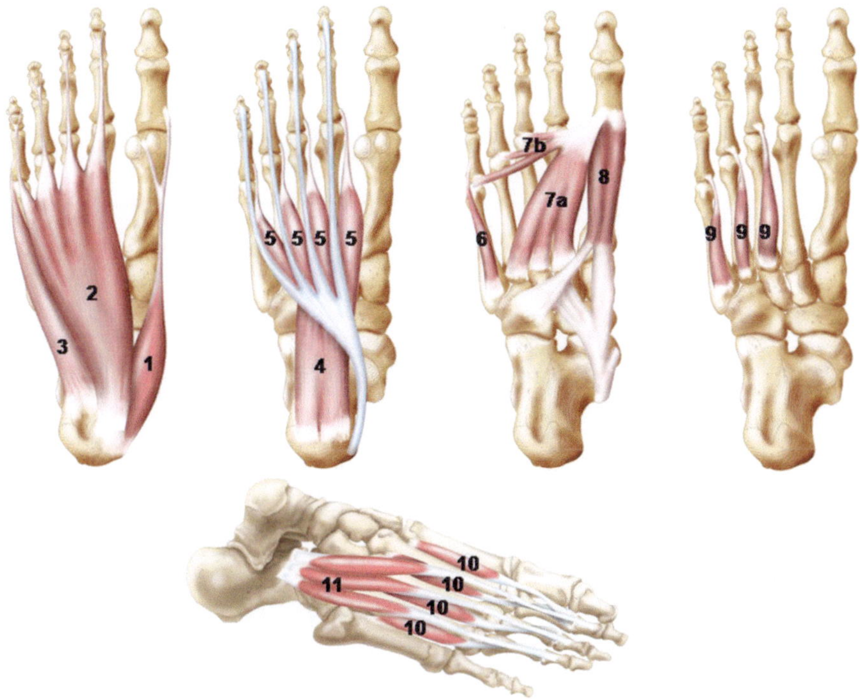

Fig. 3.19 The plantar intrinsic muscles
(1) Abductor hallucis, (2) flexor digitorum brevis, (3) abductor digiti minimi, (4) quadratus plantae (note its insertion into the flexor digitorum tendon), (5) lumbricals (note their origin from the flexor digitorum longus tendon), (6) flexor digiti minimi, (7) adductor hallucis oblique (a) and transverse (b) heads, (8) flexor hallucis brevis, (9) plantar interossei, (10) dorsal interossei, and (11) extensor digitorum brevis. (Taken from: McKeon et al. [64])

Kelly and co-workers used ultrasound-guided intramuscular EMG to study the three largest plantar intrinsic muscles, specifically the abductor hallucis, flexor digitorum brevis, and quadratus plantae [74]. Significant increase of muscle activation patterns was measured with increased postural sway suggesting that the plantar intrinsic muscles play an important role in balance and postural control. This research group subsequently studied the role of the same three plantar intrinsic muscles in stabilizing or providing stiffness to the medial longitudinal arch of the foot [75]. This study measured increased stretch and activation of the plantar intrinsic muscles with increasing load deformation of the longitudinal arch of the foot. With electrical stimulation the researchers demonstrated that the plantar intrinsic muscles act along with the plantar aponeurosis to attenuate longitudinal arch deformation during walking. Specific rotation of foot skeletal segments were measured depending on which muscle were stimulated. The abductor hallucis, the largest of all the plantar intrinsics, demonstrated kinematic effects in all three body planes resulting in calcaneal inversion as well as dorsiflexion (increased calcaneal inclination). The calcaneus was also observed to move into abduction with stimulation of the

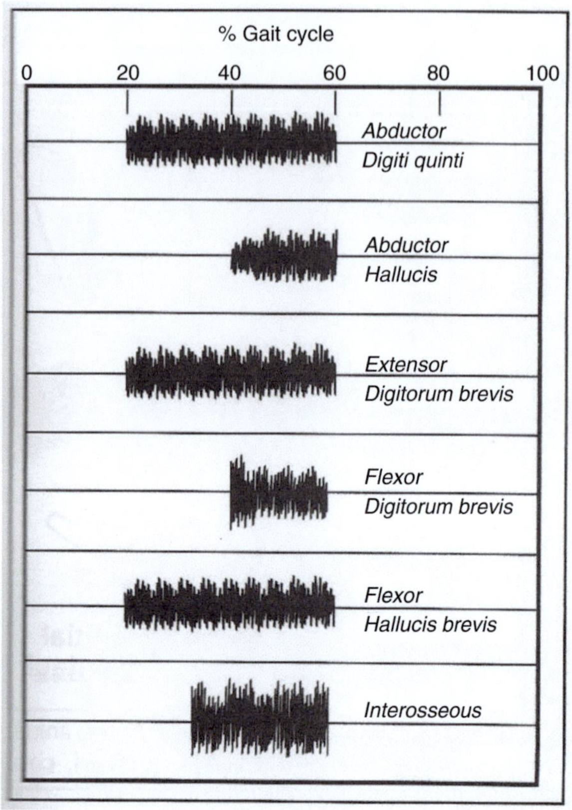

Fig. 3.20 Intrinsic foot muscle action during stance. (Adapted from Mann and Inman [71]). (From: Perry and Burnfield [52] Figs. 4–16 page 69. Reprinted with permission)

abductor hallucis indicating closed chain supination of the midtarsal joint. This was supported by the observation of flexion and adduction of the metatarsals with stimulation of the abductor hallucis muscle. The flexor digitorum brevis and quadratus plantae only exerted influence in the frontal and transverse plane. The authors speculated that the more medial location of the abductor hallucis provided better moment arm to affect the sagittal plane and height of the longitudinal arch of the foot.

Storage and Return of Energy

Other investigators have speculated that stiffness of the longitudinal arch of the foot is regulated by the active contractile elements of the plantar intrinsic muscles as well as the passive elastic elements of the plantar aponeurosis and the plantar ligaments [76, 77]. The role of the passive plantar ligament structures to provide storage of energy and elastic recoil has been dubbed the "foot spring" by Ker et al. [78]. They measured energy strain storage capacity in the plantar aponeurosis, the spring ligament, and the long and short plantar ligaments. It appears that the plantar intrinsics also respond to load with stretch and contraction to provide stability to the entire foot [76].

Decreased cross-sectional area with ultrasound study of the abductor hallucis and flexor hallucis brevis has been observed in patients with pes planus compared to normal arch feet [67, 79]. This contrasts with hypertrophy of other arch supporting muscles in the leg in pes planus such as the FHL and FDL [66, 68]. The plantar intrinsic muscles may not have to work as hard in pes planus because the plantar fascia becomes taut in pes planus and takes over plantar flexing toes, a so-called reverse windlass [80].

The critical initiation of supination motion of the rearfoot at 60% of the stance phase of gait has been validated by many kinematic studies carried out after the initial report by McPoil and Cornwall in 1994 [9]. See Figs. 3.1 and 3.3. At 60% of the stance phase of gait, which corresponds to 40% of the overall gait cycle, the swing phase foot has just passed the stance phase foot, and heel rise is initiated. This is the beginning of terminal stance.

Rearfoot supination and elevation of the calcaneus from the supportive surface rely upon elastic recoil of passive ligamentous structures in the foot as well as dynamic contraction of both extrinsic and intrinsic musculature. This point of the gait cycle marks the initiation of positive work which is carried out during the final 40% of the stance phase of gait.

During the first 60% of the stance phase of gait, certain motions of segments within the foot allow stretching, elongation, and release of energy to dissipate shock. Furthermore, the stretching of tendon and ligament structure allows storage of strain energy which is released during recoil and shortening of these structures during terminal stance and pre-swing, the final 40% of the stance phase of gait.

The longitudinal arch of the foot compresses under load during the first 60% of the stance phase of gait, resulting in lengthening and increased strain in the soft tissue supports. Elastic recoil of these ligamentous structures causes the release of stored strain energy. This compression-recoil mechanism has been called the foot-spring by Ker and is postulated to provide significant contribution to metabolic efficiency in human gait [78].

The plantar aponeurosis stretches during loading of the foot in early stance [76]. Compression and elongation of the longitudinal arch is restricted when the plantar aponeurosis stiffens at end range. In fact all the key ligamentous structures including the spring ligament and long and short plantar ligaments stiffen under load to provide overall stiffness of the foot and elastic recoil during the last half of the stance phase of gait [76]. Loading, elongation, and stiffening of key ligaments are integral to the energy storage mechanism which allows push off or propulsion during gait [77]. This loading is made possible by unique movement of the twisted plate configuration of the osteoligamentous structure of the human foot.

MacConnail observed that maximal loading and stiffening of the key ligaments of the foot occurred when the rearfoot was pronated and the forefoot was supinated [81]. These motions occur during the first 60% of the stance phase of gait. The talonavicular joint moves into dorsiflexion and abduction which causes a lowering or compression of the medial longitudinal arch. This motion also loads the spring ligament and the plantar aponeurosis. Pronation of the rearfoot lowers calcaneal pitch, increasing strain on the long and short plantar ligaments. The human foot remains stable during the first half of the stance phase of gait due to tensioning of these key ligaments. At the same time, significant strain energy is stored in these ligaments with loading. Stearne and McDonald have demonstrated that if pronation and arch compression are inhibited by over-corrective foot orthoses, the energy storage-release mechanism of key ligaments is significantly disrupted [82, 83].

What causes the recoil of the arch supportive ligaments and release of stored strain energy? It may be a combination of dynamic contraction of certain extrinsic and intrinsic muscles. This has been named the "Foot Core" mechanism by McKeon and co-workers. They show that the tendons from the lower leg muscles, as well as the plantar intrinsic muscles and the primary ligaments on the plantar surface of the foot are oriented in a crisscross network to provide optimal lever arm for support of the medial, lateral and transverse arches (Figs. 3.14, 3.19 and 3.21).

Besides muscular contraction, there are mechanical processes where supination of the foot, driven by proximal motion in the leg, releases the plantar ligaments from their maximal state of tension. Supination occurs in the stance phase foot as the contralateral swing phase limb passes by, inducing external rotation of the pelvis, femur, and tibia in the weight-bearing extremity which is fixed on the ground. This occurs at heel rise or the beginning of terminal stance which is at 60% completion of the stance phase of gait. At the start of terminal stance, the rearfoot begins to supinate, and forefoot plantarflexion occurs from dynamic muscular action as well as elastic recoil of the plantar fascia and arch ligaments. The arch of the foot will begin to raise in terminal stance reaching a maximum height after toe-off [84–86].

Twisting the Plate Stores and Releases Energy

The osteoligamentous structure of the human foot "untwists" during the first half of the stance phase of gait to absorb energy, dissipate shock, and adapt to terrain irregularity. This osteoligamentous structure then recoils and "twists" into rearfoot inversion and forefoot pronation. This results in significant increase in arch height during the final 15% of stance [84–86]. Nester showed that the first ray undergoes approximately 15° of plantar flexion from the start of terminal stance to the end of pre-swing [28]. The beginning of terminal stance with heel rise corresponds to maximal tensile strain in the plantar fascia along with peak intrinsic muscle activity [87–89]. Recoil of the medial slip of the plantar aponeurosis as well as the abductor hallucis, flexor hallucis longus, and brevis cause stiffening of the first ray, plantarflexion of the medial column, and elevation of the medial longitudinal arch during terminal stance and pre-swing (Fig. 3.22). The peroneus longus is integral to stiffening of the first ray as it abducts the first metatarsal into a close-packed position against the intermediate cuneiform. The peroneus longus also performs a critical function during terminal stance to pronate the forefoot

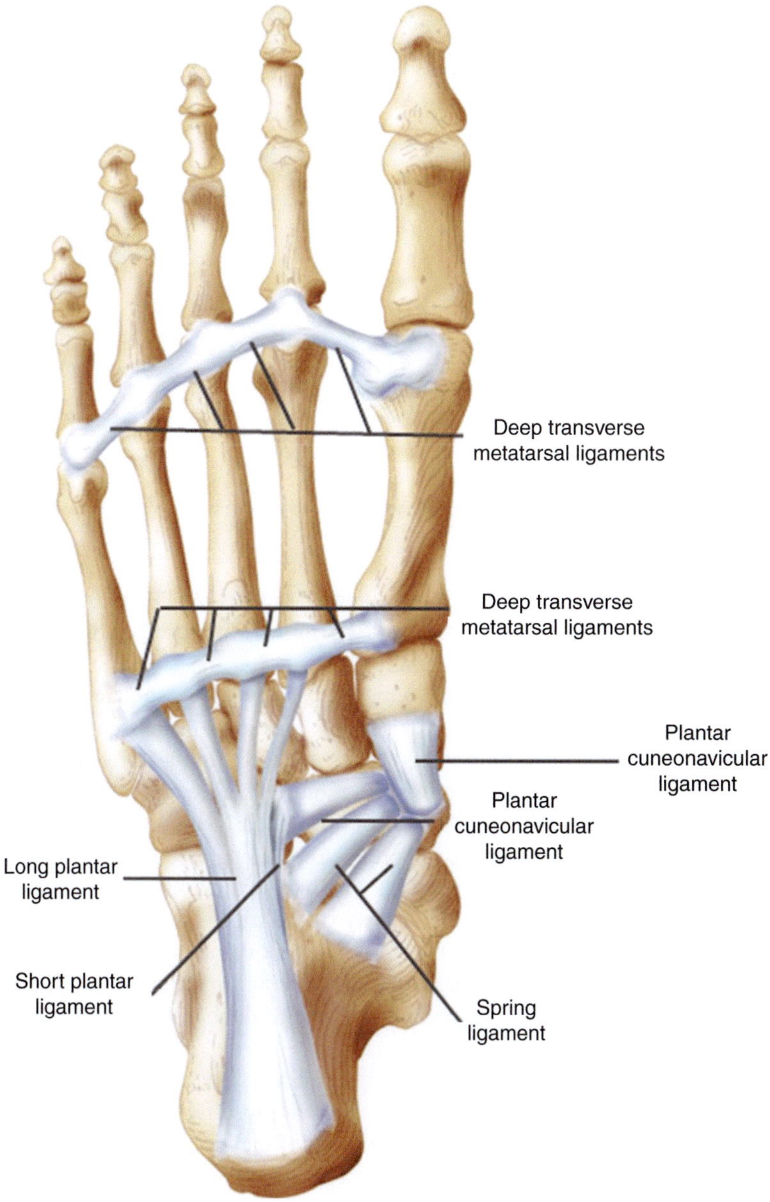

Fig. 3.21 The plantar ligaments which make up part of the arch spring mechanism. (Taken from: McKeon et al. [64])

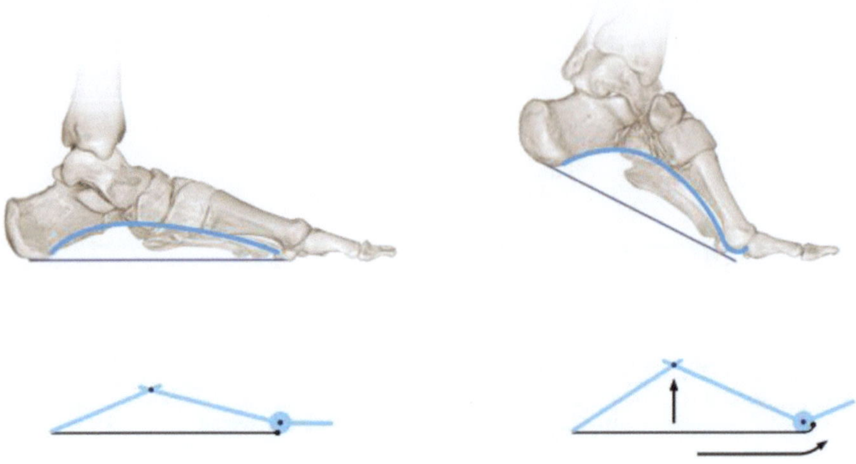

Fig. 3.22 The windlass is coupled with plantarflexion of the forefoot initiatited from elastic recoil of the long flexor tendons

which transfers load from lateral to medial across the first MTPJ to engage the windlass mechanism. This is all part of the "twisting" of the forefoot on the rearfoot as originally described by MacConnail [81].

Contrary to popular belief, the windlass mechanism by itself cannot explain elevation and shortening of the arch of the foot during terminal stance and pre-swing. The arch of the foot raises before the windlass is engaged, and a study has shown that approximately 20° of digital dorsiflexion is required before arch raising is seen during terminal stance [90]. While tightening of the plantar fascia and activity of the plantar intrinsics in terminal stance are important arch raising mechanisms, Wearing proposes that other factors are also operating which raise the arch prior to push off [91]. During terminal stance contractile activity of the triceps drops off rapidly to become totally inactive by pre-swing [84, 92]. This removes an arch lowering force even though the ankle is plantar flexing from elastic recoil in the Achilles. There is also unweighting of the foot due the commencement of double limb support at the start of pre-swing [93, 94].

The critical motions within the foot during terminal stance and pre-swing are unique to humans and are clearly part of an energy storage and release mechanism (Fig. 3.23). Instead of a simple "rigid lever" concept of the foot in propulsion previously embraced by most authorities, the human foot is now recognized as a spring-like structure which undergoes considerable motion during push off .

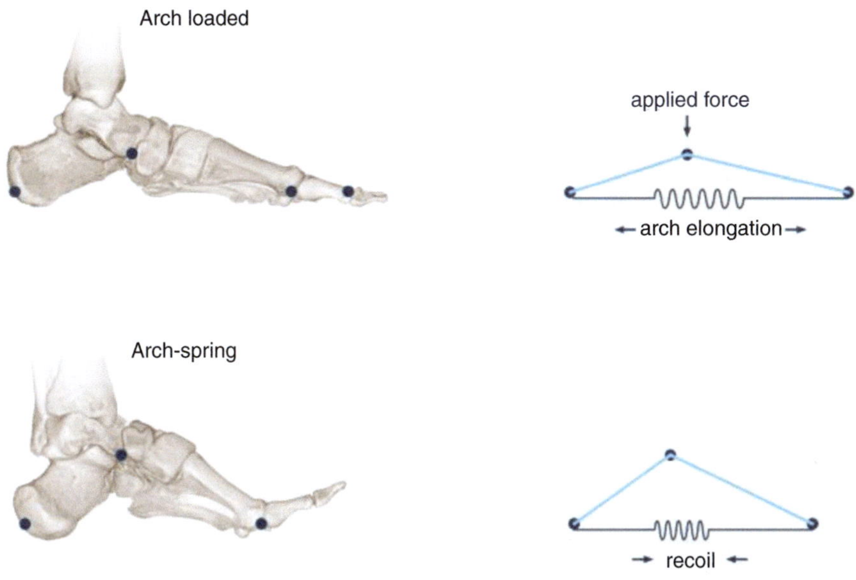

Fig. 3.23 The arch spring mechanism utilizes stored elastic energy during arch compression in midstance and releases energy in terminal stance

References

1. Root ML, Orien WP, Weed JH. Clinical biomechanics: Normal and abnormal function of the foot, vol. 2. Los Angeles: Clinical Biomechanics Corp; 1977. p. 138–9.
2. Root ML, Weed JH, Sgarlato TE, Bluth DR. Axis of motion of the subtalar joint. J Am Pod Assoc. 1966;56(4):149–55.
3. Close JR, Inman VT, Poor PM, Todd FN. The function of the subtalar joint. Clin Orthop. 1967;50(1–2):159–79.
4. Inman VT, Ralston HJ, Todd FN. Human Walking. Baltimore: Williams and Wilkins Company; 1981.
5. Wright DG, Desai ME, Henderson WH. Action of the subtalar and ankle-joint complex during the stance phase of walking. J Bone Joint Surg. 1964;46-A(2):361–82.
6. Elftman H, Manter J. The evolution of the human foot with special reference to the joints. J Anat. 1935;70:56–67.
7. Hamill J, Bates BT, Knutzen KM, Kirkpatrick GM. Relationship between selected static and dynamic lower extremity measures. Clin Biomech. 1989;4:217–25.
8. Mueller MJ, Norton BJ. Reliability of kinematic measurements of rear-foot motion. Phys Ther. 1992;72:731–7.
9. McPoil TG, Cornwall MW. Relationship between neutral subtalar joint position and pattern of rearfoot motion during walking. Foot Ankle Int. 1994;15:141–5.
10. Apkarian J, Naumann S, Cairns B. A three-dimensional kinematic and dynamic model of the lower limb. J Biomech. 1989;22(2):143.

11. Kadaba MP, Ramakrishnan HK, Wootten ME. Measurement of lower extremity kinematics during level walking. J Orthop Res. 1990;8(3):383–92.
12. Davis RB, Ounpuu S, Tyrburski D, Gage JR. A gait analysis data collection and reduction technique. Hum Mov Sci. 1991;10:575–87.
13. Deschamps K, Staes F, Roosen P, Desloovere K, Bruyninckx H, Matricall G. Body of evidence supporting the clinical use of 3D multisegment foot models: a systematic review. Gait Posture. 2011;33(3):338–49.
14. Leardini A, Benedetti MG, Catani F, Simoncini L, Giannini S. An anatomically based protocol for the description of foot segment kinematics during gait. Clin Biomech. 1999;14:528–36.
15. Cornwall MW, McPoil TG. Three-dimensional movement of the foot during the stance phase of walking. J Am Podiatr Med Assoc. 1999;89:56.
16. Cornwall MW, McPoil TG. Relative movement of the navicular bone during normal walking. Foot Ankle Int. 1999;20:507.
17. Carson MC, Harrington ME, Thompson N, O'Connor JJ, Theologis TN. Kinematic analysis of a multi-segment foot model for research and clinical applications: a repeatability analysis. J Biomech. 2001;34:1299–307.
18. Hunt AE, Smith RM, Torode M, Keenan AM. Inter-segment foot motion and ground reaction forces over the stance phase of walking. Clin Biomech. 2001;16:592–600.
19. Woodburn J, Nelson KM, Siegel KL, Kepple TM, Gerbber LH. Multisegment foot motion during gait: proof of concept in rheumatoid arthritis. J Rheumatol. 2004;31(10):1918–27.
20. Rattanaprasert U, Smith R, Sullivan M, Gilleard W. Three-dimensional kinematics of the forefoot, rearfoot, and leg without the function of tibialis posterior in comparison with normals during stance phase of walking. Clin Biomech (Bristol, Avon). 1999;14:14–23.
21. Ward ED, Smith KM, Cocheba JR, Patterson PE, Phillips RD. In vivo forces in the plantar fascia during the stance phase of gait: sequential release of the plantar fascia. J Am Podiatr Med Assoc. 2003;93(6):429–42.
22. Nester CJ, Liu AM, Ward E, Howard D, Cocheba J, Derrick T, Patterson P. Invitro study of foot kinematics using a dynamic walking cadaver model. J Biomech. 2007;40(9):1927–37.
23. Lundgren P, Nester C, Liu A, Arndt A, Jones R, Stacoff A, Wolf P. Lundberg a:invasive in vivo measurement of rear-, mid- and forefoot motion during walking. Gait Posture. 2008;28(1):93–100.
24. Nester CJ, Liu AM, Ward E, Howard D, Cocheba J, Derrick T, Patterson P. In vitro study of foot kinematics using a dynamic walking cadaver model. J Biomech. 2007;40(9):1927–37.
25. Astion DJ, Deland JT, Otis JC. Motion of the hindfoot after simulated arthrodesis. JBJS. 1997;79-A:241.
26. Wulker N, Stukenborg C, Savory KM, Alfke D. Hindfoot motion after isolated and combined arthrodeses: measurements in anatomic specimens. Foot Ankle. 2000;21:921.
27. Lundberg A, Svensson OK, Bylund C, et al. Kinematics of the ankle/foot complex. Part 2: pronation and supination. Foot Ankle. 1989;9:248.
28. Nester CJ, Jarvis HL, Jones RK, Bowden PD, Anmin L. Movement of the human foot in 100 pain free individuals aged 18–45: implications for understanding normal foot function. J Foot Ankle Res. 2014;7:51.
29. Buldt AK, Murley GS, Butterworth P, Levinger P, Menz HB, Landorf KB. The relationship between foot posture and lower limb kinematics during walking: a systematic review. Gait Posture. 2013;38(3):363–72.
30. Hillstrom HJ, Song J, Kraszewski AP, Hafer JF, Mootanah R, Dufour AB, Chow BS, Deland JT 3rd. Foot type biomechanics part 1: structure and function of the asymptomatic foot. Gait Posture. 2013;37(3):445–51.
31. Lephart SM, Riemann BL, Fu FH. Introduction to the sensorimotor system. In: Lephart SM, Fu FH, editors. Proprioception and neuromuscular control in joint stability. Champaign: Human Kinetics; 2000. p. 37–51.
32. Riemann BL, Lephart SM. The sensorimotor system, part I: the physiologic basis of functional joint stability. J Athl Train. 2002;37(1):71–9.

33. Fernando M, Crowther R, Lazzarini P, et al. Biomechanical characteristics of peripheral diabetic neuropathy: a systematic review and meta-analysis of findings from the gait cycle, muscle activity and dynamic barefoot plantar pressure. Clin Biomech. 2013;28:831–45.
34. Herb CC, Chinn L, Dicharry J, McKeon PO, Hart JM, Hertel J. Shank-rearfoot joint coupling with chronic ankle instability. J Appl Biomech. 2014;30(3):336–72.
35. Koldenhoven RM, Feger MA, Fraser JJ, Saliba S, Hertel J. Surface electromyography and plantar pressure during walking in young adults with chronic ankle instability. Knee Surg Sports Traumatol Arthrosc. 2016;24(4):1060–70.
36. Koldenhoven RM, Feger MA, Fraser JJ, Hertel J. Variability in center of pressure position and muscle activation during walking with chronic ankle instability. J Electromyogr Kinesiol. 2018;38:155–61.
37. Richie DH. Functional instability of the ankle and the role of neuromuscular control. A comprehensive review. J Foot Ankle Surg. 2001;40(4):240–51.
38. Riemann BL, Lephart SM. The sensorimotor system, part II: the role of proprioception in motor control and functional joint stability. J Athl Train. 2002;37(1):80–4.
39. Grillner S. The role of muscle stiffness in meeting the changing postural and locomotor requirements for force development by the ankle extensors. Acta Physiol Scand. 1972;86:92–108.
40. Louie JK, Mote CD Jr. Contribution of the musculature to rotatory laxity and torsional stiffness at the knee. J Biomech. 1987;20:281–300.
41. Sinkjaer T, Toft E, Andreassen S, Hornemann BC. Muscle stiffness in human ankle dorsiflexors: intrinsic and reflex components. J Neurophysiol. 1988;60:1110–21.
42. Rack PM, Ross HF, Thilmann AF, Walters DK. Reflex responses at the human ankle: the importance of tendon compliance. J Physiol. 1983;344:503–24.
43. Grabiner MD. Bioelectric characteristics of the electromechanical delay preceding concentric contraction. Med Sci Sports Exerc. 1986;18:37–43.
44. Beckman SM, Buchanan TS. Ankle inversion injury and hypermobility: effect on hip and ankle muscle electromyography onset latency. Arch Phys Med Rehabil. 1995;76(12):1138–43.
45. Munn J, Sullivan SJ, Schneiders AG. Evidence of sensorimotor deficits in functional ankle instability: a systematic review with meta-analysis. J Sci Med Sport. 2010;13(1):2–12.
46. Sefton JM, Hicks-Little CA, Hubbard TJ, et al. Segmental spinal reflex adaptations associated with chronic ankle instability. Arch Phys Med Rehabil. 2008;89(10):1991–5.
47. Ciccotti MG, Kerlan RK, Perry J, Pink M. An electromyographic analysis of the knee during functional activities, II: the anterior cruciate ligament deficient knee and reconstructed profiles. Am J Sports Med. 1994;22:651–8.
48. McKeon PO, Hertel J. Systematic review of postural control and lateral ankle instability. Part 2: is balance training clinically effective? J Athl Train. 2008;43(3):305–15.
49. Taube W, Gruber M, Beck S, Faist M, Gollhofer A, Schubert M. Cortical and spinal adaptations induced by balance training: correlation between stance stability and corticospinal activation. Acta Physiol (Oxf). 2007;189:347–58.
50. Taube W, Kullmann N, Leukel C, Kurz O, Amtage F, Gollhofer A. Differential reflex adaptations following sensorimotor and strength training in young elite athletes. Int J Sports Med. 2007;28:999–1005.
51. Beck S, Taube W, Gruber M, Amtage F, Gollhofer A, Schubert M. Task-specific changes in motor evoked potentials of lower limb muscles after different training interventions. Brain Res. 2007;1179:51–60.
52. Perry J. Gait analysis: Normal and pathological function. 2nd ed. Thorofare: SLACK inc; 2010.
53. Kirby KA. Subtalar joint axis location and rotational equilibrium theory of foot function. J Am Pod Med Assoc. 2001;91(9):465–87.
54. Silver RL, de la Garza J, Rang M. The myth of muscle balance. J Bone Joint Surg. 1985;67-B:432–7.
55. Wickiewicz TL, Roy RR, Powell PL, Edgerton VR. Muscle architecture of the human lower limb. Clin Orthop Rel Res. 1983;179:317–25.

56. Zajac FE. Muscle coordination of movement: a perspective. J Biomech. 1993;26(Suppl 1):109–24.
57. Maganaris CN. In vivo measurement-based estimations of the moment arm in the human tibialis anterior muscle-tendon unit. J Biomech. 2000;33:375–9.
58. Klein P, Mattys S, Rooze M. Moment arm length variations of selected muscles acting on talocrural and subtalar joints during movement: an in vitro study. J Biomech. 1996;29:21–30.
59. Hintermann B, Nigg BM, Sommer C. Foot movement and tendon excursion: an in vitro study. Foot Ankle Int. 1994;15:386–95. https://doi.org/10.1016/0021-9290(95)00025-9.
60. An KN, Takahashi K, Harrigan TP, Chao EY. Determination of muscle orientations and moment arms. J Biomech Eng. 1984;106:280–2.
61. Lee SM, Piazza SJ. Inversion-eversion moment arms of gastrocnemius and tibialis anterior measured in vivo. J Biomech. 2008;41:3366–70.
62. Wang R, Gutierrez-Farewik EM. The effect of subtalar inversion/eversion on the dynamic function of the tibialis anterior, soleus, and gastrocnemius during the stance phase of gait. Gait Posture. 2011;34:29–35.
63. McCullough MB, Ringleb SI, Arai K, Kitaoka HB, Kaufman KR. Moment arms of the ankle throughout the range of motion in three Planes. Foot Ankle Int. 2011;32(3):300–6.
64. McKeon PO, Hertel J, Bramble D, Davis I. The foot core system: a new paradigm for understanding intrinsic foot muscle function. Br J Sports Med. 2015;49:290.
65. Hui HJ, Beals TC, Brown NT. Influence of tendon transfer site on moment arms of the flexor Digitorum longus muscle. Foot Ankle Int. 2007;28(4):441–7.
66. Wacker J, Calder JD, Engstrom CM, Saxby TS. MR morphometry of posterior tibialis muscle in adult acquired flat foot. Foot Ankle Int. 2003;24(4):354–7.
67. Angin S, Crofts G, Mickle KJ, Nester CJ. Ultrasound evaluation of foot muscles and plantar fascia in pes planus. Gait Posture. 2014;40:48–52.
68. Murley GS, Menz HB, Landorf KB. Foot posture influences the electromyographic activity of selected lower limb muscles during gait. J Foot Ankle Res. 2009;2(1):35.
69. Hunt AE, Smith RM. Mechanics and control of the flat versus normal foot during the stance phase of walking. Clin Biomech. 2004;19(4):391–7.
70. Gray EG, Basmajian JV. Electromyography and cinematography of leg and foot ("normal" and flat) during walking. Anat Rec. 1968;161:1–15.
71. Mann R, Inman VT. Phasic activity of intrinsic muscles of the foot. J Bone Joint Surg Am. 1964;46:469–81.
72. Headlee DL, Leonard JL, Hart JM, Ingersoll CD, Hertel J. Fatigue of the plantar intrinsic foot muscles increases navicular drop. J Electromyogr Kinesiol. 2008;18:420–5.
73. Fiolkowski P, Brunt D, Bishop M, Woo R, Horodyski M. Intrinsic pedal musculature support of the medial longitudinal arch: an electromyographic study. J Foot Ankle Surg. 2003;42:327–33.
74. Kelly LA, Kuitunen S, Racinais S, Cresswell AG. 2012 recruitment of the plantar intrinsic foot muscles with increasing postural demand. J Clin Biomech. 2012;27:46–51.
75. Kelly LA, Cresswell AG, Racinais S, Whiteley R, Lichtwark G. Intrinsic foot muscles have the capacity to control deformation of the longitudinal arch. J R Soc Interface. 2014;11:20131188.
76. Caravaggi P, Pataky T, Günther M, Savage R, Crompton R. Dynamics of longitudinal arch support in relation to walking speed: contribution of the plantar aponeurosis. J Anat. 2010;217:254–61.
77. Bates KT, et al. The evolution of compliance in the human lateral mid-foot. Proc R Soc Lond Ser. 2013;B 280:1818.
78. Ker RF, Bennett MB, Bibby SR, Kester RC, Alexander RM. The spring in the arch of the human foot. Nature. 1987;325:147–9.
79. Angin S, Mickle KJ, Nester CJ. Contributions of foot muscles and plantar fascia morphology to foot posture. Gait Posture. 2018;61:238–42.
80. Aquino A, Payne C. The role of the reverse windlass mechanism in foot pathology. J Am Podiatr Med Assoc. 2000;34:32–4.
81. MacConaill MA. The postural mechanism of the human foot. Proc R Ir Acad. 1944-45;50B:265–78.

82. Stearne SM, McDonald KA, Alderson JA, North I, Oxnard CE, Rubenson J. The foot's arch and the energetics of human locomotion. Sci Rep. 2016;6:19403.
83. McDonald KA, Stearne SM, Alderson JA, North I, Pires NJ, Rubenson J. The role of arch compression and metatarsophalangeal joint dynamics in modulating plantar fascia strain in running. PLoS One. 2016;11:e0152602.
84. Hunt AE, Smith RM, Torode M. Extrinsic muscle activity, foot motion and ankle joint moments during the stance phase of walking. Foot Ankle Int. 2001;22:31–41.
85. Cashmere T, Smith R, Hunt A. Medial longitudinal arch of the foot: stationary versus walking measures. Foot Ankle Int. 1999;20:112–8.
86. Hunt AE, Smith RM, Torode M, et al. Inter-segment foot motion and ground reaction forces over the stance phase of walking. Clin Biomech. 2001;16:592–600.
87. Yang SM, Kayamo J, Norimatsu T, et al. Dynamic changes of the arches of the foot during walking. In: Winter DA, Norman RW, Wells RP, et al., editors. Biomechanics IX: international series on biomechanics. Champaign: Human Kinetics; 1985. p. 417–22.
88. Welte L, Kelly LA, Lichtwark GA, Rainbow MJ. Influence of the windlass mechanism on arch-spring mechanics during dynamic foot arch deformation. J R Soc Interface. 2018;15:20180270.
89. Farris DJ, Kelly LA, Cresswell AG, Lichtwark GA. The functional importance of human foot muscles for bipedal locomotion. PNAS. 2019;116(5):1645–50.
90. Phillips RD, Law EA, Ward ED. Functional motion of the medial column joints of the foot during propulsion. J Am Podiatr Med Assoc. 1996;86:474–86.
91. Wearing SC, Smeathers JE, Urry SR, Henning EM, Hills AP. The pathomechanics of plantar fasciitis. Sports Med. 2006;36(7):585–611.
92. Kameyama O, Ogawa R, Okamoto T, et al. Electric discharge patterns of ankle muscles during the normal gait cycle. Arch Phys Med Rehabil. 1990;71:969–74.
93. Blanc Y, Balmer C, Landis T, et al. Temporal parameters and patterns of the foot roll over during walking: normative data for healthy adults. Gait Posture. 1999;10:97–108.
94. Siegel KL, Kepple TM, O'Connell PG, Gerber LH, Stanhope SJ. A technique to evaluate foot function during the stance phase of gait. Foot Ankle. 1995;16:764–70.

Disorders of the First Ray: Part 1 Hallux Abductovalgus Deformity

4

The most debated concept about foot and ankle pathology may be the role of instability of the first ray in various foot disorders.

–Corinne Van Beek and Justin Greisberg

Key Points

1. The first ray moves about an axis which allows equal amounts of dorsiflexion with inversion and plantarflexion with eversion.
2. In pes planus foot posture, the axis of the first ray allows greater range of transverse plane adduction and abduction.
3. Hallux abductovalgus (HAV) deformity results from excessive transverse plane adduction of the first ray.
4. Constrictive footwear, causing chronic lateral deviation of the hallux, is an extrinsic risk factor for HAV deformity.
5. Intrinsic risk factors for HAV deformity include isolated weakness of the abductor hallucis, inflammatory joint disease, long first metatarsal, and metatarsus adductus deformity.
6. The appearance of a "pronated" or everted first metatarsal in HAV deformity is the result of midfoot or hindfoot pronation, not from independent eversion of the first ray.

D. H. Richie Jr, *Pathomechanics of Common Foot Disorders*,
https://doi.org/10.1007/978-3-030-54201-6_4

Introduction

Instability or dysfunction of the first ray is involved in more pathologies than any other segment of the human foot. Hallux abductovalgus, hallux rigidus, hammertoes, metatarsalgia, and flatfoot deformity are all commonly associated with first ray dysfunction. At the same time, the biomechanics of the first ray continues to be misunderstood and misrepresented in articles about hallux valgus treatment, most notably those published by foot and ankle surgeons.

Anatomy

The first ray is composed of the first metatarsal and the medial cuneiform bones (Fig. 4.1). The main articulations of the first ray include the first metatarsocuneiform joint (first TMT joint) and the medial naviculocuneiform joint. Other articulations of the first ray include those between the first metatarsal and second metatarsal, the medial cuneiform articulation with the second metatarsal, as well as the articulation between the medial and intermediate cuneiform.

Stability of the first ray is dependent upon specific anatomic structures which can provide either static or dynamic support, or both. The peroneus longus muscle provides a significant dynamic stabilizing force on the first ray both in the direction of plantarflexion and abduction. The plantar fascia provides both static and dynamic stability to the medial arch of the foot and to the first ray. Static stability to the first

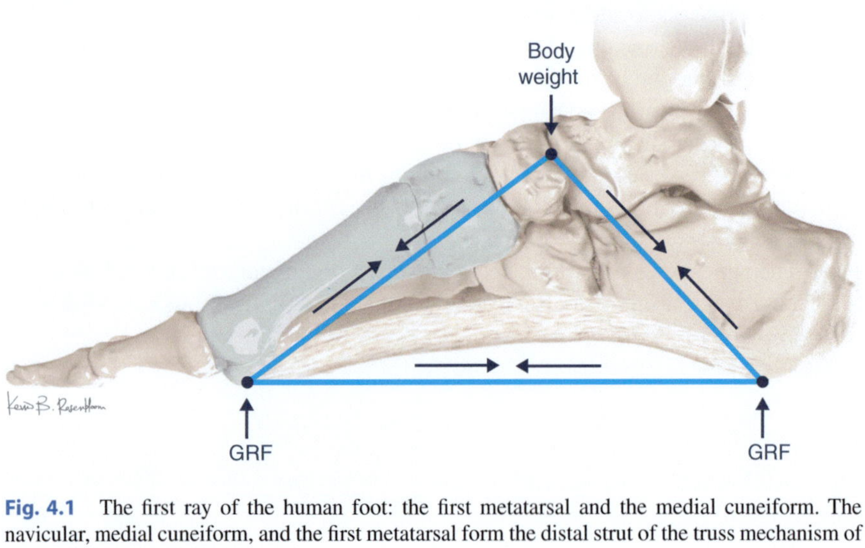

Fig. 4.1 The first ray of the human foot: the first metatarsal and the medial cuneiform. The navicular, medial cuneiform, and the first metatarsal form the distal strut of the truss mechanism of the medial longitudinal arch

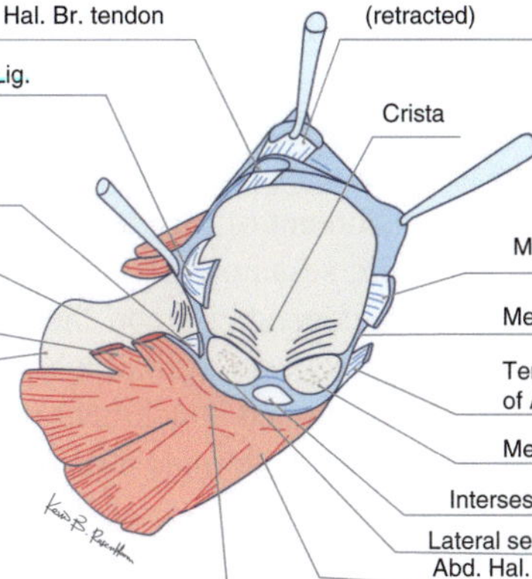

Fig. 4.2 Soft tissue structures of the first MTPJ surrounding the head of the first metatarsal

ray is provided proximally by the intermetatarsal ligaments, plantar metatarsocuneiform ligaments, and interosseous ligaments.

Stability of the first ray is also dependent upon muscles and ligaments surrounding the first metatarsophalangeal joint (first MTPJ). Failure of some of these soft tissue structures will lead to significant loss of stability of the first ray in both the transverse and sagittal planes. The soft tissucs surrounding the first MTPJ consist of seven muscles, eight ligaments, and two sesamoid bones (Fig. 4.2)

The seven muscles surrounding the 1st MTPJ are:

1. Abductor hallucis
2. Medial head of the flexor hallucis brevis
3. Lateral heads of the flexor hallucis brevis
4. Flexor hallucis longus
5. Adductor hallucis
6. Extensor hallucis brevis
7. Extensor hallucis longus

The eight ligaments are:

1. The medial sagittal hood
2. Lateral sagittal hood
3. Medial sesamoid
4. Lateral sesamoid
5. Intersesamoid
6. Medial collateral
7. Lateral collateral
8. Transverse metatarsal

Of these 15 soft tissue structures, only 4 actually attach to the first metatarsal: the medial and lateral sesamoid ligaments and the medial and lateral collateral ligaments. These ligaments attach to the medial or lateral tubercle of the head of the first metatarsal. At this relatively small point of origin, the four ligaments attach the sesamoid complex to the first metatarsal and to the proximal phalanx of the hallux providing an indirect attachment of four intrinsic muscles which influence the first MTPJ. In reality, the entire sesamoid complex including muscular attachments of the first MTPJ are totally dependent upon the integrity of attachment of the four wings of the collateral ligaments to a small tubercle on each side of the head of the first metatarsal. This small, limited attachment is vulnerable to attrition and rupture in HAV deformity. This leads to complete disruption of stability of all seven muscles acting across the first MTPJ.

A key specialization of the human foot is the presence of a well-developed plantar aponeurosis which has strong insertions into the proximal phalanx of all 5 digits. It will be shown in Chap. 4 that the plantar aponeurosis, via the sesamoid complex, provides the most important sagittal plane stabilizing influence on the first ray. Another unique feature of the human foot is the larger diameter of the first metatarsal head, compared to the lesser metatarsals. The diameter of the first metatarsal is functionally enlarged by the sesamoid bones. The combined radius of the first metatarsal and the sesamoids provides a longer lever arm for the flexor hallucis brevis tendons as well as the distal slips of the plantar aponeurosis to act on the 1st MTPJ (Fig. 4.3). These two structures invest the medial and lateral sesamoids and then extend into a strong insertion on the proximal phalanx of the hallux. The enhanced stabilizing force of these structures on the hallux facilitates load bearing of first ray and hallux during heel rise and toe off. The mechanism whereby the plantar

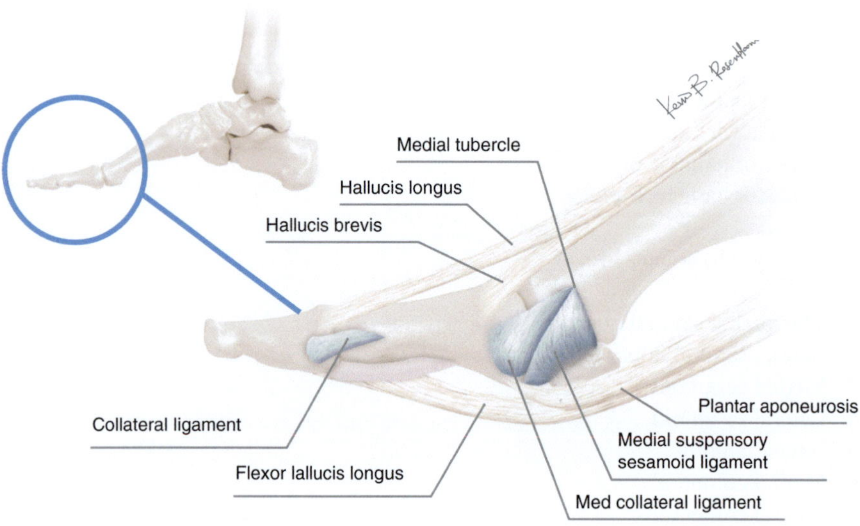

Fig. 4.3 The plantar aponeurosis, medial suspensory sesamoid, and medial collateral ligament structures of the first MTPJ

aponeurosis winds around the head of the first metatarsal with extension of the hallux has been described by Hicks as a "windlass mechanism" which is critical to stability of the medial longitudinal arch of the foot [1].

Specialization of the First Ray

Based on fossil evidence, the first ray is the last or final structure to have evolved or to have achieved specialization in the human foot. In so doing, the first ray provides the most unique feature which differentiates the human foot from all other primate feet. In studying fossil specimens, Kidd postulates that the lateral column of the foot evolved sooner than the medial column [2]. The lateral column of the human foot contains a bone enlargement or "beak" on the plantar-medial margin of the cuboid which allows a close packing or locking of the calcaneal-cuboid joint. The under hanging beak of the cuboid is seen in humans as well as fossil specimens of pre-humans, but it is not seen in apes. See Fig. 1.2 in Chap. 1. In Kidd's study of a pre-human fossil specimen, the cuboid beak was present; however the medial column showed an increased talar neck angle and increased talar head torsion angle, similar to what is seen in apes. In humans, the reduction of the neck angle and reduction of head torsion angle of the talus are features which reduced divergence of the first ray and also provided stability to the midtarsal joint.

If we look at ontogeny alone, the order of ossification of the bones of the medial and lateral column verifies the more recent specialization of the first ray. Indeed, when examining the sequence of chondrification and ossification of the bones of the human foot in-utero, this hierarchy of the lateral column over the medial column for specialization is repeated. The central metatarsals chondrify first followed by the 5th metatarsal and the cuboid. The talus and the calcaneus are next to chondrify. The first metatarsal and medial cuneiform follow, and the navicular is the last bone of the foot to chondrify. Clearly, the medial column lags behind the lateral column in the order of maturation. This may explain why the medial column is responsible for more foot pathologies in humans than the lateral column.

Metatarsus primus varus is a medial deviation of the first metatarsal and is a deformity often associated with hallux abductovalgus (HAV) deformity. The medial or varus alignment of the first metatarsal is significant in the fetus, measuring 32 degrees at 8 weeks. This medial angulation of the first metatarsal then reduces to 6–10 degrees in the adult foot. Retention of metatarsus primus varus and medial angulation of the first metatarsal- medial cuneiform joint is a hallmark of juvenile HAV deformity.

Sagittal plane motion in the medial column during gait approximates the same amount of motion which occurs in the lateral column. In the frontal and transverse planes, the talonavicular joint provides the dominant motion, while the joints of the first ray and the joints of the lateral column (calcaneal-cuboid and cuboid-5th metatarsal) contribute equally. This is an important recent finding from bone pin studies which refutes a long-standing belief that the lateral column of the foot is a relatively rigid segment compared to the medial column.

Many myths about contribution of motion of the joints of the first ray continue to be propagated in the medical literature. This will be further discussed in Chap. 6. Most notably, the so-called condition of "first ray hypermobility" is erroneously thought to occur at the first metatarsocuneiform joint. Surgeons commonly fuse this joint as a remedy for hypermobility. Not only is the concept of first ray hypermobility greatly misunderstood, the notion that it occurs in this joint is clearly refuted by many quality studies.

Motion occurring at the medial naviculocuneiform joint is twice that of the motion occurring at the first metatarsocuneiform joint [3]. This dispels the myth that the first metatarsal medial cuneiform joint is the primary contributor to "hypermobility" of the first ray in many foot pathologies. In fact, the bone pin studies verify that the first metatarsal- medial cuneiform joint moves less than any other joint of the foot. Therefore, fusion of this joint would not be expected to have any significant effect on overall motion or function of the foot.

Direction and Range of Motion of the First Ray

The first ray moves in the direction of dorsiflexion, combined with inversion and slight adduction (towards the midline). Plantarflexion is combined with eversion and slight abduction. This motion is not a "pronation-supination" motion as often erroneously reported in the surgical literature. Direction of motion of the first ray was first described by Hicks in 1953, using an external jig to track bone motion in 15 cadaver specimens [4]. Hicks described an axis of rotation of the first ray which is positioned in the transverse plane or parallel to the weight-bearing surface and runs from the navicular to base of the third metatarsal, oriented 45 degrees from the transverse and sagittal planes. In Fig. 4.4 the orientation of this axis demonstrates that the amount of dorsiflexion is equivalent to the amount of inversion; plantarflexion is equivalent to eversion, while transverse plane motion is minimal.

In 1979, the author was part of a research team headed by Merton Root who attempted to verify the findings of Hicks as they isolated pure first ray motion of the foot in 24 cadaver specimens [5]. Average range of motion measured in the specimens in the sagittal plane was 12 mm (measured with calipers), and total frontal plane motion was 8 degrees (measured with a bubble level). In every cadaver specimen, first ray dorsiflexion was accompanied by inversion, while plantarflexion was accompanied by eversion. The entire foot was fixed with an apparatus so that motion of the first ray was observed relative to the foot itself as well as to the ground, with no distortion or contribution of neighboring joint motion.

In 1999, Johnson and Christensen used a sophisticated three-dimensional tracking device placed on the bones of the first ray of seven cadaver specimens to measure direction and range of motion [6]. In each specimen, first ray dorsiflexion was combined with inversion, while plantarflexion was combined with eversion. In this study, tensioning the peroneus longus tendon contributed more motion in the frontal plane rather than the sagittal plane. The authors concluded that rotation of the first ray in the direction of eversion combined with plantarflexion is critical for stiffening this segment in the sagittal plane. Johnson and co-workers described a "locking

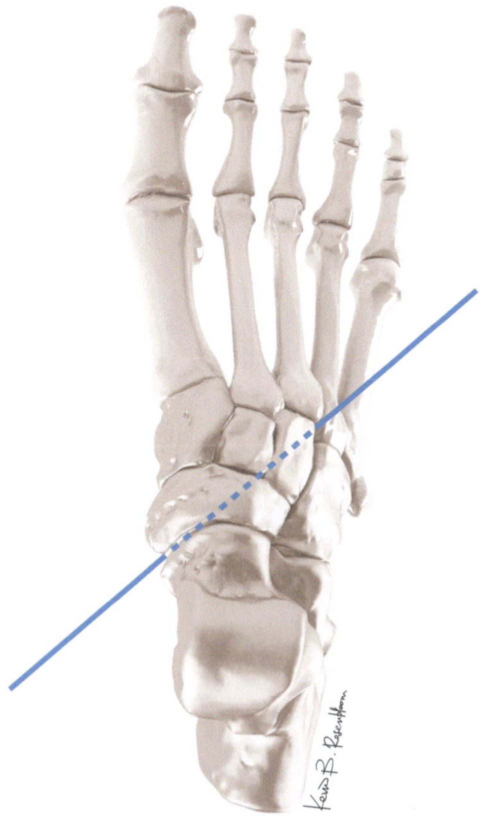

Fig. 4.4 The axis of the first ray

"mechanism initiated by the peroneus longus whereby the first ray is stabilized into a plantarflexed, abducted, and everted position.

Perez and co-workers further verified a locking mechanism of the first ray [7]. This cadaver study measured significant reduction of sagittal plane motion when the first ray was moved from a position of inversion to a position of eversion. Perez et al. propose that when the first metatarsal is positioned into plantarflexion and eversion, then the first ray becomes close packed with less range of motion available due to anatomic blocking at the interface between the medial cuneiform bone and the base of the second metatarsal (Fig. 4.5).

The axis of motion of the first ray plays a critical role in the pathomechanics of hallux abductovalgus deformity (HAV). In the human body, bone segments rotate, and this motion then allows calculation of an axis of rotation. This axis is a theoretical or imaginary line about which the bone segments of a joint rotate upon in a perpendicular plane. The axis is likened to the pin of a hinge which rotates in a perpendicular plane. Not all joints in the human body rotate like a simple hinge. Many have a helical motion around multiple joint axes.

A common misunderstanding is that an axis of joint range of motion is a real physical entity. In fact, an axis of rotation cannot be plotted and imagined unless the

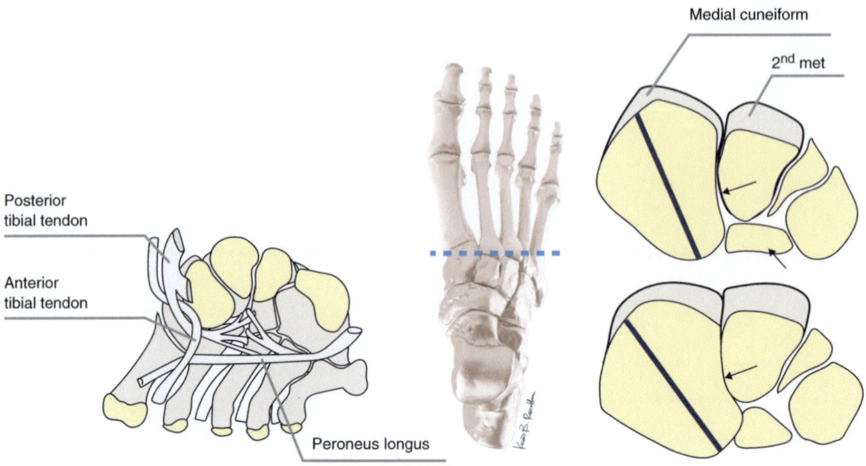

Fig. 4.5 Osseous locking of the first ray via peroneus longus eversion

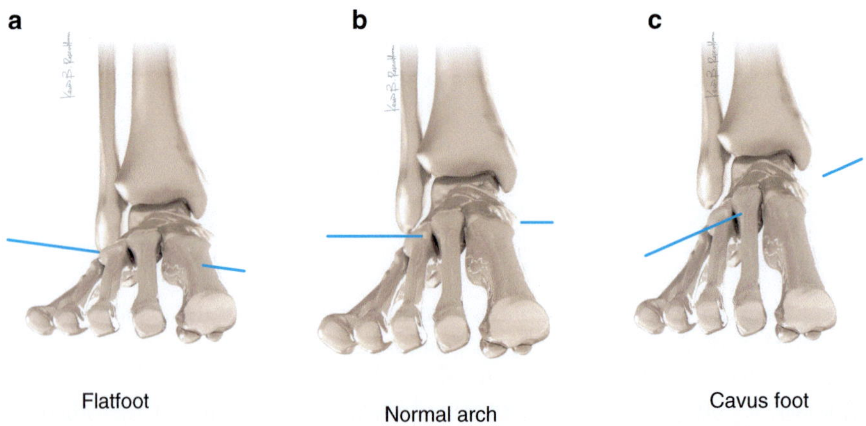

Fig. 4.6 Foot posture determines orientation of the axis of the first ray. (**a**) Pes planus orients axis more vertical to allow more adduction transverse plane motion (**b**) Normal arch allows minimal transverse plane motion (**c**) Cavus feet align first ray to favor abduction in transverse plane

joint actually moves. The joint movement defines the axis and not vice versa. An important concept with joint axes is the fact that the location and orientation of the axis are dependent upon the position of the bone segments adjacent to the joint. In other words, joint axes can "shift" depending on the position of the bone segments both proximal and distal to the joint.

The navicular, cuneiform, and first metatarsal can move closer to the ground with arch flattening which will change the orientation of the first ray axis of motion to become more vertically oriented from its usual horizontal-oriented position (Fig. 4.6). This vertical orientation will allow more transverse plane range of motion, something which is not seen in average or in normal arched feet. This variation of axis orientation of the first ray will be explored in more detail when we discuss pathomechanics of HAV deformity.

Another common misconception when evaluating first ray range of motion is determining position of the first metatarsal relative to the ground versus position of the first metatarsal relative to the rest of the foot. The first ray is capable of moving independently from the rest of the foot. When the first ray is loaded during weight bearing, ground reaction force will move the first metatarsal into dorsiflexion and inversion. If no other joints in the foot were allowed to move during this loading, the first metatarsal would be inverted *relative to the rest of the foot*. The orientation of first metatarsal *relative to the ground* is dependent upon the position of the foot itself. If the foot, moving at joints proximal to the first ray, i.e., midtarsal and subtalar joints, rotates into the direction of pronation, the first ray would be carried into the direction of pronation with the foot, *once all joint range of motion within the first ray was taken up*. Thus, while the first ray might be dorsiflexed and inverted relative the rest of the foot, global foot pronation carries the first ray into eversion *relative to the ground*. Researchers have mistakenly looked at the everted position of the first ray relative to the ground in weight-bearing subjects and have concluded that the axis of motion proposed by Hicks was incorrect [8–11]. However, if these researchers had been able to isolate pure first ray motion, *relative to the foot,* they would clearly verify that the studies by Hicks, Kelso et al., and Johnson et al. were entirely accurate [4–6].

When evaluating the intrinsic risk factors for HAV, it becomes apparent that *whole foot* pronation is a major contributor to the pathomechanics of the deformity. Pronation of the entire foot will significantly affect position of the first metatarsal on the ground. Position of the first ray relative to the ground versus relative to the rest of the foot is an important concept in hallux valgus surgery as well as flatfoot surgery.

Hallux Abductovalgus Deformity

Hallux valgus is a progressive deformity of the foot characterized by lateral deviation of the great toe, combined with medial deviation of the first metatarsal. Often, a frontal plane valgus rotation of the hallux occurs, therefore validating the fully descriptive term hallux abductovalgus (HAV). Since its first description by Carl Hueter in 1870, the cause of this condition has yet to be fully elucidated and agreed upon [12]. There are certainly a host of extrinsic and intrinsic factors which have been implicated, but no single dominant cause has been identified. As a result, it is universally accepted that hallux abductovalgus has multifactorial origins. Not surprisingly, there are many different surgical procedures used to correct HAV deformity, with no consensus about which procedure gives the best outcome.

A review of extrinsic risk factors for HAV deformity will be presented, followed by a detailed description of the pathomechanics of the condition. This will be followed by a review of proposed intrinsic risk factors which have been debated for years in terms of their overall ranking or hierarchy of importance in causing this deformity to develop.

Extrinsic Risk Factors for HAV

Two extrinsic risk factors recognized for causing HAV deformity are *improper footwear* and *overuse of the first MTPJ* due to occupation or lifestyle. In terms of the latter, the literature does not support occupation or lifestyle as a direct contributor to risk of HAV, although repetitive abnormal loading of the first MTP over life due to intrinsic biomechanical imbalance has certainly been shown to be a causative factor.

The most common extrinsic factor which has been validated in causing HAV deformity is improperly fitted footwear. Multiple studies have shown a clear relationship of HAV and patients who wear shoes with elevated heels and/or constrictive toe boxes [13–16]. The contribution of footwear to risk of HAV is illustrated in published studies demonstrating that risk of HAV increases when women who are barefoot during early life change to a shoe-wearing lifestyle [17–19].

Most of these studies of footwear and HAV focused on female subjects. Some will question whether being female is a risk factor by itself? To add more intrigue, it appears that footwear, regardless of fit or design, will negatively affect women more than men. In a classic study conducted by I.B. Shine, a large group of habitually shod vs unshod people were compared for incidence of HAV deformity [20]. The study was conducted on the island of St. Helena, where roughly half the population wear shoes on a daily basis and half remain barefoot all their life.

When comparing these two groups, Shine found that risk of developing HAV increased in an almost linear fashion with each year of wearing shoes. Hallux valgus deformity, regardless of severity, was found in only 2% of the unshod population, while it was prevalent in 16% of the men and 48% of the women who had worn shoes for over 60 years. Why women were at far greater risk is confounding in this study as the style of shoes worn by men and women on the island of St. Helena is identical and would not be described as constrictive or "inappropriate" in terms of design. This propensity for women to develop HAV at a significant higher frequency compared to men cannot be fully attributed to footwear.

In a survey of over 2000 women, Menz and co-workers found a graded increase in the risk of hallux valgus with increasing narrowness of the toe box in footwear worn between 20 and 29 years of age and, to a lesser extent, at ages 30 and 39 years [21]. Interestingly, no association with heel height of shoes and risk of developing HAV was noted.

Heredity is another contributing risk factor for HAV and was evaluated along with footwear in a study of monozygotic and dizygotic female twins published by Munteanu and co-workers [22]. Shoes with constrictive toe boxes during the fourth decade were shown to be a risk factor for HAV. This finding was specific to the monozygotic twin group rather than the dizygotic twin group, and neither group showed greater risk with higher heel shoes. This study did identify a familial risk of HAV where the twin of an affected person has a fourfold increase chance of also having the deformity. However, this shared risk among twins did not appear to be due to shared genes but due to shared environment and lifestyle, particularly with the shared preference for footwear design between the two twins.

Heredity has long been speculated as being a significant risk factor for developing HAV deformity. Many studies have identified a familial risk of HAV, but did they did not evaluate environmental or intrinsic factors as suggested by Munteanu and co-workers [24–28].

Aside from footwear preferences, simply being female is a risk factor for HAV. This relationship has been cited in multiple publications, but no quality study has been conducted to accurately verify the risk and determine why [23, 29]. Studies of surgical patients undergoing hallux valgus surgery show demographic findings which have a ratio of 1:15 male-to-female [30–32]. However, these studies of surgical populations do not necessarily document the true incidence of HAV pathology in the general population. There are many other reasons why women, compared to men, seek medical treatment. Furthermore, there are different factors which could influence women versus men to make a decision to undergo corrective foot surgery. A systematic review and meta-analysis of 78 papers reporting results of 76 surveys of the general population found that the incidence of HAV was 23% in adults age 18–65 and increased to 35.7% in elderly adults over the age of 65 [33]. The pooled estimate of HAV incidence in women was 30% and 13% in males, revealing a two-fold increased chance of females compared to males of getting a bunion deformity.

There are intrinsic differences in the anatomic features of the first ray which have been cited in the literature including a rounder head of the first metatarsal, greater degree of metatarsus primus varus, as well as overall incidence of metatarsus adductus which are seen in a greater frequency in women compared to men [34–37]. Women may have other intrinsic risk factors for developing HAV deformity during life [38]. Ligament laxity has been identified to be more common in all patients with HAV deformity, and more common in women [35, 38]. However, there are no quality studies to actually verify these anecdotal reports in the literature. More severe systemic diseases causing ligament laxity such as Marfan syndrome, Ehlers-Danlos syndrome, and rheumatoid arthritis are strongly associated with increased risk of HAV [39, 40].

A study by Uchiyama and co-workers determined a higher content of type III collagen in the medial collateral ligament of the first MTPJ in HAV patients compared to healthy patients [41]. Overall stiffness and tensile modulus of the MCL was lower in HAV patients, and these changes were proposed to be the result of, rather than the cause of, injury to the soft tissue structures around the first MTPJ.

Pathomechanics of Hallux Abductovalgus Deformity

When reviewing the scientific literature, it is clear that two basic theories of pathomechanics of HAV are proposed. One theory focuses on the failure of certain soft tissue structures at the level of the first MTPJ. The other theory focuses proximally on abnormal alignment and instability of the first metatarsocuneiform joint, also known as the first tarsal-metatarsal joint (TMT). These theories propose a mechanical mechanism causing HAV deformity which is based upon several extrinsic and intrinsic contributing factors.

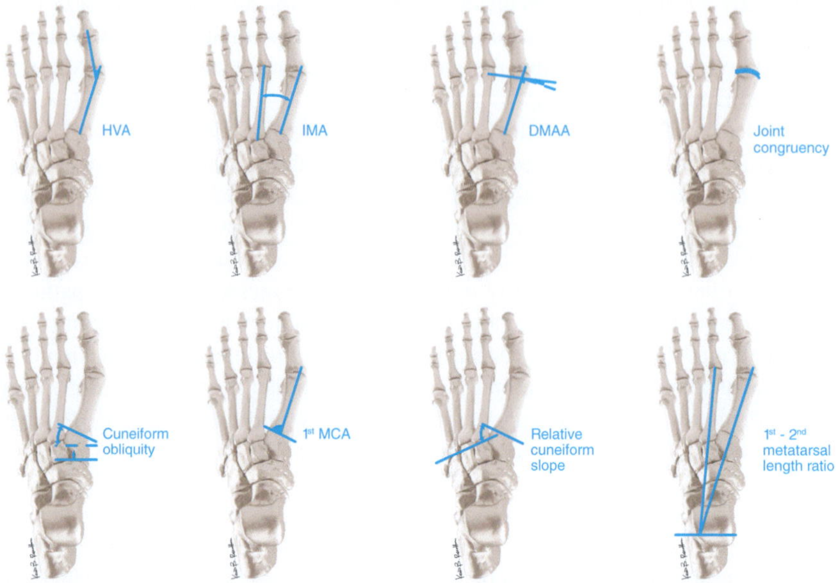

Fig. 4.7 Measurements of JVH are illustrated and include hallux valgus angle (HVA), intermetatarsal angle (IMA), distal metatarsal articular angle (DMAA), metatarsal cuneiform angle (MCA), relative first to second metatarsal length ratio (1:2 MT ratio), Meary's angle, and MTPJ congruency, which have been described and reported previously in the literature. Two novel measurements to characterize the bony morphology of the medial cuneiform-first metatarsal that we described are cuneiform obliquity (CO) and relative cuneiform slope (RCS)

HAV deformity is commonly evaluated with radiographic examination (Fig. 4.7). On weight-bearing anterior/posterior view (AP), the *hallux abductus* angle is drawn (HA angle) which measures the bisection of the hallux and the first metatarsal. When this angle exceeds 15 degrees, HAV deformity is identified. The *intermetatarsal angle* (IM angle) or the 1–2 metatarsal angle measures the bisection of the first and second metatarsals and exceeds 10 degrees in HAV deformity. Other key measurements on the AP radiograph is the angle formed by the articular cartilage on the head of the first metatarsal relative to the long axis of the bone, known as the *distal metatarsal articular angle* (DMMA) also known as the *proximal articular set angle* (PASA). Finally the alignment of the articulation between the first metatarsal and the medial cuneiform is determined by the *metatarsal cuneiform angle* (MCA) which is formed by the bisection of the first metatarsal and line connecting the medial and lateral distal extent of the medial cuneiform joint surface.

A logical question is whether deviation of the hallux laterally preceeds or causes medial deviation of the first metatarsal, or vice versa? A related controversy is, as HAV deformity progresses, whether the hallux drives the first metatarsal or whether deviation of the first metatarsal alignment determines the degree of severity of hallux valgus.

Insight into this controversy was provided by Kilmartin et al. who performed radiographic evaluation of 36 cases of unilateral and 60 cases of bilateral HAV deformity in school-age children, mean age 10 years [42]. In those cases of unilateral HAV, the unaffected foot showed increased medial angulation of the first metatarsal (IM angle) on AP radiograph compared to the feet of children without HAV deformity. Based on an assumption that most children with unilateral HAV are likely to progress to bilateral HAV later in life, the presence of medial deviation of alignment of the first metatarsal in the unaffected foot suggests that this might be the initial step in the pathomechanics of the deformity.

When HAV deformity occurs before skeletal maturity has been reached in the foot, the condition is called juvenile hallux abductovalgus. Juvenile HAV deformity may be an entirely different condition than acquired or idiopathic HAV seen in adults. Both the study by Kilmartin et al. and a more recent study by Kaiser et al. confirm malalignment of the first metatarsal, perhaps originating with deformity at the first TMT joint [43]. However, Kaiser et al. state that the first TMT joint may not be the location of the initial congenital deformity leading to juvenile HAV. They state that the distal metatarsal articular angle is the primary driving force in juvenile hallux abductovalgus deformity. In this study, the subjects were young patients (mean age 14 years), so the angulation deformity of the articular cartilage would be likely to be congenital rather than acquired.

This study also showed an obliquity, or medial angulation of the medial cuneiform as a secondary driver in juvenile HAV deformity. Caution should be taken as position of the x-ray beam, and position of the foot can greatly influence these measurements [44]. Furthermore, we still do not know whether juvenile HAV begins distal at the first MTPJ level or more proximal at the first TMT joint.

An argument against the first metatarsal being the main driver of HAV is the fact that correction of hallux valgus angle with certain surgical procedures will cause a secondary reduction of the intermetatarsal angle. Therefore, alignment of the hallux can affect alignment of the first metatarsal. Studies of arthrodesis of the first MTPJ and even with Keller arthroplasty show very impressive reduction of IM angle with simple realignment of the hallux [45–47].

Two current concept reviews of the topic of HAV have been published in the orthopedic literature in 2007 and in 2011 [48, 49]. Both papers extensively reviewed the literature, and both propose a mechanism for HAV deformity which starts with *lateral deviation of the hallux at the first MTPJ.* The extrinsic factor most likely to cause this deviation is constrictive footwear. The intrinsic factors include abnormal loading of the hallux from gait or postural alignment abnormality, ligament laxity around the first MTPJ, chronic inflammation, abnormal muscle firing, and abnormal alignment of the articular cartilage on the head of the first metatarsal.

As will be shown with the review of intrinsic risk factors, pes planus and rearfoot pronation are both strongly associated with HAV deformity. A pes planus foot posture changes alignment of the first ray axis to allow greater magnitude of transverse plane rotation. Also rearfoot pronation causes excessive medial loading of the forefoot. This will cause dorsal loading of the first ray followed by *supination of the forefoot*, untwisting the lamina pedis foot structure. Locking of the calcaneocuboid joint requires *pronation of the forefoot* and supination of the rearfoot, i.e., twisting the lamina pedis.

Medial loading of the first ray prior to osseous locking of the calcaneocuboid joint may cause inefficient engagement of the windlass mechanism. Lack of stabilization of the first ray via the peroneus longus along with excessive dorsal mobility of the first metatarsal during heel rise will cause jamming of the first MTPJ in the sagittal plane. According to Root et al., the hallux will compensate sagittal plane jamming by moving laterally, thus allowing the body to progress forward over the first MTPJ [5]. It appears that an elongated first metatarsal and/or metatarsus adductus, both proven intrinsic risk factors for HAV, can contribute to this lateral "escape mechanism" for the hallux when stability across the first MTPJ is compromised.

Primary Events in the Development of Hallux Abductovalgus Deformity

Step 1: Attenuation of Medial Structures of the First MTPJ

The medial soft tissue structures at the 1st MTPJ will attenuate under repetitive strain or load created by external forces causing lateral displacement of the hallux. The same medial soft tissue structures can be strained or attenuated from dorsal-medial migration of the head of the first metatarsal. The small discreet attachment of the medial collateral ligaments and sesamoid (suspensory) ligaments on the tubercle of the head of the first metatarsal is vulnerable to structural failure. Once the attachment of these two ligaments fails, the entire sesamoid complex and intrinsic muscular attachments to the proximal phalanx are significantly compromised.

The hallux will be forced into a lateral direction at the first MTPJ by rearfoot pronation during heel rise causing medial-to-lateral ground reaction forces directed against the hallux. Pronation of the rearfoot orients the axis of the first ray to allow greater transverse plane adduction of the first metatarsal. This mechanism occurs during terminal stance when loads on the first ray are increased. Therefore, rearfoot pronation can initiate HAV deformity by causing lateral drift of the hallux or medial deviation of the first metatarsal. Finally, lateral drift of the hallux can be induced by the extrinsic factor of constrictive footwear. Whether a lifestyle of wearing fashionable shoes alone or whether other intrinsic factors must also contribute to the mechanism of HAV has yet to be proven.

Lateral translation of the hallux or medial subluxation of the first metatarsal will create tensile strain on the following medial structures of the first MTPJ: See Fig. 4.8.

1. Tendon of the abductor hallucis
2. Tendon of the medial head of flexor hallucis brevis
3. Medial insertion of plantar aponeurosis

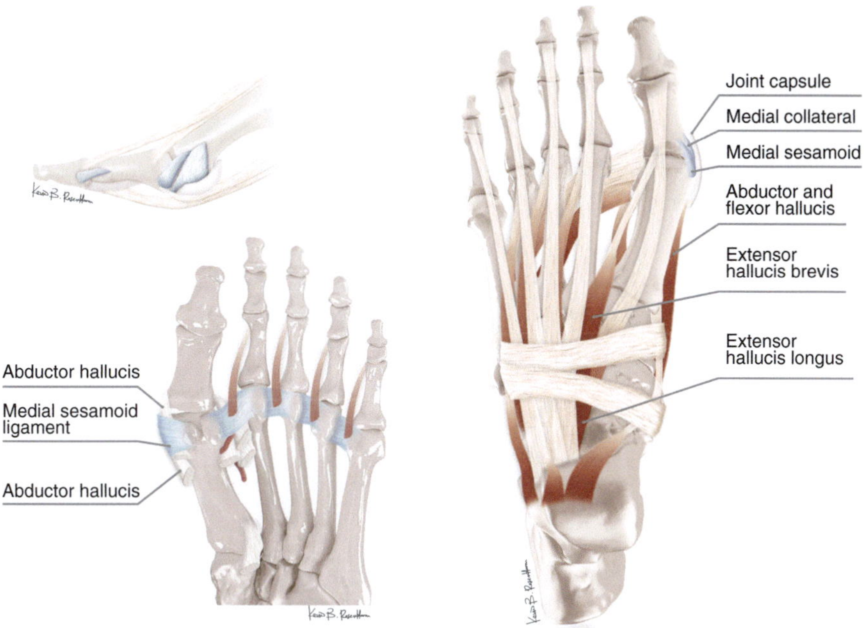

Fig. 4.8 Lateral deviation of the hallux

4. Medial collateral ligament
5. Medial sesamoid ligament
6. Medial sagittal hood ligament

Attenuation of these structures allows lateral deviation and valgus deviation of the hallux. Abnormal rotation of the hallux is initiated by external moments generated by ground reaction forces or constrictive footwear (Fig. 4.9). Also, internal moments generated by muscle forces or medial displacement of the first metatarsal can by themselves initiate the progression of HAV deformity.

The distal aspect of the first metatarsal has few ligamentous constraints and is vulnerable to retrograde force applied by the hallux. The collateral ligaments and the sesamoid ligaments are the only soft tissue structures which directly insert into the head of the first metatarsal. The medial collateral ligament controls abduction and valgus rotation of the hallux. The medial sesamoid (suspensory) ligament controls adduction or medial migration of the first metatarsal. Lateral rotation of the hallux on the head of the first metatarsal places tensile load on the medial collateral ligament. Dorsiflexion and inversion of the first metatarsal creates tensile load on the medial sesamoid (suspensory) ligament. Chronic tensile loading of the medial collateral and medial sesamoid ligaments can lead to attenuation and loss of medial stabilization of the first MTPJ. It should be noted that eversion of the first metatarsal would not place tensile load on the medial collateral ligament or the medial

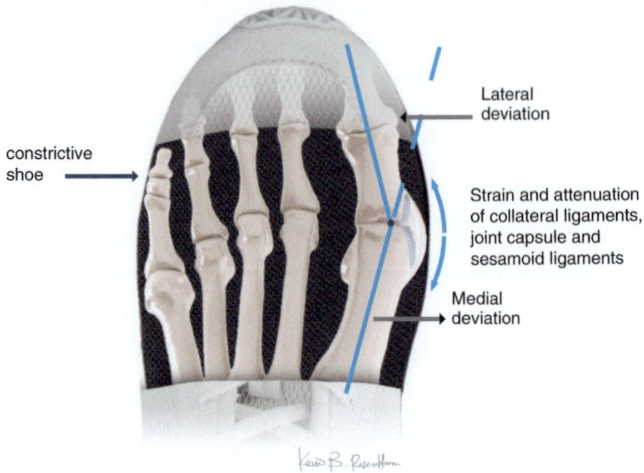

Fig. 4.9 Step 1 of the progression of HAV deformity. Lateral shift of the hallux from constrictive footwear

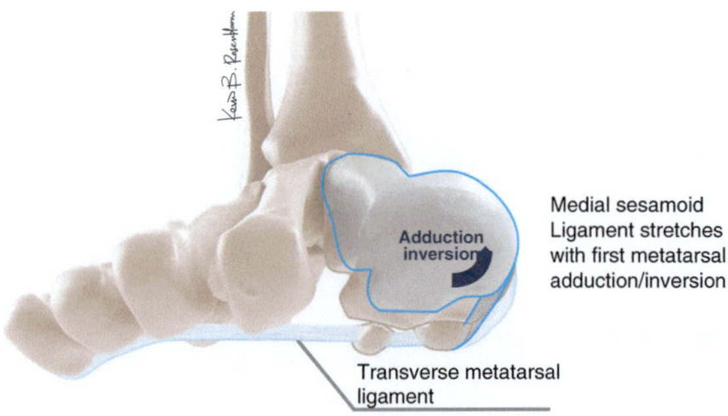

Fig. 4.10 Step 2 of HAV medial shift of the first metatarsal in HAV deformity. Note inversion as well

sesamoid ligament. If these two ligaments eventually detach from their origin on the tubercle of the head of the first metatarsal, significant instability of the entire sesamoid complex will occur. It is difficult to envision how so-called “pronation” of the first metatarsal would place strain on any of the medial structures of the 1st MTPJ. It is also important to note that the bursa which forms with chronic irritation from constrictive footwear corresponds to the location of the tubercle on the medial aspect of the head of the first metatarsal.

Step 2: Medial Shift of the First Metatarsal

At some point in the progression of HAV deformity, the first metatarsal deviates medially (Fig. 4.10). This could be caused by the lateral deviation of the hallux. However, medial shift of the first metatarsal can also be the first event of HAV deformity which occurs from the combination of pes planus and vertical alignment of the axis of the first ray. Progressive transverse plane adduction of the first ray causes pressure and attenuation of the medial capsular structures of the first MTPJ, leading to lateral drift of the hallux. Whether lateral drift of the hallux initiates HAV or whether lateral drift of the hallux follows excessive transverse plane instability of the first ray, a progressive cycle of hallux abduction along with first metatarsal adduction will ensue.

As the hallux abducts, the ground reaction force (GRF) and the action of the flexor hallucis longus resisting that force now produces a proximal and medially directed moment on the head of the first metatarsal. The magnitude of this medial force is proportional to the severity of the hallux abductus angle and increasingly works to displace the first metatarsal into adduction [50].

Step 3: Formation of the Bunion

The "bunion" deformity develops from the medial prominence of the first metatarsal head. Research has shown that there is no true bone hypertrophy to explain the medial eminence of the bunion deformity [50]. The overall width of the head of the first metatarsal is not increased in HAV deformity [51]. Clinically, the painful bunion is actually soft tissue hypertrophy around the damaged medial capsular structures as well as exposure of the medial aspect of the first metatarsal head.

Intraoperatively, what appears to be bone overgrowth is actually degeneration of the medial articular cartilage. This occurs from loss of pressure and stimulation from the base of the proximal phalanx of the hallux which has now displaced laterally. The so-called "sagittal groove" is not the demarcation of bone overgrowth but rather the point of loss of articular cartilage (Fig. 4.11). Erosion of the articular cartilage on the medial aspect of the head of the first metatarsal creates the appearance of a medial eminence of bone overgrowth, i.e. a "bunion" deformity.

The sagittal groove and medial eminence are therefore alterations of the original anatomy of the head of the first metatarsal which undergoes adaption in HAV deformity.

Step 4: Escape of First Metatarsal Away from the Sesamoid Complex

As HAV deformity progresses, the deep transverse metatarsal ligament remains intact and firmly attached the sesamoid complex. The deep transverse metatarsal ligament is not directly attached to the first metatarsal which escapes medially leaving the sesamoids behind [51, 52]. This medial escape of the first metatarsal from

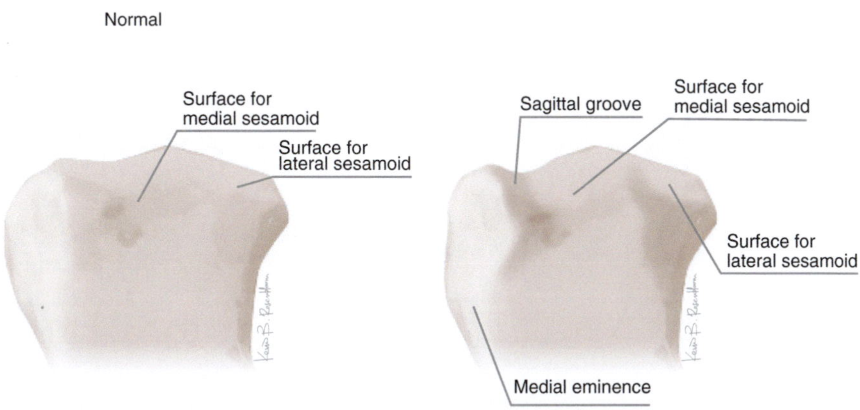

Fig. 4.11 Adaptive changes on first metatarsal head in HAV deformity

the sesamoid complex is now recognized as the primary event leading to "hypermobility" or excessive dorsal mobility of the first ray which is a hallmark of HAV deformity. The plantar aponeurosis is no longer centered under the head of the first metatarsal and the stiffening effect of the windlass mechanism is compromised. This will be discussed in detail in Chap. 4.

The axis of the first ray will dictate that the first metatarsal will move in the direction of inversion when ground reaction forces drive it dorsal, while the hallux, in retrograde fashion, drives it medial. This dorsiflexed and inverted position of the first metatarsal changes its orientation to the sesamoid apparatus, an event which is critical to the propagation of HAV deformity (Fig. 4.12).

Initially, medial drift of the first metatarsal, relative to the sesamoids, is resisted by the crista on the inferior surface of the first metatarsal. With continued medial shift of the first metatarsal combined with inversion of the first metatarsal, the crista is shifted medial, relative to the sesamoids. The crista abuts on the inferior surface of the medial sesamoid and undergoes erosion. Thus, the medial eminence, the sagittal groove, and relocation of the crista are all adaptive changes on the articular surface of the first metatarsal as HAV deformity progresses (Fig. 4.11).

While the sesamoids appear to "shift" laterally on radiographs, they actually remain tethered to the second metatarsal due the strong attachment of the deep transverse metatarsal ligament (Fig. 4.10). At the same time, the sesamoids remain firmly attached to the hallux. The soft tissue structures which invest the sesamoids and anchor to the hallux have profound influence on the deformity assumed by the hallux as HAV progresses.

Step 5: Muscle Imbalance at the First MTPJ

A muscle imbalance develops which propagates HAV deformity (Fig. 4.13). The sesamoids shift lateral, relative to the head of the first metatarsal, due to loss of integrity of the medial ligaments of the first MTPJ. The abductor hallucis shifts from a medial position to a plantar position providing a greater plantar flexion

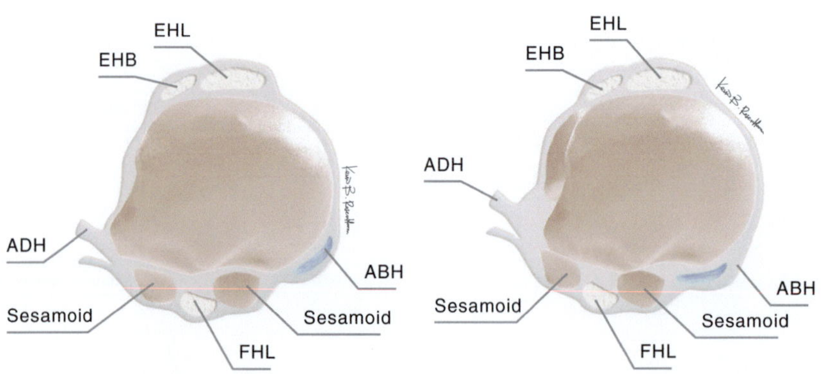

Fig. 4.12 Impingement of medial sesamoid on crista of first metatarsal head. The normal alignment (left) and valgus position of sesamoids relative to an inverted position of first metatarsal is shown on the right image

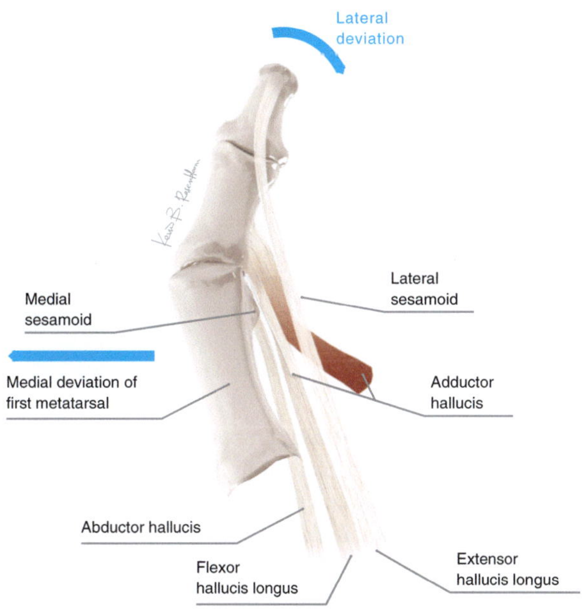

Fig. 4.13 Muscle imbalance at the first MTPJ

vector rather than an abduction vector. Keep in mind that these muscles are named according to classical anatomic terms whereby abduction refers to movement of the segment away from the second metatarsal, or in a medial direction. The adductor hallucis is now un-opposed leading to further lateral displacement of the hallux. The flexor hallucis longus, embedded between the sesamoids, is now displaced laterally and rotates with the sesamoids into a valgus position relative to the head of the first metatarsal. The two heads of the flexor hallucis brevis and the adductor hallucis also displace laterally and assume a valgus position relative to the medial and varus rotated position of the first metatarsal.

With further attenuation of the medial ligaments, the sesamoids will eventually shift lateral, relative to the first metatarsal. The lateral sesamoid may shift dorsal to the first metatarsal head, thereby pulling the sesamoid apparatus into a valgus relationship to this bone. This establishes a significant shift of vectoral alignment or direction of moment of the dynamic and static structures attaching to the sesamoid apparatus and to the hallux. The tendons of the abductor hallucis and medial head of the flexor hallucis brevis rotate in a plantar-lateral direction. This creates a muscle imbalance which causes the hallux to rotate into a valgus attitude.

In reality, the sesamoids do not displace in HAV deformity. Rather, the first metatarsal rotates relative to the sesamoids and the rest of the foot. Attenuation of the medial collateral and medial sesamoid ligaments of the first MTPJ allows the first metatarsal to escape the "grip" of the sesamoid complex. The sesamoids will appear to be "pronated" or everted on the head of the first metatarsal when viewed with a radiograph taken with an axial sesamoid view (Fig. 4.14). This everted position of the sesamoids is actually the result of first ray inversion along with errosive changes on the plantar-lateral aspect of the head of the first metatarsal.

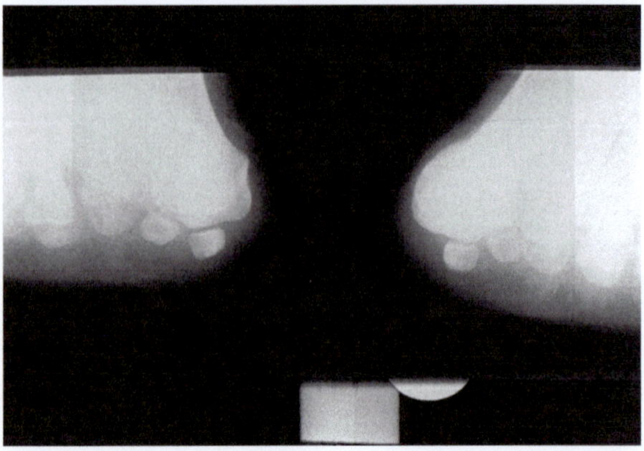

Fig. 4.14 Valgus alignment of the sesamoid complex relative to the first metatarsal in HAV deformity

Step 6: The Frontal Plane Deformity

Rotation of the first metatarsal in the direction of adduction and inversion, relative to the sesamoid apparatus, causes erosion of the articular surface between the metatarsal and the two sesamoid bones. Most notably, the crista on the inferior surface of the first metatarsal becomes eroded from the laterally displaced position of the medial sesamoid. This crista is normally critical to maintaining position of the sesamoids beneath the first metatarsal head. Loss of the crista allows further medial migration of the first metatarsal away from the sesamoids.

Increased plantar-abduction moment at the medial aspect of the proximal phalanx, without opposing plantarflexion-adduction moment applied at the lateral proximal phalanx, will cause a frontal plane imbalance at the base of the proximal phalanx, resulting in valgus rotation. Valgus rotation of the hallux may carry the sesamoids into valgus rotation away from the head of the first metatarsal, since the medial ligaments are attenuated. Thus, HAV patients will demonstrate various magnitudes of valgus or lateral rotation of the sesamoid apparatus, relative to the head of first metatarsal with plantar-axial radiographic imaging of the forefoot (Fig. 4.14). At the same time, the axis of motion of the first ray dictates that the first metatarsal will be inverted relative to the second metatarsal as it moves dorsally with loading from ground reaction forces. The more the first metatarsal rotates into a varus direction, so will the sesamoids and hallux appear deviated relative to the first metatarsal in a valgus direction.

A new trend in HAV surgery is performing a correction of alignment of the first metatarsal to reduce a perceived "pronation" deformity which has developed as part of the condition. The term pronation is inappropriate as the first metatarsal would have to dorsiflex, evert, and abduct to fulfill the requirements of this triplane direction. Investigators reporting a "pronation" deformity in HAV patients are actually describing an eversion deformity of either the first metatarsal, the sesamoid complex, or both [53–56].

The lateral and dorsal shift of alignment of the sesamoids relative to the first metatarsal can suggest a so-called "pronation" deformity of the first metatarsal in HAV deformity. Much of the evidence supporting this theory comes from studies using a plantar-axial sesamoid view with plain radiographs as well as simulated weight-bearing CT imaging. Even without first metatarsal rotation, the axial view will show an everted position of the sesamoids relative to the supportive surface in HAV deformity which is explained by the natural progression of the pathology [55]. Relative to the inverted position of the first metatarsal, the sesamoids will appear to be "pronated" or everted as they shift lateral and dorsal. As for detecting first metatarsal "pronation," the plantar axial sesamoid radiograph requires a positioning device which dorsiflexes the hallux 45 degrees causing retrograde plantarflexion and eversion of the first metatarsal. True curve beam weight-bearing CT imaging of patients with HAV deformity have refuted the erroneous reports that the first metatarsal actually rotates into eversion relative to the rest of the foot [57].

Further explanation for the apparent "pronation" of the first metatarsal position, *relative to the supportive surface*, seen in patients with HAV deformity is provided by Kimura and co-workers [58]. They compared weight-bearing CT images taken of the feet of ten patients with HAV deformity to ten healthy control patients. With loading of the foot, motion of the bones of the medial column of the foot was studied. As expected, at the first metatarsomedial cuneiform joint, the first metatarsal moved in a significant greater direction of dorsiflexion, inversion, and adduction compared to healthy controls. This follows the axis of motion of the first ray described by Hicks, Kelso, and others [4–6]. At the naviculocuneiform joint, the medial navicular actually moved in the direction of eversion in HAV patients and in the direction of inversion in healthy controls, although the magnitude was only a few degrees. At the talonavicular joint, the navicular moved on the talus in the direction of eversion in both subject groups and the magnitude of frontal plane motion at this joint exceeded the frontal plane motion at the joints of the first ray located more distal (Fig. 4.15).

To sum up the study by Kimura et al., frontal plane motion with loading in patients with HAV deformity resulted in:

- The first metatarsal moving 4.9 degrees into inversion
- The medial cuneiform moving 1.5 degrees into eversion
- The navicular moving 9.6 degrees into eversion
- The net motion of the entire medial column rotating 6.2 degrees in the direction of eversion relative to the tibia

Thus the contribution of *proximal joints* was more significant than the motion *within the first ray* itself relative to final frontal plane position. Yet surgical procedures designed to correct so called "pronation" position of the first metatarsal in HAV deformity focus on rotating the first metatarsomedial cuneiform joint [56]. This is clearly not the correct joint, and may be the wrong joint which is contributing to the overall everted position of the first metatarsal in HAV deformity [58].

In summary, any evaluation technique which shows the first metatarsal everted to the ground in HAV deformity is detecting whole foot pronation which is a hallmark of this condition. When eversion of the rearfoot and midfoot joints exceeds inversion of the first ray itself, the first metatarsal will be everted to the ground (Fig. 4.15). Remedies to correct the first metatarsal eversion should be directed to the location of the foot where this deformity originates.

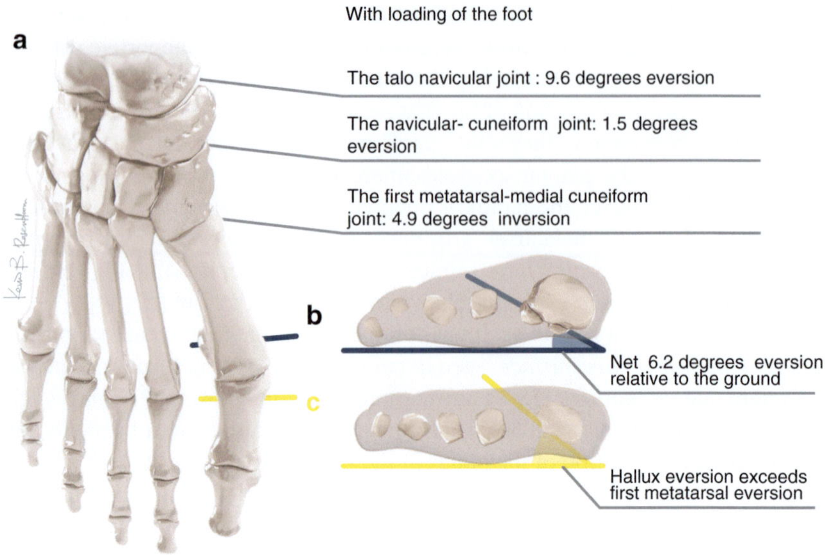

Fig. 4.15 Pronation of proximal joints causes net eversion of first metatarsal in HAV deformity. Eversion is contributed primarily by the talonavicular joint, not by the joints of the firt ray. (From: Kimura et al. [58])

Intrinsic Risk Factors for HAV Deformity

Muscle Imbalance as a Pre-disposing Factor

An interesting, and overlooked intrinsic risk factor for HAV which allows lateral drift of the hallux is weakness or abnormal activity of the abductor hallucis muscle. Studies have shown decreased bioelectric activity of the abductor hallucis in HAV patients compared to healthy controls [59, 60]. There were no differences in adductor hallucis activity when comparing the two groups.

One question is whether HAV deformity itself causes weakness or abnormal activity in the abductor hallucis muscle? In the study by Arinci et al., there was no correlation between increasing deformity of HAV with decreasing muscle activity of abductor hallucis [60]. Therefore, increased deformity did not cause decreased muscle activity. Deformity did not influence muscle activity. Therefore, diminished muscle activity of the abductor hallucis must precede the actual deformity and not be the result of it.

Besides decreased bioelectric activity, there may be intrinsic weakness of the abductor hallucis which causes HAV deformity. Stewart and co-workers studied the abductor hallucis using ultrasound and compared characteristics between patients with HAV and healthy patients [61]. They found that this muscle had decreased dorsoplantar thickness, decreased medo-lateral width, and decreased cross-sectional area in HAV patients. These differences were not related to the degree of HAV deformity, so the findings did not suggest that the muscle underwent atrophy as deformity increased. Rather, the muscle changes were consistent between mild, moderate, and severe HAV deformity suggesting that the abductor hallucis may be abnormal before the deformity begins.

Another Intrinsic Risk Factor: Chronic Inflammation

Chronic inflammatory conditions have been found to be related to HAV deformity [62, 63]. Coughlin and co-workers reported that 10% of patients undergoing HAV surgery have co-existing inflammatory joint disease, most notably rheumatoid arthritis [64]. Rheumatoid arthritis with accompanying synovitis weakens the soft tissue structures surrounding the first MTPJ leading to instability and risk of dislocation [65]. Patients with rheumatoid arthritis are susceptible to acquired pes valgus deformity which, by itself, is also a pre-disposing factor for HAV [66]. Just like juvenile HAV, patients with inflammatory joint disease may demonstrate a unique pathomechanics for the condition which is different than the more common acquired HAV deformity.

Metatarsal Length as a Risk Factor

Morton proposed that a short first metatarsal becomes hypermobile which then causes a myriad of foot problems [67, 68]. Contrary to popular teaching, Morton never attributed his theories of first ray abnormality to causing the condition known as hallux abductovalgus. More important, many studies have refuted Morton's theory that a short first metatarsal causes any foot problem, let alone HAV deformity [69–71]. In fact, a long first metatarsal has been identified as a risk factor for developing HAV deformity.

One flaw in Morton's methodology was his technique of measuring relative length of the first and second metatarsal. Hardy and Clapham reported that any medial deviation of the first metatarsal as seen in metatarsus adductus and metatarsus primus varus would distort the transverse line used as a measurement of length as proposed by Morton [72]. Grebing and Coughlin used the arc method proposed by Hardy and found that increased length of the first metatarsal was more likely in HAV patients rather than decreased length [73]. See Fig. 6.2 in Chap. 6. Grebing found an average of 2 mm increased length of the first metatarsal in HAV patients, while Hardy found an average of 4 m increased length.

Root proposed that an elongated first metatarsal was inhibited from plantarflexing below the second metatarsal [74]. This may leave the first ray in a dorsiflexed and inverted position which could initiate progressive HAV deformity.

Others have described a "buckle point" created by an elongated first metatarsal which inhibits dorsiflexion of the first MTPJ and encourages deviation of the hallux laterally to escape joint pressure [75]. Retrograde, medial deviation of the first ray can result as the hallux moves lateral from increased joint pressure. This may explain the relationship between increased metatarsal length and increased IM angle in patients with HAV [76].

Intrinsic Risk Factor: Metatarsus Adductus

Medial angulation of the metatarsals, relative to the axis of the lesser tarsus, is measured on a standing anterior/posterior radiograph and is termed the metatarsus adductus angle. Root et al. were among the first to recognize that medial angulation of the metatarsals will place the digits in an abducted position at their respective metatarsophalangeal joints [74]. This automatically places a patient in the first stage of HAV deformity and sets up a potential muscle imbalance with bowstringing of the long flexors and extensors in a lateral direction relative to their respective metatarsals. Coughlin and Jones found a statistically significant increased frequency of metatarsus adductus in their study of patients with HAV compared to the general population [32]. The presence of metatarsus adductus and increased risk of HAV has been proposed by other authors [47–50].

Intrinsic Risk Factor: Articulation Angle of the First Metatarsal

Another intrinsic deformity of the first metatarsal which could allow shift of the hallux in early HAV deformity is a deviation of the distal metatarsal articular angle (DMMA) [43] (Fig. 4.8). Podiatric literature has traditionally described this same angle as the proximal articular set angle (PASA) [77]. A laterally directed articular cartilage at the head of the first metatarsal would create a congruent joint, despite lateral deviation of the hallux at the first MTPJ. The PASA or DMMA angle is formed by drawing line which is a bisection of the long axis of the first metatarsal on an A/P radiograph and then drawing a second line which connects two dots at the medial and lateral articular margins at the head of the first metatarsal. Determining where these dots lie, or where the articular cartilage borders are on the head of the first metatarsal, is very subjective since cartilage margins are not clearly demarcated on plain radiograph.

Chi and co-workers verified previous studies demonstrating very poor inter-rater reliability in measuring DMMA [78]. Curiously, this angular deformity measured on A/P radiograph reduced with proximal metatarsal osteotomy which does not correct alignment of the articular cartilage on the head of the first metatarsal. Therefore, PASA and DMMA radiographic measurements are subject to distortion by position of the entire first ray and by the severity of hallux valgus angulation.

Extrinsic Risk Factor for HAV: Abnormal Foot Mechanics

Position of the foot during gait or static stance probably has the greatest influence on the risk for developing HAV and propagation of the deformity. Pes planus and, more importantly, excessive rearfoot pronation, has been identified to be a primary risk factor for HAV deformity [79–82]. The cause-effect relationship may revolve around excessive medial loading of the first ray and hallux. Rearfoot pronation also causes a deviation of motion of the first ray around a more vertically oriented axis, due to lowering of the medial arch of the foot. Greater excursion of the first metatarsal medially with loading would be the result of this change of first ray axis orientation.

Most reports linking pes planus with hallux valgus deformity were published decades ago and were based upon simple observation rather than valid objective data [83–86]. Later, many studies using radiographic measures were published showing no evidence supporting the theory that pes planus caused HAV deformity [87–92].

One problem with this controversy is the method of measuring or identifying a pes planus deformity in studies of patients with HAV deformity. In Coughlin's study, a Harris ink mat was used and did not find any correlation with flat arch profile and incidence or severity of HAV [32]. Only 15% of patients with HAV had moderate or severe pes planus as measured by Harris mat imprints.

In another study looking at radiographic parameters for flatfoot deformity, Coughlin and Shurnas found no correlation between hallux abductus angle and lateral talocalcaneal angle [93]. A similar finding was reported by Saragas who found no association between calcaneal pitch angle and severity of hallux valgus deformity [92]. Finally, Coughlin and Jones found no correlation between first ray mobility in HAV patients and their radiographic measures of anteroposterior talonavicular coverage, calcaneal pitch, the lateral talocalcaneal angle, and Meary's line [32].

Nix conducted a systematic review of studies which performed radiographic measures of the feet of patients with HAV deformity [13]. In this study, meta-analysis suggests that greater first intermetatarsal angle and greater first metatarsal protrusion distance, measured on dorsoplantar radiographs, are significantly associated with HAV. However meta-analysis of four studies of the flatfoot measure of calcaneal inclination angle in HAV patients and control groups showed no significant difference. Other measures of arch height including navicular height, talocalcaneal angle, talar declination angle, and first metatarsal declination angle also showed no significant differences between groups. The authors point out that inconsistent findings between studies investigating lateral radiographic measurements may be due to other factors such as differences in study samples and measurement methods.

In contrast, a more recent study using weight-bearing CT images showed a significant correlation between HAV deformity and adult acquired flatfoot [94].

Using IM angles and HA angles, the following flatfoot measurements were highly correlated in patients with AAF: talocalcaneal, talus-first metatarsal, hindfoot alignment, and talonavicular uncoverage. This was the first study to confirm an association between stage 2 AAF and two radiographic measures of HAV deformity. The authors were unable to determine a cause-effect relationship.

Flatfoot is associated with medial displacement of center of pressure of the foot upon weight bearing [95]. This may be a key factor in the development of HAV deformity, even when all radiographic parameters of flatfoot are not present.

Shibuya used center of pressure excursion index to identify flatfoot deformity and then looked at its correlation with other radiographic findings [96]. Patients with medial pressure excursion had greater tendency for radiographic measures confirming HAV deformity including intermetatarsal angle and hallux abductus angle deviation. While talar declination angle and calcaneal inclination angle was associated with HAV and flatfoot, other common radiographic angles including Meary's angle and talocalcaneal angle confirming flatfoot deformity was not correlated with the HAV group.

The most valid assessment of the role of pes planus and HAV deformity was provided by the Framingham Foot Study of 3205 individuals (mean age 66 years) providing 6393 feet for analysis [97]. Pressure mapping with the MatScan(R) allowed measurement of the Center of Pressure Excursion Index (CPEI), an indication of pronation, as well as the Modified Arch Index (MAI) which is correlated with height of the navicular. This study identified 1764 patients with hallux valgus who were compared to 3707 patients without the deformity.

The hallux valgus group had decreased loading under the hallux and increased loading under the lesser digits. Reduced loading of the hallux may be due to loss of purchase due to alteration of alignment of the plantar soft tissue restraints from the sesamoid complex.

In the Framingham Foot Study, loading under the lateral rearfoot with lower rearfoot peak pressure was reduced in the hallux valgus group compared to non-hallux valgus patients [97]. This pattern has been associated with valgus alignment of the rearfoot [95]. The CPEI values showed that patients with HAV had more pronated feet, while the MAI value showed that HAV patients had lower arch height compared to non-HAV patients. There was also significant reduced loading in the lateral half of the forefoot in HAV patients which also suggests a more pronated foot during gait.

Kinematic studies of HAV patients using three-dimensional multisegment foot models support the findings of these pressure studies. Canseco and co-workers studied 33 patients with HAV deformity and found significant gait abnormalities compared to healthy controls [98]. Patients with HAV had a slower walking velocity, decreased stride length, and prolonged stance phase compared to healthy controls. The authors also found decreased plantarflexion of the forefoot indicating flattening of the medial longitudinal arch in HAV patients. This finding as well as the loss of flexor stabilization of the hallux may explain the delay in push off seen in HAV patients, according to the authors.

Dechamps and co-workers showed similar findings in their study of HAV patients using a multisegment foot model. Patients with HAV deformity had significant eversion of the hindfoot compared to healthy controls [99]. Plantarflexion of the forefoot was diminished in both terminal stance and pre-swing in HAV patients. The authors speculated this was due to flattening (dorsiflexion) of the medial longitudinal arch as well as antalgic guarding of the first MTPJ inhibiting push off.

Pronation and HAV Deformity: How Are They Connected?

From all these studies evaluating the role of pes planus, pronation, and HAV deformity, several insights have been gained. Plain radiographic measures of flatfoot, taken in static stance, do not validate any consistent correlation with HAV deformity. Weight-bearing CT imaging may give better insight into relationship of alignment of bone segments of the foot, and early studies have shown a correlation between flatfoot deformity and HAV. Evaluation of foot function during dynamic gait has validated that pronation and medial loading of the rearfoot and forefoot is greater in HAV patients compared to healthy controls. Also, during dynamic gait, the medial arch height appears to lower more significantly in HAV patients. How these biomechanical changes can initiate HAV deformity has been elegantly explained by Glasoe in two published studies.

In their first study, Glasoe and co-workers quantified the axis of motion of the first ray in nine cadaver specimens [100]. The orientation of the first ray axis varied considerably among specimens but was directed in a more vertical direction as the arch of the foot lowered. A lower arch foot was associated with a more vertical axis orientation of the first ray. A more vertical axis of the first ray caused more abduction/adduction of the first metatarsal and increased the intermetarsal angle (IM angle) with loading.

In a subsequent paper, Glasoe et al. propose a mechanism whereby flattening of the medial longitudinal arch of the foot will contribute to HAV deformity, based upon this alteration of the first ray axis [101]. Based upon the previous work of Hicks and Kelso, Glasoe et al. suggest that the normal orientation of the axis of the first ray, which runs mediolateral in the horizontal plane, allows motion primarily in the sagittal plane and, to a lesser degree, motion in the frontal plane. The small amount of transverse plane motion available at the first ray is dependent upon the position of the navicular which is located at the most medial point of the first ray axis. With flattening of the medial longitudinal arch, the navicular will drop below the more stable third metatarsal base, which is the lateral point of the first ray axis (Fig. 4.6). This causes a shift of alignment of the axis of the first ray into a more vertical direction allowing more transverse plane motion of this skeletal segment. With loading of the first ray during midstance and terminal stance, dorsiflexion of the first metatarsal would be accompanied by adduction which would increase the 1–2 intermetatarsal angle. Hallux valgus deformity would thus develop as a secondary condition from this deviated first ray axis.

Glasoe suggests that custom foot orthotic therapy might intervene to address this primary cause of HAV deformity. He cites the two studies by Torkki and co-workers who published 1-year and 2-year follow-up trials comparing interventions of orthoses, surgery, and no treatment in 211 adult patients with mild to moderate HAV deformity [102, 103]. After 2 years, patients treated with custom foot orthoses were as satisfied as those having surgery and were more satisfied than controls. Glasoe speculated that these results suggest that custom foot orthoses fitted to the medial arch will orient the axis of the first ray to a more horizontal direction and reduce transverse plane motion [101].

Instability of the First Ray: Cause or Effect of HAV Deformity?

A vertical orientation of the first ray axis might also be interpreted in the clinical setting as a condition of laxity or loss of stability of the first metatarsal in the transverse plane. More specifically, many authors have attributed the primary factor for HAV deformity to begin with a laxity or instability of the first metatarsocuneiform joint [76, 104–106]. This instability has been described to occur in both the transverse and sagittal planes. Romash et al. noted excessive transverse plane motion of the first metatarsal at its articulation with the medial cuneiform in HAV patients.

However, this was a radiographic study evaluating the angle of articulation of the first metatarsocuneiform joint [107]. Subsequent studies have revealed considerable unreliability of radiographic measurement of the angle of articulation of this joint.

This role of the first metatarsocuneiform joint contributing excessive transverse plane motion was studied by Faber and co-workers [108]. In this cadaveric study, the first metatarsocuneiform joint contributed 82% of the transverse plane motion available in the first ray compared to only 18% contributed by the naviculocuneiform joint. As with previous studies, Faber et al. noted that sagittal plane motion of the first ray was greater at the naviculocuneiform joint than the first metatarsocuneiform joint.

In the previous cited study by Kimura, HAV patients showed greater transverse plane motion at the first metatarsocuneiform joint compared to healthy controls [58]. In this 3-D weight-bearing CT imaging study, the first metatarsomedial cuneiform joint moved in the direction of adduction with loading 1.1 degrees in healthy patients and 3.3 degrees in HAV patients. Transverse plane motion at the naviculocuneiform joint was less than 1 degree in both groups. In both the Faber and Kimura studies, excessive transverse plane motion at the first metatarsocuneiform joint was seen in HAV patients. The question is was this finding a result of hallux valgus at the first MTPJ? Or, did this transverse plane hypermobility of the first metatarsocuneiform joint precede and cause the HAV deformity?

There is ample evidence that the medial displacement of the first metatarsal in the transverse plane can be the result of and a response to increased lateral displacement of the hallux, and not vice versa [61–63, 69–71]. Coughlin, who has published many studies on this subject, is a strong proponent of this observation [32, 87, 92]. Furthermore, when hallux valgus angulation is corrected surgically, medial deviation of the first metatarsal reduces without any procedure performed directly on this bone [45–47]. Thus, so-called hypermobility of the first metatarsocuneiform joint in the transverse plane appears to be a *secondary* phenomenon in response to instability in the more distal first MTPJ. Conversely, a change in direction of range of motion of the first ray, dependent upon foot posture, will allow greater proportional range of motion in the transverse plane but will not necessarily cause excessive motion or "hypermobility" in this plane. Sagittal plane instability of the first metatarsocuneiform joint appears to be a similar *secondary result* of changes distally as HAV deformity progresses. We will discuss the phenomenon of hypermobility of the first ray in the next chapter.

References

1. Hicks JH. The mechanics of the foot: II. The plantar aponeurosis and the arch. J Anat. 1954;88:25–30.
2. Kidd R. The past is the key to the present: thoughts on the origins of human foot structure, function and dysfunction as seen from the fossil record. Foot. 1998;8:75–84.
3. Lundgren P, Nester C, Liu A, Arndt A, Jones R, Stacoff A, Wolf P, Lundberg A. Invasive in vivo measurement of rear-, mid- and forefoot motion during walking. Gait Posture. 2008;28(1):93–100.

4. Hicks JH. The mechanics of the foot. Part I: the joints. J Anat. 1953;87(4):345–57.
5. Kelso SF, Richie DH Jr, Cohen IR, Weed JH, Root ML. Direction and range of motion of the first ray. J Am Podiatr Med Assoc. 1982;72(12):600–5.
6. Johnson C, Christensen JC. Biomechanics of the first ray part 1. The effects of the peroneus longus function. A three dimensional kinematic study on a cadaver model. J Foot Ankle Surg. 1999;38(5):313–21.
7. Perez HR, Leon KR, Jeffrey CC. The effect of frontal plane position on first ray motion: forefoot locking mechanism. Foot Ankle Int. 2008;29(1):72–6.
8. D'Amico JC, Schuster RO. Motion of the first ray: clarification through investigation. J Am Podiatry Assoc. 1979;69:17–23.
9. Oldenbrook LL, Smith CE. Metatarsal head motion secondary to rearfoot pronation and supination: an anatomical investigation. J Am Podiatry Assoc. 1979;69:24–8.
10. Dayton P, Kauwe M, Feilmeier M. Clarification of the anatomic definition of the bunion deformity. J Foot Ankle Surg. 2014;53(2):160–3.
11. Cornwall MW, McPoil TG. Motion of the calcaneus, navicular, and first metatarsal during the stance phase of walking. J Am Podiatr Med Assoc. 2002;92:67–76.
12. Hueter C. [Klinik der Gelenkkrankungen mit Einschluss derOrthopaedie]. Edited, Vogel, Leipzig, 1870.
13. Nix SE, Vicenzino BT, Collins NJ, Smith MD. Characteristics of foot structure and footwear associated with hallux valgus: a systematic review. Osteoarthr Cartil. 2012;20:1059–74.
14. Thompson FM, Coughlin MJ. The high price of high-fashion footwear. J Bone Joint Surg Am. 1994;76-A:1586–93.
15. Frey C. Foot health and shoewear for women. Clin Orthop Relat Res. 2000;372:32–44.
16. Dawson J, Thorogood M, Marks SA, Juszczak E, Dodd C, Lavis G, et al. The prevalence of foot problems in older women: a cause for concern. J Public Health. 2002;24:77–84.
17. Barnicot NA, Hardy RH. The position of the hallux in West Africans. J Anat. 1955;89:355–61.
18. MacLennan R. Prevalence of hallux valgus in neolithic New Guinea population. Lancet. 1966;1:1398–400.
19. Kato T, Watanabe S. The etiology of hallux valgus in Japan. Clin Orthop Relat Res. 1981;157:78–81.
20. Shine IB. Incidence of hallux valgus in a partially shoe-wearing community. Br Med J. 1965;7(1):1648–50.
21. Menz HB, Roddy E, Marshall M, Thomas MJ, Rathod T, Peat GM, et al. Epidemiology of shoe wearing patterns over time in older women: associations with foot pain and hallux valgus. J Gerontol A Biol Sci Med Sci. 2016;71:1682–7.
22. Munteanu SE, Menz HB, Wark JD, Christie JJ, Scurrah KJ, Minh B, Erbas B, Hopper JL, Wluka A. Hallux valgus: by nature or nurture. A twin study. Arthritis Care Res (Hoboken). 2017;69(9):p1421–8.
23. Pique-Vidal C, Sole MT, Antich J. Hallux valgus inheritance: pedigree research in 350 patients with bunion deformity. J Foot Ankle Surg. 2007;46:149–54.
24. Johnston O. Further studies of the inheritance of hand and foot anomalies. Clin Orthop. 1956;8:146–60.
25. Lee CH, Lee S, Kang H, Jung DE, Song YM, Lee K, et al. Genetic influences on hallux valgus in Koreans: the healthy twin study. Twin Res Hum Genet. 2014;17:121–6.
26. Hannan MT, Menz HB, Jordan JM, Cupples LA, Cheng CH, Hsu YH. High heritability of hallux valgus and lesser toe deformities in adult men and women. Arthritis Care Res (Hoboken). 2013;65:1515–21.
27. El Haber N, Hill KD, Cassano AM, Paton LM, Macinnis RJ, Cui JS, et al. Genetic and environmental influences on variation in balance performance among female twin pairs aged 21–82 years. Am J Epidemiol. 2006;164:246–56.
28. Hsu YH, Liu Y, Hannan MT, et al. Genome-wide association meta-analyses to identify common genetic variants associated with hallux valgus in Caucasian and African Americans. J Med Genet. 2015;52:762–9.

29. Gottschalk FA, Sallis JG, Beighton PH, Solomon L. A comparison of the prevalence of hallux valgus in three South African populations. S Afr Med J. 1980;57:355–7.
30. Saro C, Andrén B, Wildemyr Z, Felländer-Tsai L. Outcome after distal metatarsal osteotomy for hallux valgus: a prospective randomized controlled trial of two methods. Foot Ankle Int. 2007;28:778–87.
31. Thordarson D, Ebramzadeh E, Moorthy M, Lee J, Rudicel S. Correlation of hallux valgus surgical outcome with AOFAS forefoot score and radiological parameters. Foot Ankle Int. 2005;26:122–7.
32. Coughlin MJ, Jones CP. Hallux valgus: demographics, etiology, and radiographic assessment. Foot Ankle Int. 2007;28:759–77.
33. Nix S, Smith M, Vicenzino B. Prevalence of hallux valgus in the general population: a systematic review and meta-analysis. J Foot Ankle Res. 2010;3:21.
34. Ferrari J, Malone-Lee J. The shape of the metatarsal head as a cause of hallux abductovalgus. Foot Ankle Int. 2002;23:236–42.
35. Ferrari J, Hopkinson DA, Linney AD. Size and shape differences between male and female foot bones: is the female foot predisposed to hallux abducto valgus deformity? J Am Podiatr Med Assoc. 2004;94:434–52.
36. Dykyj D, Ateshian GA, Trepal MJ, MacDonald LR. Articular geometry of the medial tarsometatarsal joint in the foot: comparison of metatarsus primus adductus and metatarsus primus rectus. J Foot Ankle Surg. 2001;40:357–65.
37. O'Connor K, Bragdon G, Baumhauer JF. Sexual dimorphism of the foot and ankle. Orthop Clin North Am. 2006;37:569–74.
38. McNerney JE, Johnston WB. Generalized ligamentous laxity, hallux abducto valgus and the first metatarsocuneiform joint. J Am Podiatry Assoc. 1979;69:69–82.
39. Shi K, Tomita T, Hayashida K, Owaki H, Ochi T. Foot deformities in rheumatoid arthritis and relevance of disease severity. J Rheumatol. 2000;27:84–9.
40. Beighton P, Horan F. Orthopaedic aspects of the Ehlers-Danlos syndrome. J Bone Joint Surg Br. 1969;51:444–53.
41. Uchiyama E, Kitaoka HB, Luo ZP, Grande JP, Kura H, An KN. Pathomechanics of hallux valgus: biomechanical and immunohistochemical study. Foot Ankle Int. 2005;26(9):732.
42. Kilmartin TE, Barrington RL, Wallace WA. Metatarsus primus varus. A statistical study. J Bone Joint Surg Br. 1991;73:937–40.
43. Kaiser P, Livingston K, Miller PE, May C, Mahan S. Radiographic evaluation of first metatarsal and medial cuneiform morphology in juvenile hallux valgus. Foot Ankle Int. 2018;39(10):1223–8.
44. Brage ME, Holmes JR, Sangeorzan BJ. The influence of x-ray orientation on the first metatarsocuneiform joint angle. Foot Ankle Int. 1994;15:495–7.
45. Cronin JJ, Limbers JP, Kutty S, Stephens MM. Intermetatarsal angle after first metatarsophalangeal joint arthrodesis for hallux valgus. Foot Ankle Int. 2006;27:104–9.
46. Mann RA, Katcherian DA. Relationship of metatarsophalangeal joint fusion on the intermetarsal angle. Foot Ankle. 1989;10:8–11.
47. Antrobus JN. The primary deformity in hallux valgus and metatarsus primus varus. Clin Orthop Relat Res. 1984;184:251–5.
48. Easley ME, Trnka HJ. Current concepts review: hallux valgus part 1: pathomechanics, clinical assessment and non-operative management. Foot Ankle Int. 2007;28:654–9.
49. Perera AM, Mason L, Stephens MM. The pathogenesis of hallux valgus. J Bone Joint Surg. 2011;93:1650–61.
50. Snijder CJ, Snijder JG, Philippens MM. Biomechanics of hallux valgus and spread foot. Foot Ankle. 1986;7:26–39.
51. Alvarez R, Haddad RJ, Gould N, Trevino S. The simple bunion: anatomy at the metatarsophalangeal joint of the great toe. Foot Ankle. 1984;4(5):229–40.
52. Haines RW, McDougall D. The anatomy of hallux valgus. J Bone Joint Surg. 1954;36B:272–93.
53. Scranton PE, Rutkowski R. Anatomic variations in the first ray- part 1: anatomic aspects related to bunion surgery. Clin Orthop Relat Res. 1980;151:244–55.

54. Eustace S, Obyrne J, Stack J, Stephens MM. Radiographic features that enable the assessment of first metatarsal rotation: the role of pronation in hallux valgus. Skelet Radiol. 1993;22(3):153–6.
55. Kim Y, Kim SK, Young KW, et al. A new measure of tibial sesamoid position in hallux valgus in relation to coronal rotation of the first metatarsal in CT scans. Foot Ankle Int. 2015;36(8):944–52.
56. Dayton P, Kauwe M, Feilmeier M. Is our current paradigm for evaluation and management of the bunion deformity flawed? A discussion of procedure philosophy relative to anatomy. J Foot Ankle Surg. 2015;54(1):102–11.
57. Collan L, Kankare JA, Mattila K. The biomechanics of the first metatarsal bone in hallux valgus: a preliminary study utilizing weight bearing extremity CT. Foot Ankle Int. 2013;19(3):155–61.
58. Kimura T, Kubota M, Taguchi T, Suzuki N, et al. Evaluation of first-ray mobility in patients with hallux valgus using weight-bearing CT and a 3-D analysis system. A comparison with normal feet. J Bone Joint Surg Am. 2017;99:247–55.
59. Mortka K, Lisiński P, Wiertel-Krawczuk A. The study of surface electromyography used for the assessment of abductor hallucis muscle activity in patients with hallux valgus. Physiother Theory Pract. 2018;34(11):846–51.
60. Arinci Incel N, Genç H, Erdem HR, Yorgancioglu ZR. Muscle imbalance in hallux valgus: an electromyographic study. Am J Phys Med Rehabil. 2003;82:345–9.
61. Stewart S, Ellis R, Heath M, Rome K. Ultrasonic evaluation of the abductor hallucis muscle in hallux valgus: a cross-sectional observational study. BioMed Central Musculoskelet Disord. 2013;14:45.
62. Scranton PE Jr, McDermott JE. Prognostic factors in bunion surgery. Foot Ankle Int. 1995;16:698–704.
63. Coughlin MJ, Saltzman CL, Nunley JA Jr. Angular measurements in the evaluation of hallux valgus deformities: a report of the ad hoc committee of the American Orthopaedic Foot & Ankle Society on angular measurements. Foot Ankle Int. 2002;23:68–74.
64. Coughlin MJ. Hallux valgus in men: effect of the distal metatarsal articular angle on hallux valgus correction. Foot Ankle Int. 1997;18:463–70.
65. Belt EA, Kaarela K, Lehto MU. Destruction and arthroplasties of the metatarsophalangeal joints in seropositive rheumatoids arthritis: a 20-year follow-up study. Scand J Rheumatol. 1998;27:194–6.
66. Turner D, Woodburn J, Helliwell P, et al. Pes planovalgus in rheumatoid arthritis: a descriptive and analytical study of foot function determined by gait analysis. Musculoskeletal Care. 2003;1:21–33.
67. Morton DJ. Metatarsus atavicus. J Bone Joint Surg. 1927;9:531–44.
68. Morton DJ. Structural factors in static disorders of the foot. Am J Surg. 1930;9:315–26.
69. Mancuso JE, Abramow SP, Landsman MJ, Waldman M, Carioscia M. The zeroplus first metatarsal and its relationship to bunion deformity. J Foot Ankle Surg. 2003;42:319–26.
70. Lundberg BJ, Sulja T. Skeletal parameters in the hallux valgus foot. Acta Orthop Scand. 1972;43:576–82.
71. Bryant A, Tinley P, Singer K. A comparison of radiographic measurements in normal, hallux valgus, and hallux limitus feet. J Foot Ankle Surg. 2000;39:39–43.
72. Hardy RH, Clapham JC. Observations on hallux valgus: based on a controlled series. J Bone Joint Surg. 1951;33-B:376–95.
73. Grebing BR, Coughlin MJ. Evaluation of Morton's theory of second metatarsal hypertrophy. J Bone Joint Surg. 2004;42-A:1375–86.
74. Root ML, Orien WP, Weed JH. Normal and abnormal function of the foot, vol. 2. Los Angeles: Clinical Biomechanics; 1977. Forefoot deformity caused by abnormal subtalar joint pronation; p 349–460.
75. Heden RI, Sorto LA Jr. The buckle point and the metatarsal protrusion's relationship to hallux valgus. J Am Podiatry Assoc. 1981;71:200–8.

76. Klaue K, Hansen ST, Masquelet AC. Clinical, quantitative assessment of first tarsometatarsal mobility in the sagittal plane and its relation to hallux valgus deformity. Foot Ankle Int. 1994;15:9–13.
77. Gudas CJ, Marcinko DE. The complex deformity known as hallux abductor valgus. In: Marcinko DE, editor. Comprehensive textbook of hallux valgus reconstruction. St Louis: Mosby; 1992.
78. Chi TD, Davitt J, Younger A, Holt S, Sangeorzan BJ. Intra- and inter-observer reliability of the distal metatarsal articular angle in adult hallux valgus. Foot Ankle Int. 2002;23:722–6.
79. Harris M-C, Beeson P. Generalized hypermobility: is it a predisposing factor towards the development of juvenile hallux abducto-valgus? Part 2. Foot. 1998;8:203–9.
80. Glasoe WM, Allen MK, Saltzman CL. First ray dorsal mobility in relation to hallux valgus deformity and first intermetatarsal angle. Foot Ankle Int. 2001;22:98–101.
81. Kalen V, Breecher A. Relationship between adolescent bunions and flatfeet. Foot Ankle. 1988;8:331–6.
82. Perera AM, Mason L, Stephens MM. The pathogenesis of hallux valgus. J Bone Joint Surg Am. 2011;93:1650–61.
83. Joplin RJ. Sling procedure for correction of splayfoot, metatarsus primus varus, and hallux valgus. J Bone Joint Surg. 1950;32:779–85.
84. Rogers WA, Joplin RJ. Hallux valgus, weak foot, and the Keller operations: an end-result study. Surg Clin North Am. 1947;27:1295–302.
85. Silver D. The operative treatment of hallux valgus. J Bone Joint Surg. 1923;5:225–32.
86. Schoenhaus HD, Cohen RS. Etiology of the bunion. J Foot Surg. 1992;31:25–9.
87. Canale PB, Aronsson DD, Lamont RL, Manoli A 2nd. The Mitchell procedure for the treatment of adolescent hallux valgus. A long-term study. J Bone Joint Surg. 1993;75-A:1610–8.
88. Coughlin M. Juvenile hallux valgus: etiology and treatment. Foot Ankle. 1995;16:682–97.
89. Kilmartin TE, Wallace WA. The significance of pes planus in juvenile hallux valgus. Foot Ankle. 1992;13:53–6.
90. Mann RA, Coughlin MJ. Hallux valgus: etiology, anatomy, treatment and surgical considerations. Clin Orthop. 1981;157:31–41.
91. Pouliart N, Haentjens P, Opdecam P. Clinical and radiographic evaluation of Wilson osteotomy for hallux valgus. Foot Ankle Int. 1996;17:388–94.
92. Saragas NP, Becker PJ. Comparative radiographic analysis of parameters in feet with and without hallux valgus. Foot Ankle Int. 1995;16:139–43.
93. Coughlin MJ, Shurnas PS. Hallux valgus in men. Part II: first ray mobility after bunionectomy and factors associated with hallux valgus deformity. Foot Ankle Int. 2003;24:73–8.
94. Roney A, Netto C, Sofka C, Sturnick MS et al. Instability of the first ray and hallux valgus in patients with adult acquired flatfoot deformity (AAFD). Foot and Ankle Orthopaedics. 2018;3. Sage Publishing.
95. Song J, Hillstrom HJ, Secord D, Levitt J. Foot type biomechanics. Comparison of planus and rectus foot types. J Am Podiatr Med Assoc. 1996;86:16–23.
96. Shibuya N, Kitterman RT, LaFontaine J, Jupiter DC. Demographic, physical, and radiographic factors associated with functional flatfoot deformity. J Foot Ankle Surg. 2014;53:168–72.
97. Galica A, Hagedorn TJ, Dufour AB, Riskowski JL, Hillstrom HJ, Casey VA, Hannan MT. Hallux valgus and plantar pressure loading: the Framingham foot study. J Foot Ankle Res. 2013;6:1–18.
98. Canseco K, Long J, Marks R, Harris G. Quantitative motion analysis in hallux valgus: rehabilitative insight. Arch Phys Med Rehab. 2007;88:E10.
99. DeChamps K, Birch I, Desloovere K, Matricali GA. The impact of hallux valgus on foot kinematics: a cross-sectional, comparative study. Gait Posture. 2010;32:102–6.
100. Glasoe W, Pena F, Phadke V, Ludewig PM. Arch height and first metatarsal joint axis orientation as related variables in foot structure and function. Foot Ankle Int. 2008;29:647–55.
101. Glasoe WM, Nuckley DJ, Ludewig PM. Hallux valgus and the first metatarsal arch segment: a theoretical biomechanical perspective. Phys Ther. 2010;90:110–20.

102. Torkki M, Malmivaara A, Scitsalo S, et al. Surgery vs orthosis vs watchful waiting for hallux valgus a randomized controlled trial. JAMA. 2001;285:2474–80.
103. Torkki M, Malmivaara A, Seitsalo S, et al. Hallux valgus: immediate operation versus 1 year of waiting with and without orthoses: a randomized controlled trial of 209 patients. Acta Orthop Scand. 2003;74:209–15.
104. Lapidus P. The operative correction of the metatarsus varus primus in hallux valgus. Surg Gynecol Obstet. 1934;58:183–91.
105. Myerson M. Metatarsocuneiform arthrodesis for treatment of hallux valgus and metatarsus primus varus. Orthopedics. 1990;13:1025–31.
106. Sangeorzan BJ, Hansen ST Jr. Modified Lapidus procedures for hallux valgus. Foot Ankle. 1989;9:262–6.
107. Romash MM, Fugate D, Yanklowit B. Passive motion of the first metatarsal cuneiform joint: preoperative assessment. Foot Ankle. 1990;10:293–8.
108. Faber FW, Kleinrensink GJ, Verhoog MW, et al. Mobility of the first tarsometatarsal joint in relation to hallux valgus deformity: anatomical and biomechanical aspects. Foot Ankle Int. 1999;20:651–6.

Disorders of the First Ray: Part 2 Hypermobility, Functional Hallux Limitus, and Hallux Rigidus

5

It has been shown that disorder in the longitudinal arch and also in the metatarsal portion of the foot originates as a functional deficiency of the first metatarsal: through either laxity of ligaments or shortness, this member lacks the necessary supporting contact with the ground.

–Dudley Morton

Key Points

1. Hypermobility of the first ray is a theoretical condition of excessive dorsiflexion motion of the first metatarsal during dynamic gait.
2. Clinical measures of hypermobility of the first ray are commonly performed in a static off-weight-bearing condition which will not engage multiple forces which occur during dynamic gait.
3. Static measures of first ray mobility do not predict first ray hypermobility during gait.
4. Stability of the first ray is dependent upon specific anatomic structures located at the first metatarsophalangeal joint (first MTPJ), providing static or dynamic restraint of motion or both.
5. Hallux abductovalgus (HAV) deformity is the only pathology of the human foot which demonstrates excessive dorsal mobility of the first ray with static exam, and this hypermobility reduces or vanishes when the first metatarsal is realigned over the sesamoids.
6. Metatarsus primus elevatus (MPE) is a static radiographic measure which does not predict load bearing capacity of the first metatarsal during gait.

D. H. Richie Jr, *Pathomechanics of Common Foot Disorders*,
https://doi.org/10.1007/978-3-030-54201-6_5

7. Patients with hallux rigidus and evidence of MPE demonstrate hypomobility of the first ray with static exam.
8. Metatarsus primus elevatus is the result of progressive degenerative arthritis and progressive plantar flexion contracture of the hallux seen in hallux rigidus.
9. Most people demonstrate greater range of motion of dorsiflexion of the first MTPJ off weight bearing compared to range of motion weight bearing and during dynamic gait.
10. Studies of patients with evidence of functional hallux limitus with static examination will show normal extension of the first MTPJ during dynamic gait.

Defining First Ray Hypermobility

The term "hypermobility" was first used by Dudley Morton in 1928 to describe abnormal first ray function [1]. Morton specifically described hypermobility as a condition resulting from an increase of dorsal movement of the first metatarsal bone with weight bearing. Root et al. agreed with the concept of hypermobility of the first ray, but modified the definition to state: "Hypermobility of the first ray is a state of abnormal 1st ray instability that occurs while the forefoot is bearing weight" [2]. A critical issue is the fact is that no kinematic study has ever been conducted to determine the normal range of motion of the first metatarsal segment during dynamic gait. Therefore, there are no parameters of determining the threshold of "hypermobility" as speculated by Morton and Root.

Measuring First Ray Mobility

Secondly, although Morton and Root described the state of hypermobility occurring during *weight bearing*, they proposed a patient evaluation technique which was performed in a static, *non-weight-bearing* condition. Both authors described a test to measure and detect hypermobility, whereby an examiner puts a dorsal force against the first metatarsal while stabilizing the lesser metatarsals in the off-weight-bearing foot (Fig. 5.1). Herein is the first shortcoming of the notion of first ray hypermobility: the original proponents of this concept used a non-weight-bearing test of passive dorsal excursion to theoretically predict or measure the *motion* of this skeletal segment during standing and during dynamic gait. The proponents of the concept of first ray hypermobility never actually measured motion of the first ray during gait in any human subject.

The clinical test for first ray mobility described by Root has been challenged for its validity because there is variability depending on foot position [3]. This test has also shown poor inter-rater reliability when performed by multiple practitioners [4].

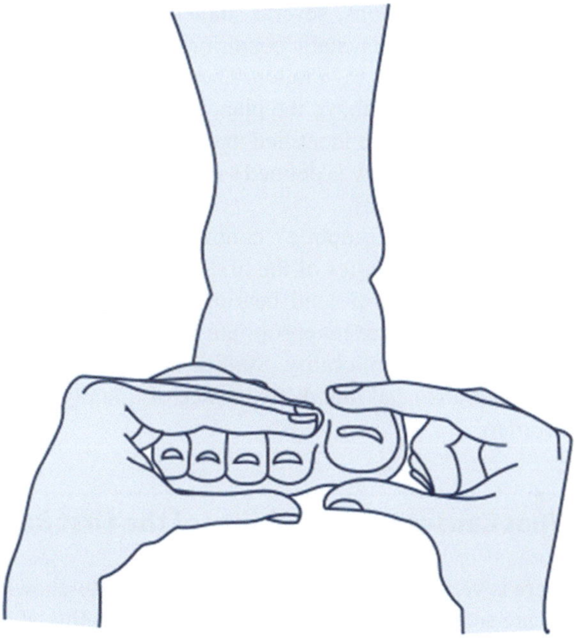

Fig. 5.1 Method to measure mobility of the first ray in static, non-weight-bearing condition as described by Morton and Root. [1, 2]

The shortcomings of this manual excursion test led to the development of mechanical devices to measure first ray mobility which have demonstrated good reliability and validity [5, 6]. Unfortunately, these devices are not practical for day-to-day use in a clinical setting. Notwithstanding, there is still a question whether a static measure of motion of any skeletal segment in the lower extremity will accurately predict performance during dynamic gait.

Allen and co-workers attempted to address this question when they used a mechanical device to measure static first ray range of motion and then performed dynamic measurement of motion of this segment during gait using optical motion capture technology [7]. The authors found no correlation between static measure of first ray range of motion and actual motion of this segment during dynamic gait. The first ray moved less during dynamic gait compared to the static condition. However, when excessive first ray motion was measured statically, it was associated with increased dynamic pronation of the midfoot and hindfoot during gait. Allen et al. concluded that *"the lack of similarity between dynamic and static measures of first ray motion in this study suggest caution is needed when interpreting both manual and mechanical static test results of first ray laxity in a healthy population."* Allen et al. point out that, compared to the static condition, dynamic gait engages multiple mechanical linkages between the hindfoot, midfoot, and forefoot. Furthermore, the sensorimotor system is engaged in dynamic gait which allows great opportunity for modification of motion and variation among individuals. This factor is important when performing static vs dynamic measures of range of motion of the first MTPJ and will be studied later in this chapter.

Despite limitations, several studies have used mechanical devices to measure first ray mobility in a static condition and have established benchmarks for normal vs hypermobile first ray motion. These standards are based upon dorsal excursion of the first metatarsal above the plane of metatarsals 2–5 rather than overall range of motion. Studies have identified that "normal" dorsal mobility averages 5 mm [5, 6, 8–11]. Hypermobility is defined as dorsal excursion of greater than 8 mm of the first metatarsal [5, 11–13].

The term "hypermobile" commonly appears in the medical literature when referring to pathologies of the first ray. In most cases, the term is used to describe a loss of stability or load bearing capacity of the first metatarsal during gait. However, even more inappropriate is the common use of static measures to identify first ray hypermobility. Studies have been undertaken to determine if these static measures can link the existence of first ray hypermobility with specific foot disorders.

What Causes Hypermobility of the First Ray?

There is very little evidence that static measures showing hypermobility of the first ray are seen in any foot disorder other than hallux abductovalgus (HAV) [14, 15]. Hypermobility of the first ray has consistently been demonstrated in studies of patients with HAV deformity [16–19].

Interestingly, none of Morton's publications mentioned any relationship between hallux valgus and first ray hypermobility. Lapidus was the first to attribute HAV deformity to first ray hypermobility [20]. He described a surgical procedure which involved arthrodesis of the first metatarsocuneiform joint to correct this underlying etiology [21]. Since that time, many authors have embraced Lapidus' explanation of the etiology of first ray hypermobility with HAV deformity and reported favorable results with his surgical procedure [22–29]. Ironically, none of these authors ever used valid methods to measure dorsal mobility of the first ray in their patients either pre-operative or post-operative. Most important, the cornerstone of the Lapidus theory is that excessive mobility of the first ray in the sagittal plane primarily occurs in the first metatarsocuneiform joint, i.e., the first TMT joint. This assumption has been disproven in numerous subsequent studies. In addition, the belief that reduction of hypermobility of the first ray can only be achieved with arthrodesis of the first metatarsocuneiform joint has also been soundly refuted.

Motion occurring at the medial naviculocuneiform joint is twice that of the motion occurring at the first metatarsocuneiform joint [30]. This dispels the myth that the first metatarsal medial cuneiform joint is the primary contributor to "hypermobility" of the first ray in many foot pathologies. In fact, the bone pin studies verify that the first metatarsal medial cuneiform joint moves less than any other joint of the foot [30]. Therefore, fusion of this joint would not be expected to have any significant effect on overall first ray motion or function of the foot.

We have learned in Chap. 5 that mobility of the first ray appears more dependent upon distal restraint of the first metatarsal at the first MTPJ rather than in the proximal location of the first metatarsocuneiform and naviculocuneiform joints.

Proximal stability of the first ray is dependent upon static ligamentous structures located on the plantar aspect of these joints [31, 32]. Faber and co-workers conducted a cadaver study and demonstrated that with HAV deformity, the first TMT joint actually has more mobility in the transverse plane than the sagittal plane [16]. Furthermore, this joint is significantly stabilized by both the peroneus longus and flexor hallucis longus muscles, probably more so than by the ligamentous structures. In particular, the flexor hallucis longus has the dual function to reduce medial displacement as well as dorsal displacement of first metatarsal.

The work of Johnson and Perez was presented in Chap. 5 demonstrating the role of the peroneus longus in plantarflexing and abducting the first metatarsal. More important, this movement engages a locking mechanism of the first ray into a close packed position with the second metatarsal [33, 34] (see Fig. 5.5, Chap. 5). Despite this finding, there is convincing evidence that these proximal structures of the first ray are not as important for sagittal plane stability or stiffness as are certain distal restraints around the first MTPJ.

Stability of the first ray is dependent upon specific anatomic structures located at its distal end, specifically at the first metatarsophalangeal joint (1st MTPJ), providing static or dynamic restraint of motion or both. The most important structure is the plantar fascia which provides both static and dynamic stability to the medial arch of the foot and to the first ray [35, 36]. Stainsby has described the relationship of the plantar fascia, the plantar plates, and the deep transverse metatarsal ligament as a "tie bar" system which not only tethers the metatarsals together in the transverse plane but also provides stability to the digits and metatarsals in the sagittal plane, resisting dorsal displacement from ground reaction forces [37]. Khaw also showed that the deep transverse metatarsal ligament provides significant restraint to dorsal motion of the first metatarsal and concluded that, combined with the plantar aponeurosis, these two structures are more important than the proximal structures to restrain dorsal displacement of the first ray [38]. As shown earlier, there is no direct attachment of the deep transverse metatarsal ligament or the plantar aponeurosis to the first metatarsal. The stability afforded to the first metatarsal by the plantar aponeurosis and the deep metatarsal tie bar ligaments is via the attachment of certain key ligaments of the sesamoid complex.

A key evolutionary advance in the human foot is the presence of a well-developed plantar aponeurosis with strong insertions into the proximal phalanx of all 5 digits [39]. In the lesser digits, the plantar aponeurosis invests the plantar plate which attaches to the base of the proximal phalanx. In the first MTPJ, there is also a plantar plate which invests the sesamoid apparatus. Another unique feature of the human foot is the larger diameter of the first metatarsal head, compared to the lesser metatarsals. The diameter of the first metatarsal is functionally enlarged by the sesamoid bones. The combined radius of the first metatarsal and the sesamoids provides a longer lever arm of plantar flexion moment for the flexor

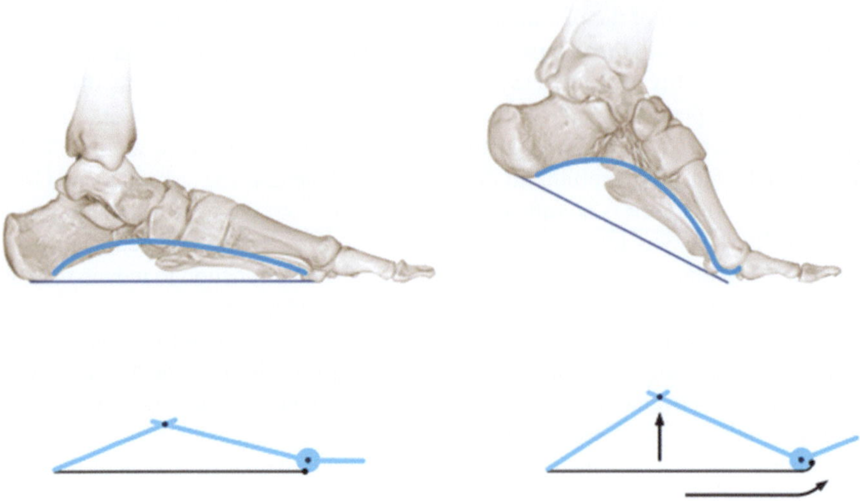

Fig. 5.2 The windlass mechanism provides stiffness to the first ray during terminal stance and pre-swing

hallucis brevis tendons as well as the distal slips of the plantar aponeurosis as these structures pass beneath the axis of rotation of the first MTPJ. The enhanced stabilizing force of these structures on the hallux favors load bearing of first ray and hallux during heel rise and toe off. The mechanism of the plantar aponeurosis "winding" around the head of the first metatarsal has been described by Hicks as a "windlass mechanism" which is critical to stability of the medial longitudinal arch of the foot [40, 41] (Fig. 5.2).

Sharkey and co-workers demonstrated that the plantar fascia can provide passive tension during static stance, as well as dynamic tension via the windlass mechanism during terminal stance and toe-off which will reduce the plantar-to-dorsal bending moments across the metatarsals [42]. In another study, these investigators demonstrated a similar offloading or shielding of bending moments of the first metatarsal by the flexor hallucis longus (FHL) [43]. Similar shielding of bending moments is provided by the flexor digitorum longus at the lesser metatarsals [44]. Proper purchase of the digits against the supportive surface will reduce plantar pressure under the heads of the metatarsals. This will offload or shield the metatarsal from dorsal displacement. Reduction of dorsal bending moment is also counteracted by retrograde plantarflexion moment placed on the metatarsal head from the proximal phalanx as the digit dorsiflexes and engages the windlass during terminal stance. The net result of both mechanisms is a reduction of dorsal bending moment or dorsal displacement of the metatarsal during standing and walking. Proper plantar purchase of the hallux prevents dorsal mobility of the first metatarsal. This plantar purchase is compromised when the sesamoids, as well as the insertion of the plantar aponeurosis and the pull of the FHL tendon are all diverted laterally from the head of the first metatarsal in HAV deformity.

Hallux Abductovalgus Causes First Ray Hypermobility

The first metatarsal loses the protective or shielding effect of the FHL, the plantar aponeurosis, and the deep transverse metatarsal ligament as it displaces medially from the sesamoid envelope during the progression of HAV deformity. Rush and co-workers demonstrated that when this displacement is corrected with an osteotomy of the first metatarsal, a 26% increase in stiffness and resistance to dorsiflexion of the first ray will occur [45]. The authors concluded that hypermobility of the first ray, which is commonly seen in HAV deformity, is due to a relative lateral shift of the sesamoids, positioning the distal slips of the plantar aponeurosis lateral and diverted away from their optimal position under the head of the first metatarsal. The FHL tendon also becomes laterally displaced from the first metatarsal and loses its proper moment arm for plantarflexion of the proximal phalanx of the hallux. Thus, increased dorsal bending moment is transmitted to the first metatarsal as the hallux loses plantar purchase to the ground. The plantar fascia, the FHL tendon, and the sesamoids therefore appear essential for providing sagittal plane stability to the first ray, particularly during terminal stance and pre-swing with engagement of the windlass mechanism.

Coughlin and co-workers noted the study by Rush et al. and undertook a number of investigations to evaluate mobility changes of the first ray in HAV deformity after surgical correction. In a cadaveric study of HAV specimens, Coughlin et al. noted a 50% reduction in dorsal excursion of the first ray when the first metatarsal was realigned with a proximal crescentic osteotomy [46]. In a follow-up clinical study, Coughlin et al. prospectively studied 103 patients with HAV deformity [47]. Using the Klaue device to measure first ray mobility, the pre-operative was 7.2 mm and diminished to 4.5 mm with after realignment of the first metatarsal with proximal crescentic osteotomy. Mobility of the first ray correlated with magnitude of metatarsus primus varus deformity [47]. This verified previous observations by Hardy and Clapham [48].

There have been many published studies citing a cause-effect relationship between hypermobility of the first ray and HAV deformity. However, few studies have used reliable mechanical measurement devices to compare mobility of the first ray in patients with HAV deformity to healthy controls. Shibuya et al. conducted a systematic review of those studies [49]. Meta-analysis of the data from those studies revealed that patients with HAV deformity have a significant 3.6 mm greater range of dorsal excursion of the first ray compared to healthy controls. They also noted that some of these studies showed that hypermobility reduces to normal when the first metatarsal is realigned in HAV surgery.

Doty and Coughlin reviewed the results of all studies of mobility of the first ray in HAV deformity and also looked at their own studies showing disappearance of "hypermobility" after the first metatarsal was realigned with osteotomy alone [50]. They concluded:

> "*While the metatarsocuneiform joint is critical to the development of hallux valgus, it is for different reasons than those envisioned by Lapidus and others. These historical theories were based on unreliable methods by which to quantify mobility. Metatarsocuneiform joint mobility may be necessary for a hallux valgus deformity to develop; however, "hypermobility" may not be the cause but rather the result of a hallux valgus deformity.*"

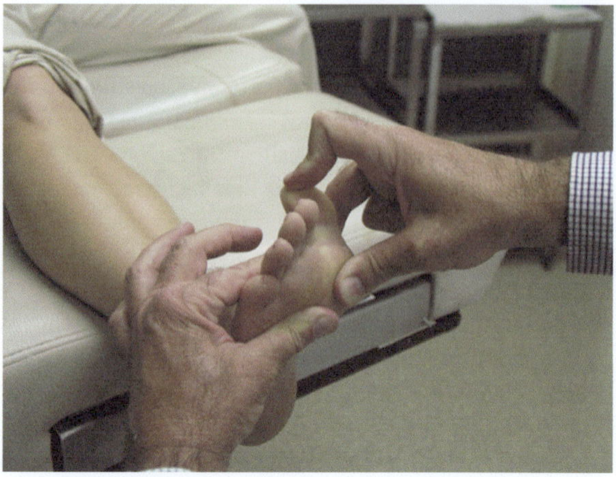

Fig. 5.3 The modified hicks test engages the windlass mechanism to mimic stiffness during pre-swing

In summary, hypermobility of the first ray is commonly found in patients with HAV deformity and rarely seen in any other foot pathology. This is because mobility of the first ray in the sagittal plane is primarily dependent upon the sesamoid complex and the flexor hallucis longus tendon located at the distal boundary of this skeletal segment. When examining the influence of these structures on stiffness of the first ray, the "modified Hick's test" will mimic pre-swing and tension the soft tissue restraints of the first MTPJ (Fig. 5.3). When these structures become laterally displaced relative to the first metatarsal in HAV deformity, significant dorsal excursion of the first ray occurs. When medial deviation of the first metatarsal in HAV deformity is relocated to a rectus alignment relative to the sesamoid complex and the flexor hallucis longus tendon, mobility of the first ray is significantly reduced. This realignment of the first metatarsal with reduction of dorsal mobility can be accomplished with an osteotomy alone without the need for arthrodesis of the metatarsocuneiform joint.

The First Ray and Limitation of Motion of the First MTPJ

Range of Motion of the First MTPJ

One of the most significant areas of misunderstanding in lower extremity biomechanics is the controversy regarding normal versus restricted range of motion of the first metatarsophalangeal joint (first MTPJ). Clinicians and surgeons will commonly assess excursion of the hallux in a dorsal and plantar direction at the first MTPJ in an off-weight-bearing exam and make a diagnosis of hallux limitus or hallux rigidus without any recognition of the futility of this measurement in actually predicting motion of this joint in dynamic gait. Furthermore, there is little information and thus no agreement about how much range of motion is necessary at the first MTPJ during walking gait. There is even less understanding about the motion requirements of this joint during running.

There is a clear difference between available range of motion in the first MTPJ measured in a non-weight-bearing position, compared to a static weight-bearing position. And, there is a difference between measurements taken in both these positions and the actual measurement of motion of the first MTPJ which takes place during dynamic gait. Static off weight bearing measures of first MTPJ motion show more range of motion than what is measured during dynamic gait.

Most healthy individuals will demonstrate up to 70 degrees of dorsiflexion of the first MTPJ in a static, off-weight-bearing examination [51]. However, when these same individuals place their feet on the ground in a weight-bearing position, this range of motion dramatically reduces. With the foot flat on the ground, dorsiflexion of the first MTPJ is limited to approximately 20 degrees, as further motion in this direction requires plantarflexion of the first metatarsal [51]. In the non-weight-bearing condition, the first ray is free to plantarflex when the hallux is moved into dorsiflexion. In the weight-bearing condition, ground reaction forces limit or block plantarflexion of the first metatarsal. During dynamic gait, the hallux does not move dorsally over a fixed foot. Rather, the first metatarsal plantarflexes on the hallux which is fixed on the ground (Fig. 5.4). Not only does the first metatarsal plantarflex, it actually moves proximal on the fixed sesamoids as the windlass mechanism is engaged. This movement changes the instantaneous center of rotation of the first MTPJ [52]. The spiral-shaped nature of the metatarsal head forces a translational sliding motion as the proximal phalanx rotates [53]. Thus, the locus of the axis of rotation has to move in an arc that necessitates metatarsal motion proximally and plantarward in order to avoid compression at the metatarsophalangeal joint [52, 53] (Fig. 5.5). No wonder it is impossible to replicate these intricate movements of skeletal segments with an off-weight-bearing examination!

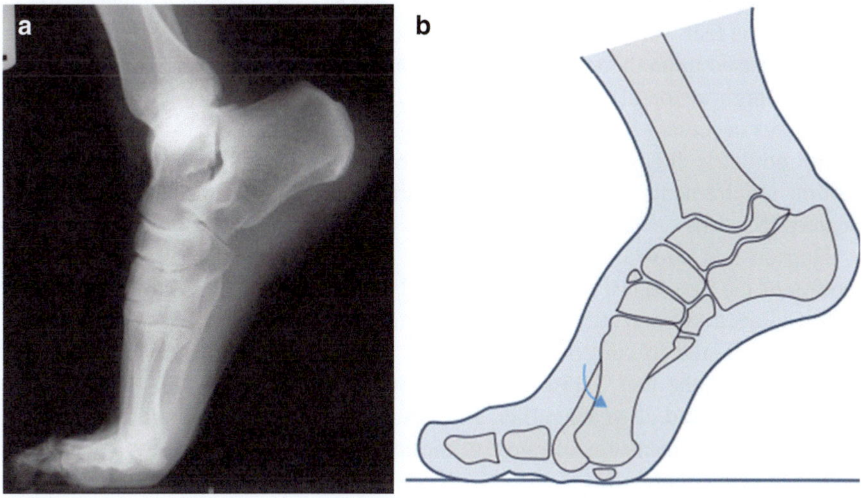

Fig. 5.4 (**a**) During terminal stance and pre-swing, the first metatarsal rotates on the fixed hallux and fixed sesamoids. (**b**) During terminal stance and pre-swing, the metatarsals plantarflex on the digits. The first metatarsal rotates proximal on the fixed sesamoids and plantarflexes relative to the lesser metatarsals

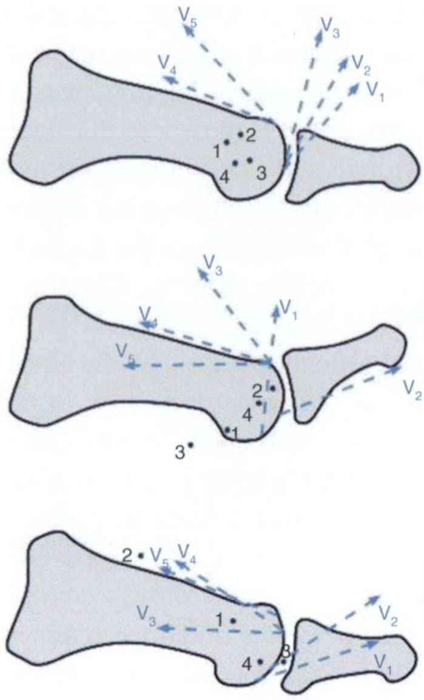

Fig. 5.5 Instant centers of rotation at the first MTPJ shift in HAV and hallux rigidus. (Adapted from: Shereff et al. [52]. 1-A, 1-B, and 1-C) Typical instant centers of rotation are marked 1-4; and surfacevelocity vectors are marked V1-5 and are plotted through the entire range of motion for the first metatarsophalangeal joint.In HAV and hallux rigidus, the surface velocity vectors are no longer tangent to the joint surface demonstrating compression and jamming

What Is the Best Way to Estimate ROM of the First MTPJ in Dynamic Gait?

Joseph was among the first to attempt to measure range of motion of the first MTPJ in a weight-bearing position [54]. He determined, using active and passive measures with weight-bearing radiographs, that the first MTPJ dorsiflexes in a range of 50–75 degrees. The feet were in a static stance position, however.

Root, based upon theoretical analysis of angular relationships of the foot, lower leg and thigh at toe-off proposed that 65–70 degrees of dorsiflexion of the first MTPJ was necessary for normal gait [2] (Fig. 5.6). Observational studies published in the podiatric literature generally concurred with Root's definition of normal motion requirements of the first MTPJ during gait, ranging between 50 and 65 degrees [55, 56].

Hopson and co-workers performed four different static measures of range of motion of the first MTPJ and then measured dynamic range of motion of this joint in the subjects during walking [57]. Their motion capture system was somewhat archaic compared to current standards, but their findings showed that maximum dorsiflexion of the first MTPJ during gait averaged 65 degrees, identical to what Root had proposed 20 years earlier. Hobson et al. showed that all static tests demonstrated greater measures of range of motion than what was used by the subjects during dynamic gait.

This same discrepancy between static and dynamic measures of first MTPJ range of motion in 33 healthy subjects was demonstrated in a study by Nawoczenski and co-workers [58]. They used four clinical tests for range of motion of the first MTPJ

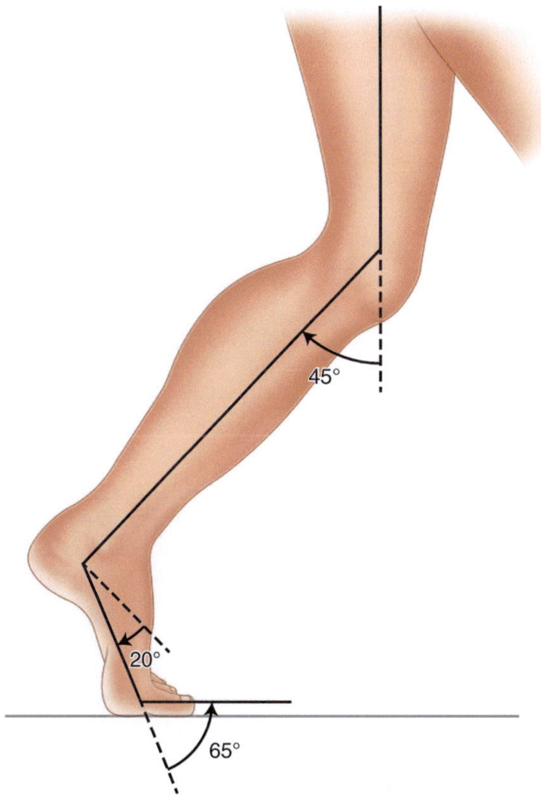

Fig. 5.6 The common accepted range of 65 degrees extension of the first MTPJ in pre-swing

and then measured actual motion used by the subjects during walking gait. The range of dorsiflexion determined by static tests were:

- Passive, non-weight-bearing extension: 57 degrees
- Passive, weight-bearing extension: 37 degrees
- Active weight-bearing extension: 44 degrees
- Heel rise, weight-bearing extension: 58 degrees

This study used an accurate electromagnetic tracking device to measure range of motion during all the tests. During dynamic gait, the subjects averaged 44 degrees of dorsiflexion at the first MTPJ. Thus, healthy subjects used less dorsiflexion range of motion of the first MTPJ than previously believed (44 degrees vs 65 degrees) (Fig. 5.7). Also, the static test which most accurately predicted range of motion during dynamic gait was the *standing active extension test* (Fig. 5.8). Finally, the off-weight-bearing static range of dorsiflexion motion of the first MTPJ in these healthy subjects was 57 degrees, yet it reduced to 37 degrees in a static weight-bearing exam.

This finding of "normal" range of motion of the first MTPJ off weight bearing, with reduction of motion in the weight-bearing condition in otherwise healthy

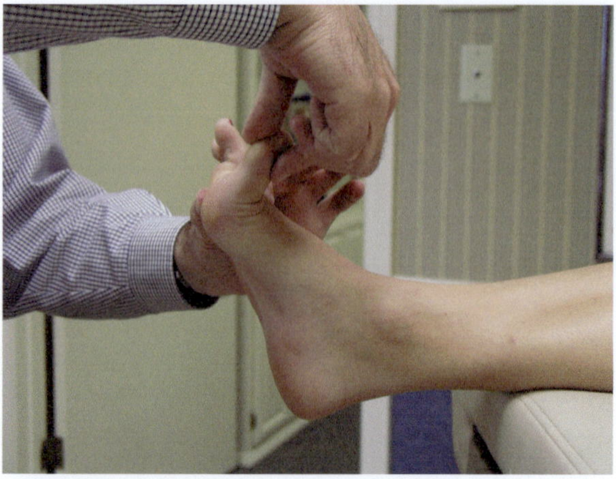

Fig. 5.7 Commonly used measure of first MTPJ extension off weight bearing will overestimate the actual range of motion used by the patient during gait

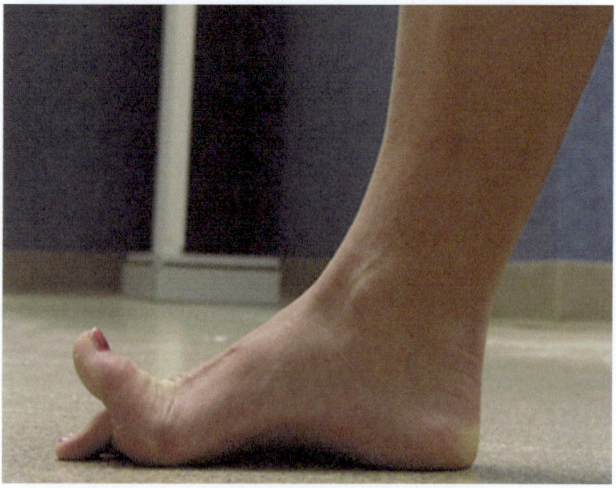

Fig. 5.8 Active extension, weight bearing will simulate the actual range of motion at the first MTPJ used by the patient during gait. This averages 45 degrees extension

subjects, is significant. According to accepted definition, these healthy subjects would all be diagnosed with "functional hallux limitus." In other words, these healthy subjects had "normal" range of motion of the first MTPJ off weight bearing but significant decreased motion while weight bearing. It turns out that this is a normal, routine finding among most healthy subjects. Thus, the condition known as functional hallux limitus remains a controversial subject because it has not been validated with accurate dynamic kinematic studies.

Pathologies Associated with Restricted Motion of the First MTPJ

A number of terms and descriptions have been used to describe various clinical conditions associated with restricted range of motion of the first MTPJ. There is a lack of agreement for use of these terms which are often used interchangeably to describe the same condition. Yet, each of these conditions are actually unique and not necessarily related to each other in terms of pathomechanics or treatment.

Restricted range of motion of the first MTPJ has been labeled with various terms including *hallux limitus, functional hallux limitus, structural hallux limitus, and hallux rigidus* [7, 56, 57]. In podiatric education, these terms describe a continuum of the same condition which starts as a functional restriction of dorsiflexion range of motion of the first MTPJ (hallux limitus) and then progresses to end-stage degenerative arthritis of the joint (hallux rigidus) [1, 55, 56].

Functional Hallux Limitus

Laird was the first to describe the term *functional hallux limitus* [59]. This condition has been proposed to exist when a patient demonstrates normal or adequate range of dorsiflexion motion of the first MTPJ off weight bearing but shows decreased range of motion in a weight-bearing condition. Curiously, this is a normal finding in healthy individuals (Figs. 5.7, 5.8, and 5.9). In the off-weight-bearing examination where the hallux is passively dorsiflexed to end range of motion, the first metatarsal is free to plantarflex, and there is no tension in the plantar aponeurosis. In static stance, the first metatarsal is prevented from plantarflexing due to ground reaction force. Also, the plantar aponeurosis is placed upon significant tension which limits extension of the first MTPJ via a reverse windlass mechanism.

In his definition, Laird stated that functional hallux limitus is present when there is greater than 50 degrees of dorsiflexion of the hallux at the first MTPJ in a non-weight-bearing condition compared to less than 12 degrees during standing. Others have stated that with greater than 50 degrees of dorsiflexion off weight bearing and anything less than 40 degrees in static stance identifies the state of functional hallux limitus [60]. To this day, there is lack of agreement about how these ranges of motion would be tested and what ranges would be considered adequate vs restricted.

Dananberg, who expanded upon the notion of functional hallux limitus, described a non-weight-bearing test whereby the hallux was dorsiflexed while the first metatarsal was loaded [61]. No specific range of motion was required to make a diagnosis of functional hallux limitus other than an observation of immediate plantarflexion motion of the first metatarsal when the hallux was dorsiflexed. Later Payne et al. determined that this test was sensitive in predicting late midstance pronation of the

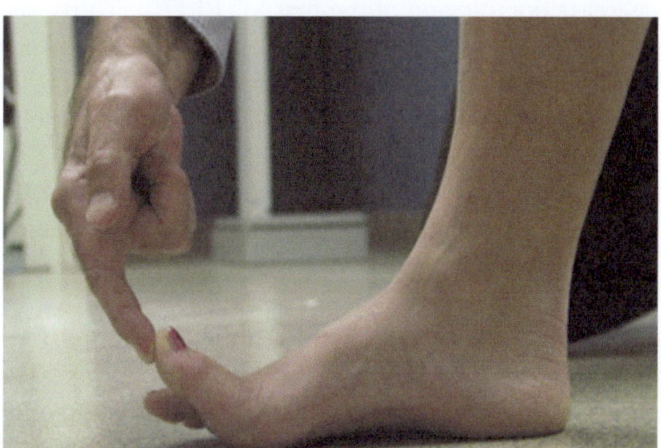

Fig. 5.9 Passive extension range of motion of the hallux is always less in a weight-bearing position than non-weight bearing as shown in same subject in Fig. 5.8

midtarsal joint during gait [62]. However, this excessive pronation determination was based on subjective observation.

Another popular method of detecting presence of hallux limitus is use of the Hubscher maneuver in a static relaxed stance to measure passive dorsiflexion of the first MTPJ [63]. This test, however, has shown poor accuracy in actually predicting the range of dorsiflexion motion of the first MTPJ during dynamic gait [60]. In this study, conducted by Halstead and Redmond, patients identified with functional hallux limitus showed 55 degrees of passive dorsiflexion of the first MTPJ off weight bearing but only 19 degrees in static stance with the Hubscher maneuver. However, these patients achieved 35 degrees of dorsiflexion of the first MTPJ during dynamic gait which was identical to healthy controls [60]. Therefore, some change occurred during dynamic gait which actually *increased* the range of dorsiflexion of the first MTPJ compared to static stance.

This raises a valid question: Can a patient with normal range of motion of the first MTPJ off weight bearing have significant restriction of motion during dynamic gait? Furthermore, does this dynamic restriction of motion at the first MTPJ lead to compensation and development of further pathologies of the lower extremity as proposed by other investigators? [64, 65].

Dorsiflexion of the hallux at the first MTPJ is coupled with plantarflexion of the first ray [66]. Reciprocally, dorsiflexion of the first ray inhibits dorsiflexion of the hallux at the first MTPJ [67]. Foot pronation causes lowering of the medial longitudinal arch placing elongation strain along the medial slips of the plantar aponeurosis. Increased tension in the plantar fascia increases plantarflexion moment exerted across the first MTPJ during weight bearing, restricting dorsiflexion via the so-called "reverse windlass" mechanism [68]. This condition would be seen only during the weight-bearing condition, whereas the patient could demonstrate normal dorsiflexion of the first MTPJ in an non-weight-bearing examination.

Another role of foot pronation is the effect of increased ground reaction forces on the first ray, leading to dorsiflexion and jamming of the first MTPJ [69]. This will be reviewed in the discussion of the condition known as hallux rigidus.

Gatt and co-workers tested the notion that increased foot pronation would decrease dorsiflexion of the first MTPJ [70]. Thirty subjects who showed less than 12 degrees of first MTPJ dorsiflexion in static stance were subdivided into two groups depending on the severity of foot pronation. As severity of foot pronation increased, the magnitude of first MTPJ dorsiflexion diminished. The authors speculated that a reverse windlass mechanism could explain the results.

An additional finding of this investigation by Gatt and co-workers determined that the static weight-bearing measure of hallux extension at the first MTPJ did not correlate with the actual range of extension during dynamic gait. Pronation did cause restriction of motion at the 1st MTPJ measured during dynamic gait. Gatt et al. showed that the severity of a pronated foot posture was asociated with greater restriction of dorsiflexion of the 1st MTPJ during dynamic gait. However, as seen with the Halstead study, the total motion during dynamic gait was far greater than motion available in static stance. Also, all evidence of hallux limitus disappeared during dynamic gait with both groups exhibiting over 45 degrees of dorsiflexion of the first MTPJ [60, 70].

To summarize, there are two studies of functional hallux limitus demonstrating reduced dorsiflexion of the first MTPJ in static stance which then improves during dynamic gait. One study showed less than 12 degrees of dorsiflexion of the first MTPJ of subjects in static stance [70], and another showed less than 19 degrees [60]. Both groups of subjects had normal range of dorsiflexion exceeding 55 degrees off weight bearing, so they met the criteria of "functional hallux limitus." This clinical evidence of hallux limitus disappeared in both groups during dynamic gait as they averaged over 45 degrees of dorsiflexion at the first MTPJ.

Thus far, the evidence for the existence of functional hallux limitus has been shown in studies where subjects demonstrate decreased dorsiflexion of the first MTPJ in static stance compared to the off-weight-bearing condition. However, the assumption that functional hallux limitus occurs in these same subjects during dynamic gait has yet to be proven by high-quality 3-D kinematic studies using multisegment foot models. The fact is that all healthy feet will demonstrate less passive extension of the first MTPJ weight bearing compared to the non-weight-bearing condition.

During dynamic gait, certain mechanisms are engaged which facilitate plantarflexion of the first metatarsal which are absent during static stance. Two such mechanisms are the role of forward momentum engaging the windlass as well as the action of the peroneus longus acting on the first ray during terminal stance.

What is needed are more dynamic gait studies to determine the relationship between motion of rearfoot, the midfoot joints, and motion the first MTPJ. These studies could verify a mechanism proposed by Finn Bojsen-Moller in 1979 in his classic paper "Calcaneocuboid Joint and Stability of the Longitudinal Arch of the Foot at High and Low Gear Push Off" [71]. In this paper, Bojsen-Moller explains how optimal dorsiflexion of the first MTPJ is contingent upon stability of the calcaneocuboid joint. Stability and alignment of this joint facilitate the peroneus longus to pronate the forefoot on the rearfoot. Forefoot pronation promotes medial transfer of load to the first MTPJ and engages the windlass mechanism. Optimally, the human foot will push off across the transverse axis of the first and second MTPJs, engaging a high-gear mechanism for the windlass rather than using a less efficient oblique axis or low-gear windlass mechanism of the lesser MTPJs [71] See Chap. 1, Fig. 1.19

This description of dynamic foot function by Bojsen-Moller is based in part upon the previous work of MacConnail describing the favorable "twisting" of the forefoot upon the rearfoot, providing stability and facilitating push off across the transverse axis of the first and second metatarsophalangeal joints [72]. This specialization of the human foot, compared to the great apes, was later studied by Griffin and coworkers verifying the transverse axis, high-gear mechanism of push off which exists in the human foot as originally proposed by Bojsen-Moller [73]. This topic will be explored in more detail in Chap. 6 when the role of first ray stability contributing to metatarsalgia is studied.

The work of Bojsen-Moller has been often cited as an explanation for the etiology of functional hallux limitus and the secondary effects of this condition on the proximal joints of the foot and leg. In his paper, Bojsen-Moller clearly states that pronation of the forefoot on the rearfoot at the midtarsal joint "pre-tightens" the plantar aponeurosis [71]. He further states that tension is developed in the plantar aponeurosis as soon as the heel rises off the ground. The larger diameter of the head of the first metatarsal combined with the additional circumference of the sesamoids favors greater mechanical advantage of the first MTPJ (a high-gear mechanism) over the lesser MTPJs to engage the windlass mechanism to raise the arch and stiffen the foot for propulsion.

Yet, the theory of functional hallux limitus describes a distal-to-proximal compensation which takes place within the foot and the entire lower extremity when "sagittal plane blockade" occurs at the first MTPJ [64]. Durrant and Chocklingham, in their review of the topic of functional hallux limitus, describe this distal-to-proximal compensation when loss of dorsiflexion of the first MTPJ *causes* instability of the midtarsal joint, specifically the calcaneocuboid joint [74]. Durrant et al. review the work of Bojsen-Moller but incorrectly state that in the healthy foot, the windlass mechanism pre-tightens the plantar fascia which *subsequently* provides close packing and locking of the calcaneocuboid joint, and not vice versa. This reliance of tightening of the plantar fascia to lock the calcaneocuboid joint was originally proposed by Dannanberg [75]. On the contrary, Bojsen-Moller proposes that pronation of the forefoot pre-tightens the plantar fascia, *before* engagement of the windlass, and this process is dependent upon an *already locked, close packed calcaneocuboid joint* [71]. Bojsen-Moller describes this pre-tensioning of the plantar fascia which occurs before heel rise based upon film analysis of a walking subject. Further tensioning of the plantar aponeurosis occurs during heel rise with engagement of the windlass, but as shown in Chap. 1, the windlass does not appear to fully engage until the first MTPJ has achieved 30 degrees of extension.

Bojsen-Moller never proposed that high-gear push off through the first MTPJ was required to lock the calcaneocuboid joint. He also never speculated about how range of motion at the first MTPJ would affect stability of the midtarsal joints. His description of optimal foot function starts with a description of how ground reaction forces cause pronation and close packing of the calcaneocuboid joint and how this sets up the eventual transfer of load to the first MTPJ during the stance phase of gait. Thus, there is nothing in Bojsen-Moller's paper to suggest a retrograde proximal compensation at the midtarsal joint which could occur with loss of dorsiflexion motion of the first MTPJ. To this date, there are no published kinematic studies verifying compensatory motion in the midfoot and hindfoot joints *after* normal dorsiflexion motion is diminished at the first MTPJ. Yet, this distal-to-proximal compensation mechanism due to loss of dorsiflexion of the first MTPJ remains a cornerstone of the functional hallux limitus theory of abnormal lower extremity function [61–65].

When there is anatomic abnormality in the first MTPJ, causing loss of motion, gait studies have shown compensatory changes. These anatomic pathologies include HAV deformity and hallux rigidus and do not have the criteria to meet the definition of functional hallux limitus. As we will see in this next section, hallux rigidus and also HAV deformity cause spatiotemporal changes in gait with increased stance duration, decreased stride length, and decreased walking velocity–all due to loss of push off thru the first MTPJ.

Hallux Rigidus

While the clinical condition known as functional hallux limitus has yet to be fully verified in quality biomechanical studies, the presence of progressive degenerative arthritis at the first MTPJ has been extensively studied. This condition has become known as hallux rigidus, a term first proposed by Cotterill in 1988 [76]. Curiously, there are still many authors who use the term hallux limitus to describe an arthritic condition of the first MTPJ, while others use the two terms hallux limitus and hallux rigidus interchangeably [77, 78]. When reviewing the medical literature, the term hallux rigidus is the preferred term to describe a state of stiffness and degenerative arthritis of the first MTPJ [79, 80].

Hallux rigidus is important to review in this chapter because it has traditionally had a strong association with hypermobility or dysfunction of the first ray. It is also a significant condition of the human foot because it does not have many good treatment options.

The symptoms of hallux rigidus include pain with range of motion of the first MTPJ and usually a dorsal "bunion" or osteophyte causing painful irritation from footwear [81]. Physical examination reveals bone enlargement, swelling, and restricted motion of the first MTPJ [82]. Radiographic findings of hallux rigidus include joint space narrowing of the first MTPJ, osteophyte formation, subchondral sclerosis, flattening of the first metatarsal head, trumpeting of the base of the proximal phalanx, and loose bodies [80, 83]. All of these clinical findings are consistent with degenerative arthritis of the first MTPJ.

Staging or grading the severity of hallux rigidus has been described by many different authors [80, 83–85]. Beeson and co-authors identified 18 classifications systems for hallux rigidus and concluded that there is a need for more objective criteria based upon sound scientific research [86]. This is evidenced by the fact that the most commonly used systems for grading severity of hallux rigidus rely on terms such as "mild or moderate pain" or "mild or moderate" restriction of joint range of motion (Table 5.1). Clinicians would disagree in their assessment of "minor" or "moderate" joint space narrowing of the first MTPJ (Fig. 5.10).

Most published classification systems for hallux rigidus rely on a clinical assessment of range of motion of the first MTPJ in an off-weight-bearing examination. In the most commonly accepted classification systems of hallux rigidus, the first stage or Grade I demonstrates diminished range of motion of the first MTPJ off weight bearing. Therefore, based upon definition, the condition known as functional hallux limitus would be an entirely different condition than hallux rigidus. Yet some speculate that hallux rigidus is preceded by functional hallux limitus. However, there are no longitudinal studies showing that functional hallux limitus causes degenerative arthritis of the first MTPJ. As will be shown, hallux rigidus does not appear to be associated with hypermobility or overload of the first ray of the foot.

A common recognized etiology for hallux rigidus is trauma [87, 88]. However, a study by study by Coughlin and Shurnas showed no association between trauma and hallux rigidus [15]. That same study showed that when hallux rigidus occurs, it is found bilateral 79% of the time. When trauma was reported as the primary cause of

Table 5.1 Classification of hallux rigidus

Grade 0	Dorsiflexion	Radiographic findings[a]	Clinical findings
	40° to 60° and/or 10% to 20% loss compared with normal side on examination	Normal	No pain; only stiffness and loss of motion
	30° to 40° and/or 20% to 50% loss compared with normal side	Dorsal osteophyte is main finding, minimal joint space narrowing, minimal flattening of metatarsal head, periarticular sclerosis	Mild or occasional pain and stiffness, pain at extremes of dorsiflexion/plantarflexion on examination
2	10° to 30° and/or 50% to 75% loss compared with normal side	Dorsal, lateral, and possibly medial osteophytes, flattened appearance to metatarsal head. No more than 1/4 of dorsal joint space involved on lateral radiograph, mild to moderate joint space narrowing and sclerosis, sesamoids not usually involved	Moderate to severe pain and stiffness that may be constant which occurs just before end range dorsi/plantar flexion
3	≤10° and/or 75% to 100% loss also ≤10° PF	Same as in Grade 2 but with substantial narrowing, with normal possibly periarticular cystic changes, more than 1/4 of dorsal joint space involved on lateral radiograph sesamoids enlarged and/or cystic and/or irregular	Nearly constant pain and stiffness at extremes of range of motion, but not at mid-range
4	Same as in Grade 3	Same as in Grade 3	Same criteria as Grade 3 but there is definite pain at mid-range of passive motion

From: Coughlin and Shurnas [15]
[a]Weight-bearing and anteroposterior and lateral radiographs are used

symptoms by patients, hallux rigidus was unilateral most of the time. This same study revealed that family history was a strong determinant of hallux rigidus; reported in 67% of patients, those with a positive family history had bilateral symptoms 95% of the time [15]. Equinus and pes planus have been speculated as predisposing to hallux rigidus, but no studies have verified this cause-effect relationship [82, 87].

A flattened metatarsal head seems to be associated with hallux rigidus. This was demonstrated in the Coughlin and Shurnas study [15] as well as in a study by Stein et al. [89]. Hunt and Anderson found a flat or chevron-shaped first metatarsal head 79% of the time in patients with hallux rigidus [90]. It is not known if this is a predisposing anatomic feature of the head of the first metatarsal to cause hallux rigidus or if it is adaptive.

Shereff showed that the kinematics of the first MTPJ are abnormal in patients with hallux valgus and hallux rigidus with displaced centers of rotation compared to intact specimens [52] (Fig. 5.5). Hallux rigidus showed jamming throughout the arc of motion which resulted in centers of rotation falling outside the metatarsal head indicating adaptation as the condition progresses. Shereff speculated that articular

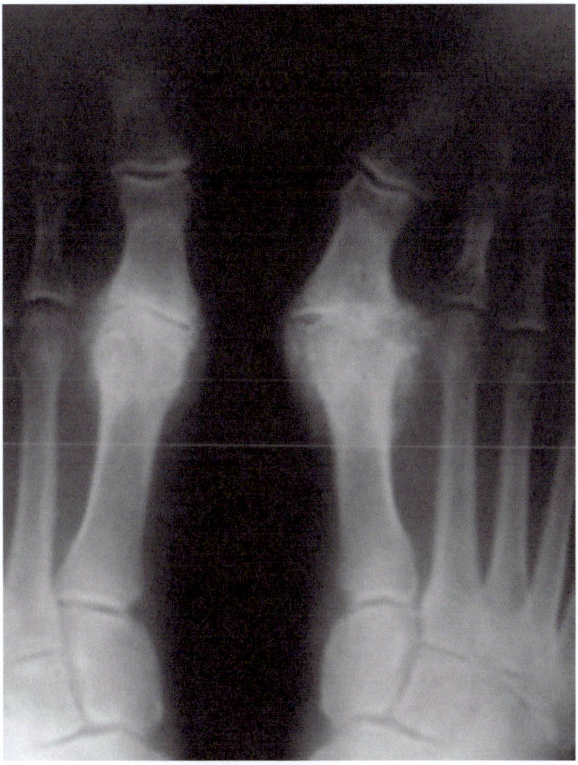

Fig. 5.10 A patient with Grade 2 hallux rigidus, left foot, and Grade 3 hallux rigidus, right foot

erosion, marginal osteophytes, and stretching and contracture of periarticular structures were probably responsible for the mechanical alterations seen in the first MTPJ in hallux rigidus. These changes are seen in middle and later stages of hallux rigidus and would likely be the result rather than the cause of the condition.

Other intrinsic risk factors for hallux rigidus were investigated by Coughlin and Shurnas in a study of 114 patients diagnosed with hallux rigidus [15]. Despite widespread belief, this study showed no association between hallux rigidus and a pes planus arch profile determined by a Harris mat imprint. Also, there was no evidence of equinus or restricted dorsiflexion range of motion of the ankle joint in this large patient pool. There was, however, a significant increase in the hallux interphalangeus angle, indicating a possible adaptation as sagittal plane motion became restricted at the first MTPJ. The incidence of a long first metatarsal was no greater in hallux rigidus patients than in the regular population, and there was no correlation with metatarsal length and severity of hallux rigidus.

With all these previous held notions about the cause of hallux rigidus, the most common accepted etiology is a dorsiflexed alignment of the first ray. This condition was termed metatarsus primus elevatus (MPE) by Lambrinudi in 1938 and has been recognized by many authors as a primary cause of hallux rigidus [91]. *Metatarsus primus elevatus is also mistakenly considered to be a sign of hypermobility of the first ray.* MPE is a static radiographic and is commonly used by foot surgeons around the world to justify plantarflexion osteotomies of the first metatarsal in the surgical correction of hallux rigidus [92–94]. The finding of MPE is also used as a criteria for performing a Lapidus fusion of the first metatarsocuneiform joint as a remedy for hypermobility of the first ray [22–29].

At first glance, it is easy to see why many people accept that MPE is a cause of hallux rigidus. The human foot will show immediate restriction of dorsiflexion motion across the first MTPJ when the first metatarsal is elevated. This was verified in a study by Roukis and Scherer where 8 mm of elevation of the first metatarsal reduced dorsiflexion of the first MTPJ by over 35% [67]. Elevation of the first ray relative to the rest of the foot in a lateral standing radiograph suggests excessive dorsal mobility during gait and arguably predicts repetitive jamming of the first MTPJ.

This raises two questions: During what part of the gait cycle will MPE become important and how much elevation of the first metatarsal is required to cause hallux rigidus? A standing lateral radiograph captures the foot in a static position which somewhat mimics the midstance period of the gait cycle. In reality, the lateral standing radiograph is not actually representative of midstance where there is only one foot supporting the body and there is tibial progression over the fixed foot, i.e., the ankle rocker. When positioning for a lateral standing radiograph, the patient's pelvis and hips are centered in the frontal plane while the two feet are next to each other, bearing weight equally on the supportive surface. This position is never achieved during any stage of the walking gait cycle.

Whether a standing lateral radiograph actually depicts midstance or not, it is more important to recognize that dorsiflexion of the first MTPJ does not occur until heel rise or terminal stance. Accurate dynamic measure of alignment of the first metatarsal relative to the hallux during heel rise is not possible in a typical clinical setting. Measurement of total range of motion used by the patient at the first MTPJ during gait would be the best way to detect functional hallux limitus or hallux rigidus, but there have been no published studies which have used this methodology on these two patient populations.

Despite these shortcomings, MPE has been studied with various methods of measurement on a standing lateral radiograph. The angular relationships between the first metatarsal to the ground and the second metatarsal to the ground can detect MPE [95] (Fig. 5.11). Also, a lateral sagittal intermetatarsal angle formed by the lines along the dorsal cortex of metatarsals 1 and 2 can be measured [80]. These

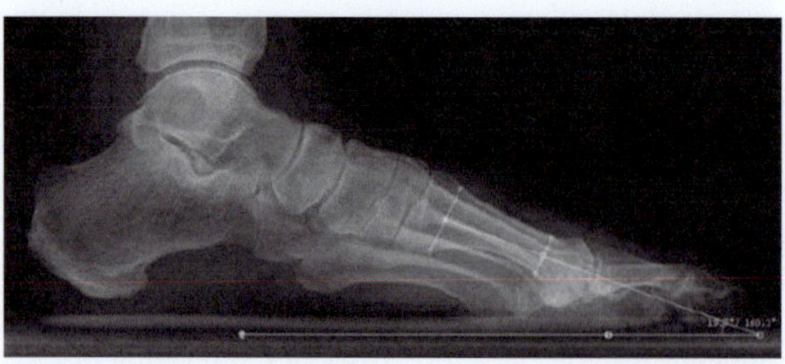

Fig. 5.11 Measuring angle of declination of the first ray. (From: Bouaicha et al. [95]). The figure describes how the MT-1-Declination angle was measured. The MT-I-declination angle is mean angle between longitudinal axis of the first metatarsal bone and the ground. The longitudinal axis of MT-1 was defined as centerline between the dorsal and plantar cortical bone. Copyright © 2010 by the American Orthopaedic Foot & Ankle Society

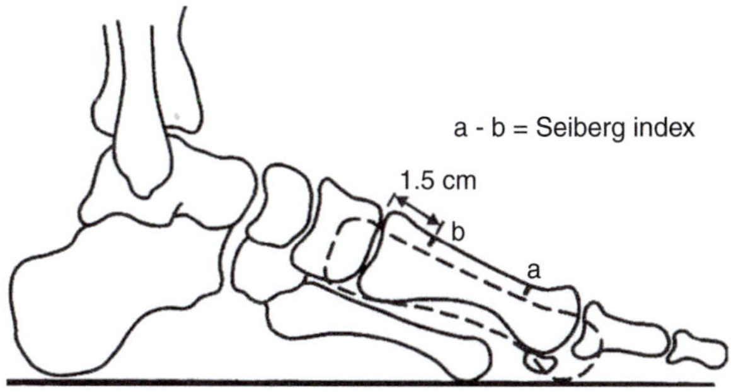

Fig. 5.12 The Seiberg index is determined by measuring the perpendicular distance from the dorsal aspect of the second metatarsal shaft to the dorsal aspect of the first metatarsal shaft at the first metatarsal neck (i.e., the metaphyseal-diaphyseal junction) and 1.5 cm from the first metatarsal base. The proximal measurement is subtracted from the distal measurement to obtain the Seiberg index. [96]

angles are difficult to draw and measure, making them prone to error. A more reliable and easy method to measure MPE was described by Seiberg and co-workers [96] (Fig. 5.12). This is a linear measure of the perpendicular distance from the dorsal aspect of the second metatarsal shaft to the dorsal aspect of the first metatarsal shaft at the first metatarsal neck (i.e., the metaphyseal-diaphyseal junction) and 1.5 cm from the first metatarsal base. The proximal measurement is subtracted from the distal measurement to obtain the Seiberg index. With a plantarly declinated first metatarsal, the Seiberg index is negative; with dorsal angulation, it is positive.

Using the Seiberg index, Roukis observed that the mean first to second metatarsal elevation was 5.8 mm in a hallux rigidus group, which was higher than what was measured in patients with other foot pathologies [97]. Bouaicha modified the Seiberg index to measure first to second metatarsal height on lateral radiographs and found that any elevation of the first metatarsal greater than 5 mm was predictive of hallux rigidus [95]. Indeed, MPE seems to be strongly associated with hallux rigidus, but is this cause or effect?

Horton and co-workers studied radiographs of patients with and without hallux rigidus and found no difference in alignment of the first ray, measuring the distance between the dorsal cortex of metatarsals 1 and 2 [98]. This confirmed findings from a previous study by Meyer et al. who warned that MPE was not a radiographic sign peculiar to hallux rigidus [99]. In cases of severe hallux rigidus, however, Horton et al. noted that the magnitude of MPE increased significantly, suggesting to the authors that *elevation of the first metatarsal occurs as a result of hallux rigidus and not* vice versa. In other words, extreme magnitude of MPE is not seen in early stage hallux rigidus, only late stage.

Coughlin and Shurmas studied 110 patients with hallux rigidus and found metatarsus primus elevatus in only 10 cases [15]. They also concluded that elevation of the first metatarsal is the *result of* and not the *cause of* hallux rigidus. Previously, Coughlin had made a similar observation in patients with HAV deformity: hypermobility of the first ray was an end result of intermetatarsal angle deformity and not the cause of hallux abductovalgus [47].

Metatarsus primus elevatus is often confused with hypermobility of the first ray, yet no studies have documented that these two clinical findings are found in the same patients. Also, hypermobility of the first ray has attributed to be a cause of hallux rigidus. If hypermobility does cause hallux rigidus, it should be a consistent finding with manual examination of these patients. Surprisingly, clinical studies fail to show any evidence of hypermobility of the first ray in patients with hallux rigidus. Coughlin and Shurmas used the Klaue device to measure dorsal mobility of the first metatarsal in 110 patients with hallux rigidus [15]. The dorsal excursion of the first metatarsal was 5.8 mm, which was far *below* the previous 8 mm threshold for hypermobility [5]. Greisberg and co-workers measured dorsal mobility of the first ray in 34 patients with hallux rigidus compared to 341 patients without the condition [100]. Patients with hallux rigidus had a significant *reduced* dorsal mobility of the first ray compared to control subjects. If anything, patients with hallux rigidus demonstrate *hypomobility* of the first ray.

Therefore, while patients with hallux rigidus often show MPE on lateral standing radiographs, they do not demonstrate true first ray hypermobility with clinical testing. The first metatarsal may appear to be dorsally positioned in static stance, yet the entire first ray does not have excessive dorsal mobility when measured off weight bearing. It is more reasonable to accept the notion that MPE is actually the *result of* the pathologic process of hallux rigidus, i.e., degenerative arthritis of the first MTPJ.

How would hallux rigidus cause metatarsus primus elevatus? A likely mechanism is the ability of the hallux to drive the first metatarsal into abnormal alignment, similar to what occurs in HAV deformity. Camasta was among the first to recognize that a plantar-flexed hallux was almost always associated with a dorsally displaced first metatarsal in hallux rigidus patients [81]. Normally, the hallux is positioned 16–18 degrees dorsiflexed relative to the first metatarsal in static stance. A condition known as *hallux equinus* is present when the hallux lies in a rectus position or a plantar-flexed position relative to the first metatarsal in static stance [101] (Fig. 5.13).

Camasta proposes that *hallux equinus* is found in a "co-dependent relationship" with MPE due to two mechanisms: a fixed forefoot varus, elevating the first MTPJ which causes the hallux to plantarflex and find the ground, and/or an unstable medial column associated with pes planus which activates the flexors of the hallux to "grip" and provide medial stability to the forefoot. According to Camasta, this "flexor stabilization syndrome of the hallux" is also seen in the lesser digits in response to

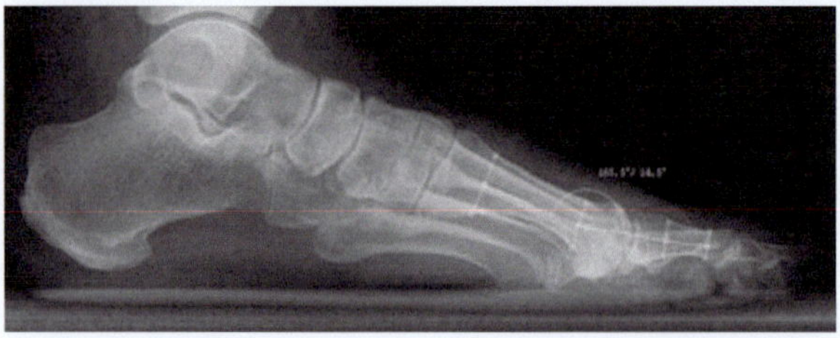

Fig. 5.13 Measurement of hallux equinus. (From: Bouaicha et al. [95]). MT-1-Dorsiflexion-angle (DFA): Angle between the longitudinal axis of the MT-1-shaft and the longitudinal axis of the proximal phalanx. In a normal foot, the hallux is dorsiflexed 16-18 degrees relative to the first metatarsal

pronation of the rearfoot and is a common cause of hammertoe deformity. Either way, Camasta proposes that flexor stabilization creates overactivity of the FHL and FHB which restricts dorsiflexion of the first MTPJ. Abnormal positioning of the hallux in an equinus attitude leads to abnormal compressive force against the head of the first metatarsal and initiates the degenerative process of hallux rigidus.

Further evidence of the role of plantarflexion contracture and hallux rigidus is provided by Roukis who found significant incidence of hallux equinus in patients with hallux rigidus, and this flexion deformity increased with increased grade of hallux rigidus [80, 97]. Bouaicha and co-workers evaluated hallux equinus, using a different term called the "first metatarsal dorsiflexion angle," and showed that the hallux was significantly more plantarflexed on the first metatarsal in patients with hallux rigidus [95]. Furthermore, there was a significant correlation with plantarflexion of the hallux and greater amount of MPE deformity.

The authors speculated that tensioning of the flexor hallucis longus tendon could be responsible for retrograde metatarsus primus elevatus (Fig. 5.14). This provides a third proposed mechanism to explain the association between hallux equinus and hallux rigidus: pain from degenerative arthritis causes soft tissue contraction or "splinting" around the first MTPJ. This attempt to restrict painful range of motion places the hallux in a plantar-flexed position causing retrograde dorsiflexion of the first metatarsal and increasing MPE deformity (Fig. 5.15). In this scenario, hallux equinus would be the result, not the cause of degenerative arthritis of the first MTPJ

Other investigators have described the role of the flexor hallucis longus tendon in the pathogenesis of hallux rigidus. In a cadaveric study, Kirane and co-workers showed that reduced excursion of the flexor hallucis longus tendon would

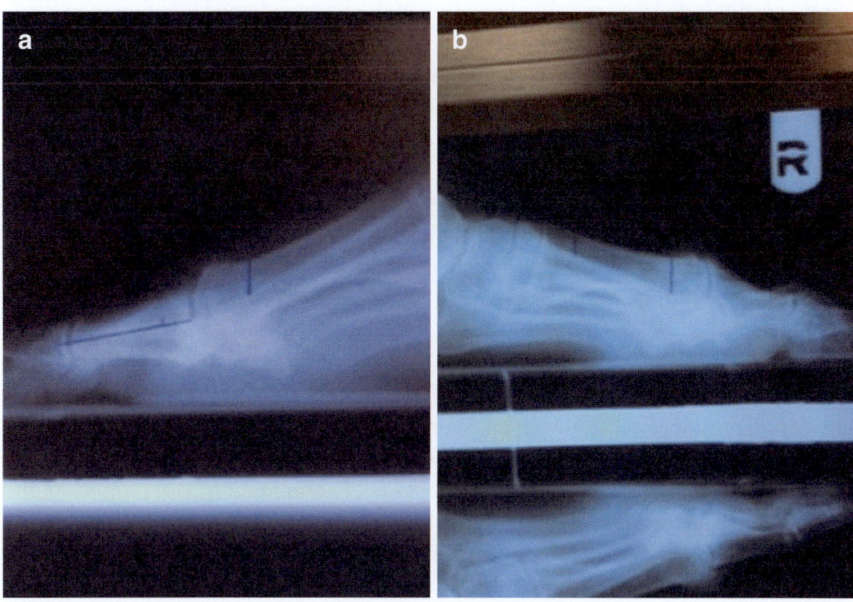

Fig. 5.14 (**a**) Metatarsus primus elevatus (MPE) associated with hallux equinus. (**b**) Increased MPE and hallux equinus with increased severity of hallux rigidus

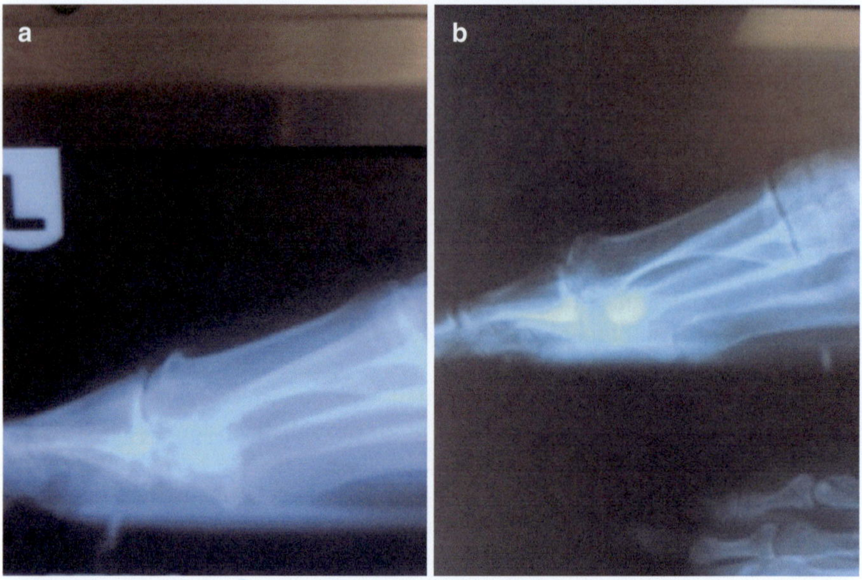

Fig. 5.15 Progression of hallux equinus and MPE in same patient with hallux rigidus: hallux drives the first metatarsal dorsal. (**a**) Stage 2 hallux rigidus. (**b**) 3 years later, same patient: stage 3 hallux rigidus with increased hallux equinus

significantly alter dorsiflexion mechanics of the first MTPJ [43]. They recommended evaluating the flexor hallucis longus for fibrosis or impingement along its length, particularly in its proximal location at the myotendinous junction in patients with hallux rigidus. This same restriction of motion at the retrotalar pulley of the FHL tendon was proposed by Valotton and co-workers as a mechanism for causing restriction of motion at the first MTPJ [102]. Their paper was primarily speculative and was intended to show a novel treatment intervention using manipulation to release the tenodesis of the FHL tendon.

Researchers have long implicated the plantar aponeurosis as a contributor to reduced dorsiflexion of the first MTPJ [36, 38, 42]. Phillips and co-authors suggested that pronation of the rearfoot leading to instability of the medial column increases tension in the plantar fascia with subsequent limitation of dorsiflexion of the hallux [66].

The potential role of the plantar fascia to limit range of motion in the first MTPJ was shown by Harton and co-workers [103]. Patients undergoing a plantar fasciotomy for plantar heel pain showed an improvement of first MTPJ dorsiflexion by 10 degrees. While few would recommend this procedure as a primary intervention for hallux rigidus, it adds validity to the concept of "decompression" osteotomies of the first metatarsal which indirectly lengthen the soft tissue structures which are limiting motion of the first MTPJ. In a study of 20 patients undergoing surgery for hallux rigidus, with a 15-year follow-up, Eliott and co-workers have shown that osteotomies intended to plantarflex the first metatarsal actually end up shortening the bone which decompresses the first MTPJ [104]. Decompression was also achieved with joint destruction procedures in this study, and both surgical procedures resulted in a reduction of metatarsus primus elevatus over time [104]. This finding supports the notion that soft tissue contractures at the first MTPJ are responsible for MPE in hallux rigidus.

In summary, the flexor hallucis longus tendon as well as the distal insertion of the plantar aponeurosis into the proximal phalanx of the hallux are important stabilizers of the first ray. Abnormal mechanics of the rearfoot, i.e. excessive pronation, can cause increased tension in these structures which will cause limitation of motion around the first MTPJ. Tensioning of the FHL, FHB will plantarflex the hallux which causes MPE and hypomobility of the first ray. Conversely, medial subluxation of the first metatarsal will cause malalignment of the FHL and plantar aponeurosis, resulting in laxity of these structures at the first MTPJ. This laxity will compromise flexion and purchase of the hallux which results in increased dorsiflexion bending moments at the first metatarsal. Laxity from lateral shift in the insertion of the plantar aponeurosis at the first MTPJ compromises the windlass mechanism which diminishes the plantarflexion moment applied to the first metatarsal. These two changes result in a loss of stiffness of the first ray, particularly during the heel rise phase of gait. This displacement of soft tissue alignment in HAV, versus contracture of soft tissues around the first MTPJ in hallux rigidus, explains how hypermobility of the first ray is seen in HAV while hypomobility is seen in hallux rigidus.

Kinematic studies of patients with hallux rigidus show a compromise of the windlass mechanism leading to compensation which is similar to what has been seen in patients with HAV deformity. Canseco and co-workers studied 22 patients with hallux rigidus and found that they had significant longer stance duration compared to healthy controls [105]. In another study of patients with HAV deformity, these same authors found that three temporal-spatial parameters of gait were abnormal with not only increased stance duration but also decreased stride length and slower walking speed [106]. In terms of these parameters, HAV appears to cause more disruption of gait than hallux rigidus.

Increased stance phase duration indicates a loss of propulsion. This can occur with antalgic guarding of the first MTPJ or dysfunction of the windlass mechanism, with both features seen in HAV and hallux rigidus. Canseco showed that both HAV and hallux rigidus demonstrate a lack of forefoot plantarflexion during terminal stance and pre-swing, reflected by a flattening of the medial longitudinal arch. Instability of the arch may compromise plantarflexion of the first ray to facilitate dorsiflexion of the hallux and engagement of the windlass mechanism.

Pressure studies have shown a loss of load bearing of the first metatarsal in patients with hallux rigidus [107]. These studies as well as observational reports show a consistent compensation during gait in patients with hallux rigidus characterized by internal rotation of the foot to direct center of pressure and load to the lateral toes [108, 109]. Pressure studies of patients with HAV deformity show inconsistent results and do not always show loss of load bearing capacity of the first metatarsal. This will be studied further in Chap. 6.

Loss of propulsion thru the first MTPJ during late stance and loss of engagement of the windlass mechanism cause compensation which is similar in both HAV patients and hallux rigidus patients. Load is shifted laterally, and stance phase is prolonged. In HAV deformity, loss of plantar purchase of the hallux could explain the shift of load laterally. In hallux rigidus, pain and lack of engagement of the windlass could explain the lateral shift to the oblique axis of the MTPJs during push off. This loss of medial shift of load bearing during propulsion and loss of the windlass may explain why patients with HAV and hallux rigidus are commonly identified to suffer from digital deformities and metatarsalgia.

References

1. Morton DJ. Hypermobility of the first metatarsal bone. J Bone Joint Surg Am. 1928;10:187–96.
2. Root ML, Orien WP, Weed JH. Motion of the joints of the foot: the first ray. In: Normal and abnormal function of the foot. Los Angeles: Clinical Biomechanics Corporation; 1977.
3. Shirk C, Sandrey MA, Erickson M. Reliability of first ray position and mobility measurements in experienced and inexperienced examiners. J Athl Train. 2006;41:93–101.
4. Grebing BR, Coughlin MJ. The effect of ankle position on the exam for first ray mobility. Foot Ankle Int. 2004;25:467–75.
5. Klaue K, Hansen ST, Masquelet AC. Clinical, quantitative assessment of first tarsometatarsal mobility in the sagittal plane and its relation to hallux valgus deformity. Foot Ankle Int. 1994;15(1):9–13.
6. Glasoe WM, Yack HJ, Saltzman CL. The reliability and validity of a first ray measurement device. Foot Ankle Int. 2000;21:240246.
7. Allen MK, Cuddeford TJ, Glasoe WM, et al. Relationship between static mobility of the first ray and first ray, midfoot, and hindfoot motion during gait. Foot Ankle Int. 2004;25:391–6.
8. Couglin MJ, Shurnas PS. Hallux valgus in men. Part II: first ray mobility after bunionectomy and factors associated with hallux valgus deformity. Foot Ankle Int. 2003;24:73–8.
9. Glasoe WM, Allen MK, Saltzman CL, Ludewig PM, Sublett SH. Comparison of two methods used to assess first ray mobility. Foot Ankle Int. 2002;23:248–52.
10. Glasoe WM, Getsoian S, Myers M, Komnick M, Kolkebeck D, Oswald W, Liakos P. Criterion-related validity of a clinical measure of dorsal first ray mobility. J Orthop Sport Phys Ther. 2005;35:589–93.
11. Glasoe WM, Grebing BR, Beck S, Coughlin MJ, Saltzman CL. A comparison of device measures of dorsal first ray mobility. Foot Ankle Int. 2005;26:957–61.
12. Faber FW, Mulder PG, Verhaar JA. Role of first ray hypermobility in the outcome of the Hohmann and the Lapidus procedure. J Bone Joint Surg. 2004;86-A:486–95.
13. Voellmicke KV, Deland JT. Manual examination technique to assess dorsal instability of the first ray. Foot Ankle Int. 2002;23:1040–1.
14. Grebing BR, Coughlin MJ. Evaluation of Morton's theory of second metatarsal hypertrophy. J Bone Joint Surg. 2004;42-A:1375–86.
15. Coughlin MJ, Shurnas PS. Hallux rigidus: demographics, etiology, and radiographic assessment. Foot Ankle Int. 2003;24(10):731–43.
16. Faber FW, Kleinrensink GJ, Verhoog MW, et al. Mobility of the first tarsometatarsal joint in relation to hallux valgus deformity: anatomical and biomechanical aspects. Foot Ankle Int. 1999;20:651–6.
17. Glasoe WM, Allen MK, Saltzman CL. First ray dorsal mobility in relation to hallux valgus deformity and first intermetatarsal angle. Foot Ankle Int. 2001;22:98–101.
18. Jones CP, Coughlin MJ, Pierce-Villadot R, Galano P, Kennedy MJ, Shurnas P, Grebing BR, Teachout L. The validity and reliability of the Klaue device. Foot Ankle Int. 2005;26:951–6.

19. Lee KT, Young K. Measurement of first-ray mobility vs. hallux valgus patients. Foot Ankle Int. 2001;22:960–4.
20. Lapidus PW. The operative correction of the metatarsus varus primus in hallux valgus. Surg Gynecol Obstet. 1934;58:183–91.
21. Lapidus PW. The author's bunion operation from 1931 to 1959. Clin Orthop. 1960;16:119–35.
22. Bacardi BE, Boyson TJ. Considerations for the Lapidus operation. J Foot Surg. 1986;25:133–8.
23. Hansen ST. Hallux valgus surgery—Morton and Lapidus were right! Clin Podiatr Med Surg. 1996;13:347–54.
24. Myerson MS. Metatarsocuneiform arthrodesis for treatment of hallux valgus and metatarsus primus varus. Orthopedics. 1990;13:1025–31.
25. Myerson MS, Badekas A. Hypermobility of the first ray. Foot Ankle Clin. 2000;5:469–84.
26. Sangeorzan BJ, Hansen ST. Modified Lapidus procedures for hallux valgus. Foot Ankle Int. 1989;9:262–6.
27. Bednarz PA, Manoli A. Modified Lapidus procedure for the treatment of hypermobile hallux valgus. Foot Ankle Int. 2000;21(10):816–21.
28. Hofbauer MH, Grossman JP. The Lapidus procedure. Clin Podiatr Med Surg. 1996;13(3):485–96.
29. Johnson KA, Kile TA. Hallux valgus due to cuneiform metatarsal instability. J South Orthop Assoc. 1994;3(4):273–82.
30. Lundgren P, Nester C, Liu A, Arndt A, Jones R, Stacoff A, Wolf P, Lundberg A. Invasive in vivo measurement of rear-, mid- and forefoot motion during walking. Gait Posture. 2008;28(1):93–100.
31. Mizel MS. The role of the plantar first metatarsal first cuneiform ligament in weight bearing on the first metatarsal. Foot Ankle Int. 1993;14:82–4.
32. Wanivenhaus A, Pretterklieber M. First tarsometatarsal joint: anatomical biomechanical study. Foot Ankle. 1989;9:153–7.
33. Johnson C, Christensen JC. Biomechanics of the first ray part 1. The effects of the peroneus longus function. A three dimensional kinematic study on a cadaver model. J Foot Ankle Surg. 1999;38(5):313–21.
34. Perez HR, Leon KR, Jeffrey CC. The effect of frontal plane position on first ray motion: forefoot locking mechanism. Foot Ankle Int. 2008;29(1):72–6.
35. Sarrafian SK. Functional characteristics of the foot and plantar aponeurosis under tibiotalar loading. Foot Ankle. 1987;8:4–18.
36. Sharkey NA, Ferris L, Donahue SW. Biomechanical consequences of plantar fascial release or rupture during gait. Part I: disruptions in longitudinal arch-conformation. Foot Ankle Int. 1998;19:812–20.
37. Stainsby GD. Pathological anatomy and dynamic effect of the displaced plantar plate and the importance of the integrity of the plantar plate–deep transverse metatarsal ligament tie-bar. Ann R Coll Surg Engl. 1997;79:58–68.
38. Khaw FM, Mak P, Johnson GR, et al. Distal ligamentous restraints of the first metatarsal. An in vitro biomechanical study. Clin Biomech (Bristol, Avon). 2005;20:653–8.
39. Sarrafian SK. Anatomy of the foot and ankle. 1st ed. Philadelphia: J.B. Lippincott Company; 1983.
40. Hicks JH. The mechanics of the foot: II. The plantar aponeurosis and the arch. J Anat. 1954;88:25–30.
41. Griffin NL, Miller CE, Schmitt D, D'Aout K. Understanding the evolution of the windlass mechanism of the human foot from comparative anatomy: insights, obstacles, and future directions. Am J Phys Anthropol. 2015;156:1–10.
42. Sharkey NA, Donahue SW, Ferris L. Biomechanical consequences of plantar fascial release or rupture during gait - Part II: Alterations in forefoot loading. Foot Ankle Int. 1999;20(2):86–96.
43. Kirane YM, Michelson JD, Sharkey NA. Contribution of the flexor hallucis longus to loading of the first metatarsal and first metatarsophalangeal joint. Foot Ankle Int. 2008;29(4):123–34.
44. Sharkey NA, Ferris L, Smith TS, et al. Strain and loading of the second metatarsal during heel-lift. J Bone Joint Surg Am. 1995;77:1050–7.

45. Rush SM, Christensen JC, Johnson CH. Biomechanics of the first ray. Part 2: metatarsus primus varus as a cause of hyper- mobility. A three-dimensional kinematic analysis in a cadaver model. J Foot Ankle Surg. 2000;39(2):68–77.
46. Coughlin MJ, Jones CP, Viladot R, Glano P, Grebing BR, Kennedy MJ, Shurnas PS, Alvarez F. Hallux valgus and first ray mobility: a cadaveric study. Foot Ankle Int. 2004;25(8):537–44.
47. Coughlin MJ, Jones CP. Hallux valgus and first ray mobility. A prospective study. J Bone Joint Surg Am. 2007;89(9):1887–98.
48. Hardy RH, Clapham JC. Observations on hallux valgus; based on a controlled series. J Bone Joint Surg Br. 1951;33-B(3):376–91.
49. Shibuya N, Roukis T, Jupiter DC. Mobility of the first ray in patients with or without hallux valgus deformity: systematic review and meta-analysis. J Foot Ankle Surg. 2017;56:1070–5.
50. Doty J, Coughlin M. Hallux valgus and hypermobility of the first ray: facts and fiction. Int Orthop. 2013;37:1655–60.
51. Hetherington VJ, Carnett J, Patterson BA. Motion of the first metatarsophalangeal joint. J Foot Surg. 1989;28:13–9.
52. Shereff MJ, Bejjani FJ, Kummer FJ. Kinematics of the first metatarsophalangeal joint. J Bone Joint Surg Am. 1986;68:392–8. 31.
53. Yoshioka Y, Siu DW, Cooke TD, Bryant JT, Wyss U. Geometry of the first metatarsophalangeal joint. J Orthop Res. 1998;6:878–85.
54. Joseph J. Range of movement in the great toe in men. J Bone Joint Surg Br. 1954;36:450.
55. Gerbert J, editor. Textbook of Bunion surgery. Mount Kisco, New York: Futura Publishing; 1989.
56. Heatherington V, Johnson R, Albritton J, et al. Necessary dorsiflexion of the first metatarsophalangeal joint during gait. J Foot Surg. 1990;29:218.
57. Hopson M, McPoil T, Cornwall M. Motion at the first metatarsophalangeal joint, reliability and validity of four measurement techniques. J Am Podiatr Med Assoc. 1995;85:198.
58. Nawoczenski D, Baumhauer J, Umberger B, et al. Relationship between clinical measurements and motion of the first metatarsophalangeal joint during gait. J Bone Joint Surg. 1999;81:370.
59. Laird PO. Functional hallux limitus. The Illinois Podiatrist. 1972;9:4.
60. Halstead J, Redmond A. Weight-bearing passive dorsiflexion of the hallux in standing is not related to hallux dorsiflexion during walking. J Orthop Sports Phys Ther. 2006;36:550.
61. Dananberg HJ. Functional hallux limitus and its relationship to gait efficiency. J Am Podiatr Med Assoc. 1986;76:648.
62. Payne C, Chutter V, Miller K. Sensitivity and specificity of the functional hallux limitus test to predict foot function. J Am Podiatr Med Assoc. 2002;92:269.
63. Carlson RE, Fleming LL, Hutton WC. The biomechanical relationship between the tendo-achilles, plantar fascia and metatarsophalangeal joint dorsiflexion angle. Foot Ankle Int. 2000;21:18–25.
64. Dananberg HL. Gait style as an etiology to chronic postural pain. Part 1. Functional hallux limitus. J Am Podiatr Med Assoc. 1993;83:433–41.
65. Payne C, Turner D, Miller K. Determinants of plantar pressures in the diabetic foot. J Diabet Complications. 2002;16:277–83.
66. Phillips RD, Law EA, Ward ED. Functional motion of the medial column joints of the foot during propulsion. J Am Podiatr Med Assoc. 1996;86(10):474–86.
67. Roukis TS, Scherer PR, Anderson CF. Position of the first ray and motion of the first metatarsophalangeal joint. J Am Podiatr Med Assoc. 1996;86:538–46.
68. Aquino A, Payne C. The role of the reverse windlass mechanism in foot pathology. J Am Podiatr Med Assoc. 2000;34:32–4.
69. Harradine P, Bevan L. The effect of rearfoot eversion on maximal hallux dorsiflexion. J Am Podiatr Med Assoc. 2000;90:390.
70. Gatt A, Mifsud T, Chockalingam N. Severity of pronation and classification of first metatarsophalangeal joint dorsiflexion increases the validity of the Hubscher manoeuvre for the diagnosis of functional hallux limitus. Foot. 2014;24:62–5.

71. Bojsen-Moller F. Calcaneocuboid joint and stability of the longitudinal arch of the foot at high and low gear push off. J Anat. 1979;129:165.
72. MacConaill MA. The postural mechanism of the human foot. Proc R Ir Acad. 1944–1945;50B:265–78.
73. Griffin NL, D'Aout K, Richmond B, Gordon F, Aerts P. Comparative in vivo forefoot kinematics of Homo sapiens and Pan paniscus. J Hum Evol. 2010;59:608–19.
74. Durrant B, Chockalingam N. Functional hallux limitus: a review. J Am Podiatr Med Assoc. 2009;99(3):236–43.
75. Dananberg HL. Gait style as an etiology to chronic postural pain. Part II. Postural compensatory process. J Am Podiatr Med Assoc. 1993;83(11):615–24.
76. Cotterill JM. Condition of stiff great toe in adolescents. Edinb Med J. 1887;33:459.
77. Harrison M. Hallux limitus. In: Proceedings of the Dewar Orthopaedic Club. /. Bone and Joint Surg. 1971;53-B(4):772.
78. Kravitz SR, LaPorta GA, Lawson JH. KLL progressive staging classification of hallux limitus and hallux rigidus. Lower Extrem. 1994;1:55–66.
79. Coughlin MJ. Hallux rigidus: demographics, etiology and radiographic assessment. Foot Ankle Int. 2003;24:731–43.
80. Roukis TS, Jacobs PM, Dawson DM, Erdmann BB, Ringstrom JB. A prospective comparison of clinical, radiographic, and intraoperative features of hallux rigidus: soft-term follow-up and analysis. J Foot Ankle Surg. 2002;41:158–65.
81. Camasta C. Hallux limitus and rigidus. Clinical examination, radiographic findings, and natural history. Clin Podiatr Med Surg. 1996;13:423–48.
82. Nilsonne H. Hallux rigidus and its treatment. Acta Orthop Scand. 1930;1:295–303.
83. Coughlin MJ, et al. Hallux rigidus: grading and long term results of operative treatment. J Bone Joint Surg Am. 2003;85:2072–88.
84. Regnauld B. Disorders of the great toe. In: Elson R, editor. The foot: pathology, etiology, epidemiology, clinical investigation and treatment. Berlin: Springer; 1986. p. 269–81.
85. Hattrup SJ, et al. Subjective results of hallux rigidus following treatment with cheilectomy. Clin Orthop Relat Res. 1988;226:182–91.
86. Beeson P, Phillips C, Corr S, et al. Classification systems for hallux rigidus: a review of the literature. Foot Ankle Int. 2008;29:407.
87. McMaster M. The pathogenesis of hallux rigidus. J Bone Joint Surg Br. 1978;60(1):82–7.
88. Mann RA, Coughlin MJ, Duvries HL. Hallux rigidus: a review of the literature and a method of treatment. Clin Orthop Relat Res. 1979;142:57–63.
89. Stein G, Pawel A, Koebke J, et al. Morphology of the first metatarsal head and hallux rigidus: a cadaveric study. Surg Radiol Anat. 2012;34(7):589–92.
90. Hunt K, Anderson R. Biplanar proximal phalanx closing wedge osteotomy for hallux rigidus. Foot Ankle Int. 2012;33(12):1043–50.
91. Lambrinudi C. Metatarsus primus elevatus. Proc R Soc Med. 1938;31(11):1273.
92. Youngswick F. Modifications of the Austin bunionectomy for treatment of metatarsus primus elevatus associated with hallux limitus. J Foot Surg. 1982;21:114–6.
93. Ronconi P, Monachino P, Baleanu P, Favilli G. Distal oblique osteotomy of the first metatarsal for the correction of hallux limitus and rigidus deformity. J Foot Ankle Surg. 2000;39:154–60.
94. McNeil DS, Baumhauer JF. Glazebrook. Evidence-based analysis of the efficacy for operative treatment of hallux rigidus. Foot Ankle Int. 2013;34(1):15–32.
95. Bouaicha S, Ehrmann C, Moor BK, et al. Radiographic analysis of metatarsus primus elevatus and hallux rigidus. Foot Ankle Int. 2010;31(9):807–14.
96. Seiberg M, Felson S, Colson JP, et al. Closing base wedge versus Austin bunionectomies for metatarsus primus adductus. J Am Podiatr Med Assoc. 1994;84:548.
97. Roukis TS. Metatarsus primus elevatus in hallux rigidus: fact or fiction? J Am Podiatr Med Assoc. 2005;95(3):221–8.
98. Horton GA, Park YW, Myerson MS. Role of metatarsus primus elevatus in the pathogenesis of hallux rigidus. Foot Ankle Int. 1999;20(12):777–80.

99. Meyer JO, Nishon LR, Weiss L, Docks G. Metatarsus primus elevatus and the etiology of hallux rigidus. J Foot Surg. 1987;26(3):237–41.
100. Greisberg J, Sperber L, Prince DE. Mobility of the first ray in various foot disorders. Foot Ankle Int. 2012;33:44–9.
101. Rzonca E, Levitz S, Lue B. Hallux equinus. The stages of hallux limitus and hallux rigidus. J Am Podiatry Assoc. 1984;74:390–3.
102. Vallotton J, Echeverri S, Dobbelaere-Nicolas V. Functional hallux limitus or rigidus caused by a tenodesis effect at the retrotalar pulley. J Am Podiatr Med Assoc. 2010;100(3):220–9.
103. Harton FM, Weiskopf SA, Goecker RM. Sectioning the plantar fascia. Effect on first metatarsophalangeal joint motion. J Am Podiatr Med Assoc. 2002;92(10):532–6.
104. Elliott AD, Borgert AJ, Roukis TS. A prospective comparison of clinical, radiographic and intraoperative features of hallux rigidus: long-term follow up and analysis. J Foot Ankle Surg. 2016;55:547–61.
105. Canseco K, Long J, Marks R, Khazzam M, Harris G. Quantitative characterization of gait kinematics in patients with hallux rigidus using the Milwaukee Foot Model. J Orthop Res. 2008;26(4):419–27.
106. Canseco K, Long J, Marks R, Harris G. Quantitative motion analysis in hallux valgus: rehabilitative insight. Arch Phys Med Rehab. 2007;88:E10.
107. Betts RP, Franks CI, Duckworth T. Foot pressure studies: normal and pathologic gait analyses. In: Jahss MH, editor. Disorders of the foot and ankle. 2nd ed. Philadelphia: W.B. Saunders; 1991. p. 484–519.
108. DeFrino PF, Brodsky JW, Pollo FE, et al. First metatarsophalangeal arthrodesis: a clinical, pedobarographic and gait analysis study. Foot Ankle Int. 2002;23:496–502.
109. Giannini S, Ceccarelli F, Faldini C, et al. What's new in surgical options for hallux rigidus? J Bone Joint Surg Am. 2004;86-A(Suppl 2):72–83.

Metatarsalgia and Digital Deformities

6

"After all, it is the toe that goes forward first and foremost, and the toe that helps to tell us if our bodies are hot or cold—in other words, the toe experiences far more than we give it credit for."

–Meia Geddes, Love Letters to the World

Key Points

1. The term metatarsalgia describes pain in the forefoot which can be the result of many different pathologies.
2. Contrary to popular belief, elongated metatarsals are not associated with elevated plantar pressures or with metatarsalgia.
3. Plantar-flexed metatarsals will increase plantar pressure and can cause localized metatarsalgia.
4. Elevated plantar pressure in the forefoot is more closely associated with digital deformities than any other condition.
5. The plantar plate of the lesser MTPJ is subjected to compressive, tensile, and shear forces which contribute to degeneration and rupture.
6. Claw toes have a different etiology and pathomechanics than plantar plate injuries.
7. The interdigital neuroma is a common cause of metatarsalgia; however specific risk factors and pathomechanics of this condition have not been validated.

Terminology and Classification

The lesser metatarsal region of the forefoot is the most common location of foot pain in older people. An examination of 784 community based adults aged 65 and older revealed that 37% had pain to palpation of the forefoot, with pain divided equally between the metatarsal heads and the intermetatarsal spaces [1]. In contrast,

D. H. Richie Jr, *Pathomechanics of Common Foot Disorders*,
https://doi.org/10.1007/978-3-030-54201-6_6

only 11% had pain on the plantar heel or along the plantar fascia. In this same group of patients, 60% had digital deformities, while 37% had bunion deformities. Together, pain on the plantar aspect of the metatarsals and digital deformities represent the two most common foot problems seen in older adults. This raises the question: Are these two conditions related to each other?

The term metatarsalgia is commonly used to describe pain on the plantar aspect of the forefoot [2, 3]. Metatarsalgia is a descriptive term and not a specific diagnosis. A study by Scranton, reporting on 98 patients presenting with metatarsalgia, revealed 23 different diagnoses accounting for pain in the forefoot [4]. These diagnoses included Morton's neuroma, plantar plate pathology, stress fracture, avascular necrosis/Freiberg's disease, brachymetatarsia, benign skin lesion, bursitis, tendinitis, capsulitis, and arthritis. That study divided diagnoses and causes of metatarsalgia into three categories:

- *Primary*, which is due to excessive mechanical loading of one or more metatarsals
- *Secondary* which is due to systemic disease affecting the MTPJs
- *Other* forefoot pain not localized to the MTPJs seen in nerve and circulatory conditions

Espinosa proposed a newer classification system of causative factors of metatarsalgia [5]. He clarified that the term "metatarsalgia" has traditionally described pain in the area of second thru fourth metatarsals, which excludes pathologies in the 1st and 5th MTPJ. A review of the literature verifies that the terms "metatarsalgia" and "central metatarsalgia" are used interchangeably and both describe pain in the central forefoot. For this chapter, the two terms are synonymous.

In the Espinosa classification, structural deformities within the foot, such as long or plantar-flexed metatarsals, are classified as a *primary* cause of metatarsalgia. Also included in the structural category is the condition known as first ray insufficiency. Primary metatarsalgia includes all conditions which are mechanically induced. Metabolic, systemic, and traumatic conditions are *secondary* causes of metatarsalgia. Finally, an *iatrogenic* category of metatarsalgia is proposed which can result from failed hallux valgus surgery or lesser metatarsal surgery.

We will focus our discussion on primary metatarsalgia which is thought to be due to abnormal mechanical overload of certain anatomic structures in the forefoot. The two most commonly accepted causes of this overload, causing primary metatarsalgia, are long central metatarsals or plantar-flexed central metatarsals. As this topic is explored, it will become clear that digital deformities are a closely related condition to metatarsalgia. The same forces which cause metatarsalgia can also cause digital deformities. Digital deformities can also cause metatarsalgia.

What Is the Role of an Elongated Metatarsal Causing Metatarsalgia?

Increased length of one or more of the central metatarsals is believed to be a factor in the development of primary metatarsalgia, a concept first proposed by Morton in 1927 [6]. In fact, the surgical management of central metatarsalgia most commonly focuses on shortening an elongated or plantar-flexed metatarsal. [2, 7] Despite this widespread surgical practice, there is no evidence showing that elongated metatarsals can cause primary metatarsalgia.

Fundamental to this misunderstanding is the method used to measure the length of the metatarsals of the human foot. Traditionally, this measurement is performed using a weight-bearing A/P radiograph Fig. 6.1b. This method involves drawing a transverse line from the distal most point of the first metatarsal and connecting to the distal most point of the head of the third metatarsal [6]. A long second metatarsal would protrude distal to this line. There can be a major error with this measurement technique as deviation of the first metatarsal in a medial direction, away from the second metatarsal, as seen in HAV deformity will automatically shorten its position from the transverse axis of the second and third metatarsals. However, the actual length of the first metatarsal has not changed.

To account for the effects of angulation of the first metatarsal relative to the second metatarsal, Hardy and Clapham developed a technique which uses a compass to draw an arc of the distal margins of the metatarsals on an A/P radiograph [8]. With this method, a true parabola is drawn, and relative elongation of one or more metatarsals can be determined (Fig. 6.2).

There are four other popular methods of examining forefoot metatarsal length patterns (Fig. 6.1):

A. The metatarsal protrusion index as described by Nilsonne [9] takes into account only the relative lengths of the first and second metatarsals. The accuracy of this measurement is compromised by medial deviation of the first metatarsal in HAV deformity
B. The metatarsal protrusion distance as described by Coughlin [10] considers the relative lengths of the first, second, and third metatarsals and describes the distance from the apex of the second metatarsal to a line connecting the apex of the first and third metatarsals. The accuracy of this measurement is compromised by medial deviation of the first metatarsal in HAV deformity
C. Meschan's metatarsal break angle describes the angle created by connecting the distal most aspects of the first, second, and fifth metatarsals [11]. The angle generally increases with long first metatarsals and decreases with short first metatarsals. The accuracy of this measurement is compromised by medial deviation of the first metatarsal in HAV deformity
D. Maestro's method uses a transmetatarsal line perpendicular to the medial border of the second metatarsal and centered on the lateral sesamoid. [12] The distance from the transmetatarsal line to the apex of each metatarsal is measured. When metatarsals are longer than the transmetatarsal line, they are assigned a positive value, and when they are negative, they are assigned a negative value. Sesamoid position can alter this measurement. The sesamoid position does not change with HAV deformity but does migrate proximal in hallux rigidus conditions as the hallux moves into equinus.

To illustrate how important angular deformity of the first metatarsal must be considered when measuring the metatarsal parabola, consider the results of two early studies by Coughlin [10, 13]. Using Morton's method of drawing a line across

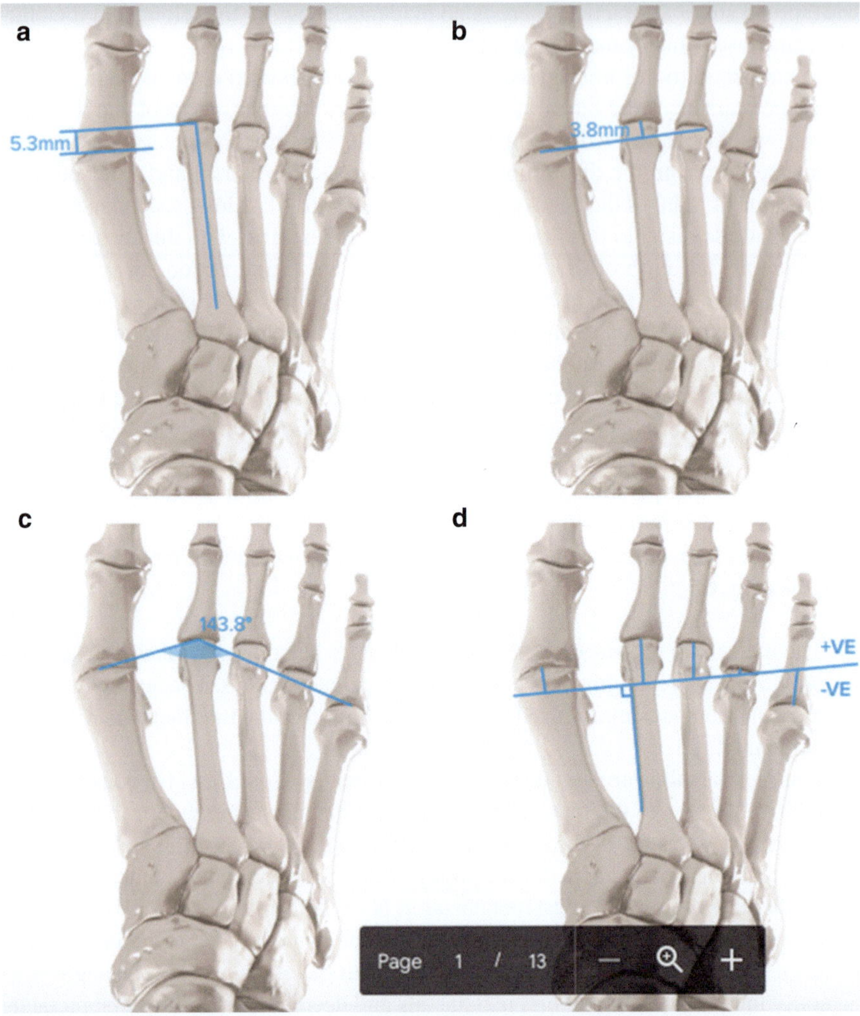

Fig. 6.1 Methods of measuring relative lengths of the metatarsals. (**a**) The metatarsal protrusion index as described by Nilsonne takes into account only the relative lengths of the first and second metatarsals. The longitudinal axis of the second metatarsal is drawn. A perpendicular line is drawn to this. The distance is measured between this line and the distal margin of the first metatarsal. (**b**) The metatarsal protrusion distance as described by Coughlin considers the relative lengths of the first, second, and third metatarsals. A line is drawn connecting the apex of the first and third metatarsals. The distance is measured from the apex of the second metatarsal. (**c**) Meschan's metatarsal break angle describes the angle created by connecting the distal most aspects of the first, second, and fifth metatarsals. The angle generally increases with long first metatarsals and decreases with short first metatarsals. (**d**) Maestro's method uses a transmetatarsal line which is perpendicular to the medial border of the second metatarsal and centered in the lateral sesamoid. The distance from the transmetatarsal line to the apex of each metatarsal is measured, determining which metatarsal is theoretically elongated

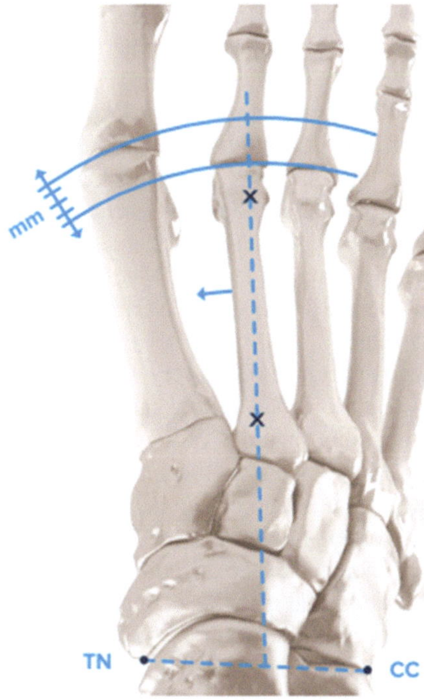

Fig. 6.2 Hardy Clapham measurement. A transverse reference line is made by bisecting two points (one at the most lateral aspect of the calcaneocuboid joint and one at the most medial aspect of the talonavicular joint). The longitudinal axis of the second metatarsal is drawn using two metaphyseal-diaphyseal reference points. The intersection of these two lines acts as a center of rotation for metatarsal length measurements. Using a compass, two arcs are drawn: one at the distal extent of the articular surface of the first metatarsal and one at distal extent of the articular surface of the second metatarsal. A perpendicular line drawn between the two arcs is measured in millimeters, with a positive value indicating a second metatarsal that is relatively longer than the first. A similar perpendicular line is drawn from the midpoint of the head of the third metatarsal to the arc of the first metatarsal

the distal margin of the 1st and 3rd metatarsal heads (see Fig. 6.1b), Coughlin showed that an elongated 2nd metatarsal was found in 80–90% of patients with crossover 2nd toe deformity.

However, later Kaz and Coughlin went back and re-measured the same radiographs using the Hardy Clapham method of a metatarsal parabola, and a long second metatarsal was found in only 21% of the patients [14]. A significant number of patients with HAV deformity in this study accounts for the initial error in measuring length of the first and second metatarsals.

In another study, Fleisher et al. studied several radiographic measures of patients with documented tear of the plantar plate at the second metatarsophalangeal joint (MTPJ) [15]. They measured second metatarsal protrusion using a technique described by Nilsonne which measures length of the second and first metatarsals only using the second metatarsal as a reference [13] (Fig. 6.1). Here, angulation of the first metatarsal into adduction would result in shortening the relative length of this segment. Since half the patients in their study with plantar plate injuries had HAV deformity with IM angle deviation of greater than 15 degrees, it is not surprising that an elongated second metatarsal was found in a similar proportion.

One of the most elaborate descriptions of variations in metatarsal parabola is provided by Maestro, who places great emphasis on metatarsal length for justification of metatarsal surgery to relieve metatarsalgia [12]. He speculates, like many others, that increased length of a metatarsal causes increased plantar pressure and metatarsalgia. In his paper, Maestro uses a standard A/P radiograph to measure metatarsal length but relies on the position of the fibular sesamoid as the point of reference (Fig. 6.1). Sesamoids can migrate proximal or distal depending on passive and dynamic contracture of the flexor hallucis brevis providing ample opportunity for error in measuring metatarsal length. Notwithstanding, Maestro identifies four different "morphotypes" of length patterns. Of these four, one morphotype is determined to be "harmonious" or "normal," yet this was only found in 31% of the feet they studied. Shortening osteotomies were recommended to restore the harmonious pattern of metatarsal lengths. Maestro used no other objective data such as plantar pressure measurements to verify the role of metatarsal length in creating metatarsalgia.

Surprisingly, there are very few quality studies or publications validating the theory that elongated central metatarsals are associated with metatarsalgia. Most studies on abnormalities of the metatarsal parabola have focused on plantar plate injuries, hallux abductovalgus, and hallux rigidus. One of the few studies focusing on primary metatarsalgia and radiographic measurements was published by Slullitel and co-workers [16]. All of the 121 subjects in this study had HAV deformity, and 45% of them also had symptoms of metatarsalgia. Using the Hardy Clapham method, abnormal metatarsal length had no statistical correlation with incidence of metatarsalgia. In fact, patients with a minus index (i.e., short first metatarsal) were less likely to have metatarsalgia. Also, the magnitude of HAV deformity measured by hallux abductus angle and intermetatarsal angle was not correlated with incidence or of metatarsalgia. Metatarsalgia was associated with increased age, increased body weight, gastrocnemius contracture, and presence of digital deformity [16].

Elongated metatarsals have been associated with digital deformities, particularly crossover toe and plantar plate injuries [10, 13, 18, 19]. With little proof, many authors have speculated that elongated metatarsals will increase plantar pressure which leads to attrition and eventual rupture of the plantar plate. A review of the medical literature on this subject shows conflicting evidence linking elongated metatarsals to plantar plate injury. We will explore the pathomechanics of plantar plate injuries later in this chapter.

Katz and Coughlin use the term "crossover toe" to essentially describe a plantar plate tear since the majority of patients in their study had a positive drawer test, consistent with this injury [14] (Fig. 6.3). In this study of 169 patients with crossover toe deformity, 44% had an elongated 2nd metatarsal. In fact, among all patients in this study, the overall length of the second metatarsal was 0.2 mm *less* than the first metatarsal.

In this same study [14], Katz and Coughlin specifically describe the error in measurement of metatarsal length which they had used in previous studies [10, 13]. Those studies simply drew a line across the distal margin of the three medial metatarsal heads, starting at the first metatarsal and ending at the third metatarsal (see Fig. 6.1b). This method was originally described by Morton [6]. With a true parabolic arc measure, an apparent elongated second metatarsal disappears and is no

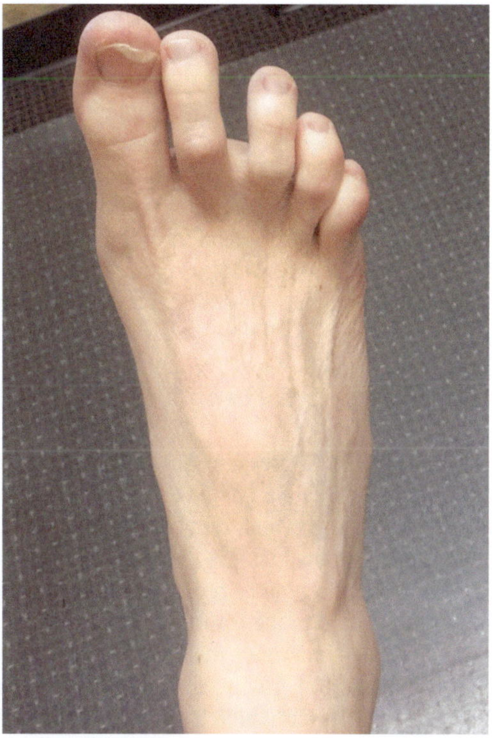

Fig. 6.3 Crossover toe deformity

longer a significant finding in patients with plantar plate injuries and crossover second toe [14]. Notwithstanding, Bhutta used the erroneous linear measure of metatarsal length in a study published after Katz and Coughlin [19]. As expected this technique found a correlation between increased 2nd metatarsal length and presence of crossover 2nd digit deformity.

In summary, metatarsal length does not appear to be correlated with primary metatarsalgia. Methods of measuring metatarsal length show conflicting results in patients with crossover toe deformity. Metatarsal length has more relevance in another form of metatarsalgia which occurs after iatrogenic shortening of a single ray or metatarsal.

The term *transfer metatarsalgia* refers to the onset of pain at a different ray than that which is mechanically impaired [20]. Transfer metatarsalgia is most likely the result of shortening of the first metatarsal after hallux valgus surgery or hallux rigidus surgery [21, 22]. Metatarsal osteotomies for HAV surgery result in shortening by an average 2–4 mm [23]. A Mitchell osteotomy, performed for HAV surgery, causes an average shortening of the first metatarsal of 4 mm and causes transfer metatarsalgia in 20% of patients undergoing this procedure [24, 25].

In a systematic review of 4 studies of 93 patients with hallux rigidus who underwent plantarflexion-shortening osteotomy of the first metatarsal, Roukis found that 30% of the patients developed post-operative metatarsalgia [26]. A similar proportion (22.3%) of all the patients who had a first metatarsal osteotomy for hallux rigidus required some type of revisional surgery.

It appears that in the healthy intact foot, relative length of the metatarsals has little association with the development of metatarsalgia. However, excessive shortening of any metatarsal in a previous healthy foot can have devastating consequences, causing intractable transfer metatarsalgia. Transfer metatarsalgia may not be purely the result of iatrogenic shortening of a metatarsal as mal-union can also cause elevation or dorsiflexion of a metatarsal. The sagittal plane orientation of the distal metatarsals may be more important than length. This has been demonstrated in studies of plantar pressure in patients with metatarsalgia.

Role of Metatarsal Length and Plantar Pressure

Abnormal plantar pressure has been attributed to be a primary mechanical force causing metatarsalgia [27]. This is validated to some degree with studies showing that interventions which reduce plantar pressure under the metatarsals improve the symptoms of metatarsalgia [28, 29]. In static stance, metatarsal length would have less influence on plantar pressure than during terminal stance and pre-swing when weight is shifted to the forefoot.

An equally important mechanical force in the pathomechanics of metatarsalgia is shear stress [30]. Yavuz et al. showed that patients with HAV deformity have significant shear force which is shifted laterally in the forefoot [31]. Plantar callous patterns in patients with metatarsalgia often show a linear shape along the line of progression suggesting a skin-shear effect caused by the metatarsals gliding in an anterior-posterior direction [32]. Unfortunately, studies of shear stress on patients with metatarsalgia are lacking, but the potential contribution of this force must be considered in the pathomechanics of all forefoot pathologies.

It is logical to assume that elongated metatarsals elevate plantar pressure under the MTPJs during dynamic gait. However, if we look at investigations which studied plantar pressure in patients with metatarsalgia, there is little evidence that elongated metatarsals actually increase plantar pressure, let alone cause metatarsalgia.

There are surprisingly few studies of plantar pressure measurement in patients with metatarsalgia. Dreeben studied 37 patients (45 feet) with primary metatarsalgia under the second metatarsal head [33]. The contralateral symptom-free feet of the patients were used as controls. Metatarsal length and height of the metatarsal head from the supportive surface were measured and then correlated with plantar pressure measurements obtained from a pedobarograph. *In this study, plantar pressure did not correlate with metatarsal length but did closely correlate with metatarsal head position or height from the ground.* When a second metatarsal osteotomy was performed, decrease in plantar pressure occurred when the metatarsal head was elevated, while shortening of the metatarsal had minimal effect on plantar pressure. In order to reduce plantar pressure and pain, the second metatarsal head required elevation by at least 3.5 mm; however the reduction in plantar pressure was not predictable with the amount of elevation. Decreased plantar pressure was noted under the first metatarsal in patients with metatarsalgia. The authors concluded that plantar pressure under the metatarsal heads is multifactorial and might be affected by the quality of the soft tissues and plantar fat pad.

The proposal that a plantar-flexed metatarsal, rather than an elongated metatarsal, is a contributor to metatarsalgia was supported by Jung et al. [34]. In this cadaver study, osteotomies were performed on the first metatarsal to determine the effect of dorsal elevation versus shortening of the bone on plantar pressure. Dorsal elevation had a greater effect on reducing pressure under the first metatarsal than shortening. Both dorsal elevation and shortening of the first metatarsal caused significant increased plantar pressure under the second metatarsal.

Kaipel and co-workers prospectively studied 46 patients with metatarsalgia compared to 45 asymptomatic patients and used radiographic and pedobarographic measures to determine the role of metatarsal length and plantar pressures in causing symptoms of metatarsalgia [35]. These researchers used a simple linear reference from the head of the second metatarsal, extending across to the first and third metatarsals to measure relative length of the bones. As shown previously, a deviated first metatarsal can alter this measurement. This probably was not a factor, as patients with HAV deformity were excluded from the Kaipel study. The relative lengths of the metatarsals did not correlate with the pressure or force under the respective metatarsal. Also, there was no difference in peak pressure under the central metatarsals comparing the group with metatarsalgia with the control group. There were no differences in metatarsal length patterns when comparing the groups with metatarsalgia to those without metatarsalgia. The only significant finding was a decrease in peak force under the first metatarsal in the metatarsalgia group, similar to what was found in the study by Dreeben et al. This validates a long accepted notion that instability of the first ray is the primary cause of metatarsalgia.

To summarize, studies do not support the notion that elongated metatarsals increase plantar pressures or correlate with the incidence of metatarsalgia. Instead, sagittal plane position of the metatarsal appears to play a more important role in plantar pressure. The position of the heads of the metatarsals relative to each other is important during terminal stance and pre-swing when plantar pressures peak in the forefoot. During these phases of gait, load should be shifting medially to the first ray. Therefore, loss of stability of the first ray during dynamic gait may be the most significant factor causing central metatarsalgia.

The Role of First Ray Hypermobility and Metatarsalgia

While HAV has demonstrated a close association with first ray instability, few investigations have documented the true incidence of metatarsalgia in patients with this deformity. Coughlin and Jones have published the most comprehensive study of patients with hallux valgus and did find a significant incidence of metatarsalgia which was found in 48% of patients with HAV deformity [36]. Yet, only 11% of all patients with HAV in this study demonstrated plantar callosities in the forefoot.

Interestingly, metatarsalgia was the primary reason for patients to seek surgery for HAV deformity in the Coughlin study.

Hypermobility or instability of the first ray has been speculated to cause metatarsalgia due to a lateral shift of pressure away from the first metatarsal to adjacent metatarsals [37]. Lateral shift of load in the forefoot has been speculated to cause synovitis of the lesser MTPJ's, plantar plate injury, neuroma, and metatarsal stress fracture [38–40].

The concept of first ray hypermobility and its association with HAV has been discussed extensively in Chap. 6. Hypermobility appears to be the *result*, rather than the *cause*, of HAV deformity. Excessive dorsal mobility of the first ray measured by static exam is reduced with realignment osteotomy of the first metatarsal [41, 42].

Griesberg et al. measured first ray translation and correlated the findings in patients with various foot disorders [43]. Increased mobility of the first ray was most likely found in patients with hallux valgus and patients with metatarsalgia. However, the authors did not separate the two patient groups, and the statistics suggest that most patients with metatarsalgia also had hallux valgus. There are no studies which show that hypermobility of the first ray is directly linked to metatarsalgia in any patient group other than those with hallux valgus.

Many clinicians attribute second metatarsal hypertrophy, seen on radiographic exam as an adaptive change resulting from first ray instability. However, this radiographic sign may be misleading. Coughlin measured second metatarsal thickness on x-ray and then measured the mobility of the first metatarsal and found no correlation between the measurements [44]. In a retrospective case control study of women athletes with a history of lesser metatarsal stress fractures, Glasoe and co-workers found no correlation with increased first ray elevation and incidence of fracture [40].

In conclusion, the concept of first ray hypermobility, measured with static exam, and its actual existence in any foot disorder other than HAV has not been validated. Hypermobility of the first ray in the sagittal plane seems to occur only when the first metatarsal shifts medially away from the sesamoid envelope and the stabilizing effects of the plantar aponeurosis and flexor hallucis longus are compromised.

Pressure Studies of HAV Patients

Pressure studies have confirmed the critical role of the hallux in accepting load during gait. *In fact, the hallux bears almost as much weight as any of the metatarsals and more than the third, fourth, and fifth metatarsals* [45]. The load carried by the great toe is more than twice that carried by all the lesser digits, combined. Only the first and second metatarsals bear more weight than the hallux [45].

While it has been well established that patients with hallux valgus have excessive motion of the first ray in static exam and a significant number of these patients have metatarsalgia, few studies have documented increased pressure under the central metatarsal heads in these patients during dynamic gait. It is assumed that there will be increased pressure under the second metatarsal resulting from simple load transfer from a hypermobile first ray. Yet, not all studies show a significant load transfer to the second metatarsal in HAV patients. What is often overlooked is the effect of deviated alignment of the hallux in HAV deformity. This not only affects load bearing of the hallux but also significantly affects loading across the entire forefoot.

When measuring plantar pressures in the forefoot, there are two major obstacles affecting clinical interpretation. Plantar pressure can be altered by variation in thickness of the integument and the plantar fat pad which often occurs in HAV patients who have a diminished fat pad under the first MTP [46]. Secondly, depending on the technology used to measure plantar pressure, discreet measurement of loading under a specific metatarsal is challenging. Finally, patients with metatarsalgia and HAV deformity may compensate with an altered gait pattern to avoid pain. Commonly, this will show a lateral shift of loading across the forefoot as the patient supinates their foot during terminal stance [35].

Some studies have demonstrated *decreased pressure under the hallux* in patients with HAV and the expected shift of pressure to the central metatarsals [47–49] (Fig. 6.4). Another pattern was measured by Kernozek showing *increased plantar pressure under the hallux* in patients with HAV and a lateral shift of pressure to the central metatarsals [50] (Fig. 6.5). One study has showed no significant lateral shift of metatarsal pressure [51]. Finally, some studies show increased pressure under both the first and second metatarsals in patients with moderate to severe HAV deformity [52, 53]. Despite a high prevalence of hypermobility shown in the off-weight-bearing exam of patients with HAV deformity, only one study has demonstrated that these patients have reduced loading of the first metatarsal during dynamic gait [54].

Fig. 6.4 HAV deformity with loss of pressure under hallux and 1st metatarsal. The lateral shift of pressure is isolated to the 2nd metatarsal

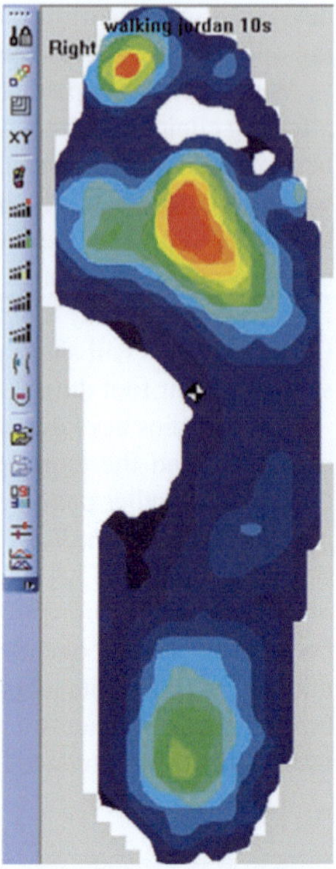

Fig. 6.5 HAV deformity with intact pressure under hallux, loss of pressure under 1st metatarsal and transfer of load to the central metatarsals

The variation of findings in plantar pressure studies of HAV patients may be due to changes which occur as HAV deformity increases. Pain can cause compensated gait patterns. Stratification of patients according to stage of HAV in future pressure studies could give insight into this variable.

What is clear is that not all patients with HAV, despite off-weight-bearing measurement of first ray hypermobility, demonstrate the expected loss of weight bearing under the hallux and first metatarsal during dynamic gait and therefore do not transfer load to the lesser metatarsals. This was seen in the study conducted by Bryant et al. where load actually increased under the first metatarsal and no lateral shift of plantar pressure was noted [51]. Conversely, Waldecker studied HAV patients, all of whom had metatarsalgia, and documented a loss of weight bearing under the hallux resulting in a lateral shift of load to the central metatarsals [48]. This shift or transfer of load was speculated to be due to a loss of plantarflexion stabilization of the hallux due to malalignment of the static and dynamic soft tissue structures which insert on the hallux. Valgus rotation of the sesamoids compromises the plantarflexion moment arm of both the plantar aponeurosis and the flexor hallucis brevis. Interestingly, Waldecker showed that the major load transfer was from the hallux to the lateral forefoot, while loading of the first metatarsal remained intact in HAV patients.

A more recent study by Hoffman et al. evaluated 36 patients with unilateral HAV deformity and compared to the contralateral healthy foot as well as 30 healthy controls [55]. A novel method of pressure measurement was used using a pedographic instrumented treadmill. A significant increased pressure was measured under the second and third metatarsals in the HAV feet. While pressure decreased under the hallux and first metatarsal in HAV feet, this change did not achieve statistical significance. Also, significant pressure was shifted to digits 4 and 5 in the HAV feet, indicating a supinated forefoot with push off thru the oblique axis of the MTPJs.

Higher forces thru the lateral digits of HAV patients were also found in the Framingham Foot Study [56]. This study of an extraordinary large cohort of human subjects allowed comparison of plantar pressure and force among a group of patients with hallux valgus (1123 subjects), hallux valgus and other foot disorders (641 subjects), and people without hallux valgus (3707 subjects). This high-powered study evaluated foot posture, center of pressure excursion during gait, and distribution of peak pressure and force in eight regions of the foot. Subjects with hallux valgus demonstrated higher peak pressure under the medial rearfoot, as well as a smaller center of pressure excursion index and a lower arch profile than healthy subjects. All of these findings link hallux valgus to a pronated rearfoot alignment during gait. Furthermore, the subjects with hallux valgus demonstrated decreased loading of the hallux along with increased loading of the lesser digits. This loading pattern shows a shift from a push off thru the hallux to a push off thru the lateral digits in hallux valgus patients. The pronated position of the rearfoot appears to be the pre-disposing factor for this altered push off seen in HAV patients compared to healthy subjects.

In summary, studies of plantar pressure in patients with primary metatarsalgia do not show consistent elevation of pressure under the central metatarsals, and elongated metatarsals do not cause elevated plantar pressure. Patients with central metatarsalgia do show decreased loading of the first metatarsal. However, the plantar pressure studies discussed in this chapter did not evaluate subjects for evidence of excessive dorsal mobility of the first ray off weight bearing. While hypermobility of the first ray is assumed to be a common finding in patients with HAV deformity, plantar pressure studies do not consistently show decreased loading of the first ray in these patients during dynamic gait. Instead, loss of loading under the hallux is seen more consistently in HAV patients than loss of loading under the first metatarsal. This may cause a lateral shift of load to the central metatarsals, and this has been verified in several studies. Finally, not all patients with HAV deformity demonstrate a lateral shift of load in the forefoot which explains why not all patients with HAV develop metatarsalgia.

Optimal Load Sharing in the Human Foot

A more common finding, in terms of plantar pressure measurement in HAV patients, is a shift of load away from the hallux to the lateral digits [56]. Loss of weight bearing under the hallux with dysfunction of the windlass mechanism is one cause of this lateral shift. This shift from a "high-gear" push off, which engages the transverse axis of the first and second MTPJ, to a "low-gear" push off engaging the oblique distal orientation of the third, fourth, and fifth MTPJs was originally described by Finn Bojsen-Moller in his classic paper published in 1979 entitled "Calcaneocuboid joint and stability of the longitudinal arch of the foot at high and low gear push off" [57]. This paper is discussed in detail in Chap. 1 of this book.

Bojsen-Moller observes that the metatarsal parabola provides two axes of motion which are available across the forefoot during push off. The relative equal lengths of the first and second metatarsals establish a transverse axis of motion across the evenly aligned first and second MTPJs which is oriented perpendicular to the line of progression. An oblique axis is aligned along the descending lengths of the third, fourth, and fifth metatarsals which is angulated from the line of progression.

Depending upon which axis is favored during push off, two "gears" of leverage are proposed by Bojsen-Moller which differ in their ability to tension the plantar fascia and engage the windlass mechanism. The "high gear push off," across the transverse axis of the first and second MTPJs, improves tensioning of the plantar aponeurosis compared to the "low-gear push off" across the oblique axis. The first metatarsal, with a larger radius compared to the lesser metatarsals, has a better mechanical advantage to increase tension and "shortening" of the plantar aponeurosis as the hallux is driven into extension during toe off. The sesamoids provide a radius of curvature of the first MTP which is twice that of the third metatarsal. When the sesamoids are shifted laterally in HAV deformity, this mechanical advantage of the windlass is compromised by shortening the radius of curvature of the first MTPJ.

If load is directed to the first and second MTPJs during terminal stance, i.e., the third rocker, a "high-gear" push off occurs across a transverse axis alignment of these joints relative to the line of progression. Conversely, if load is directed laterally to the MTPJs 3–5 during the third rocker, a low-gear push off occurs across an oblique axis relative to the line of progression (Fig. 6.6).

A kinematic study was carried out by Griffin and co-workers to verify Bojsen-Moller's observations. Griffin et al. [59] attempted to verify that medial transfer of weight to the forefoot and engagement of the windlass thru the first MTPJ was more efficient in humans compared to great apes (bonobos) during ambulation. Overall, dorsiflexion of the MTPJs was greater in humans compared to the great apes and occurred in a greater magnitude at the first MTPJ.

Pressure data in this study showed a medial transfer from midstance thru toe-off in the human subjects with a peak pressure under the first metatarsal during late stance [59]. This medial shift did not occur in the bonobos, but instead pressure increased in the lateral metatarsals during toe-off (Fig. 6.7). The authors concluded that the greater excursion of dorsiflexion of all the MTPJs in humans was due to specializations of the human foot which includes:

- A stable calcaneocuboid joint
- An adducted first metatarsal (MT) that is substantially larger than the lateral metatarsals
- Distal articular regions of the MTs that are oriented in a way to allow greater metatarsophalangeal joint excursion in the dorsoplantar plane
- Relatively short pedal phalanges

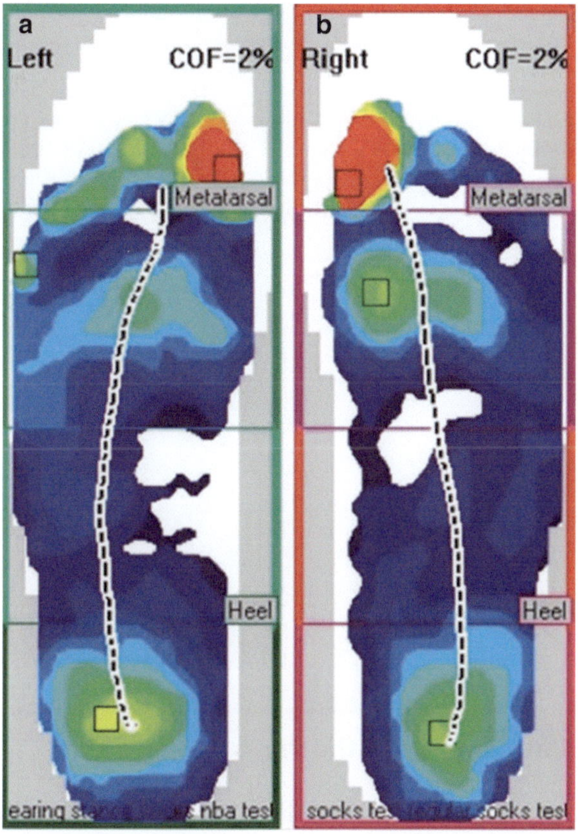

Fig. 6.6 Push off thru the oblique axis left foot (**a**) and transverse axis right foot (**b**)

In summary, this study by Griffin et al. [59] verified the observations which had been made by Bojsen-Moller 30 years earlier regarding the role of lateral column stability, medial transfer of load during terminal stance, and propulsion thru the first MTPJ unique to the human foot [57].

Pathomechanics of Digital Deformity

When load fails to shift medial during terminal stance and pre-swing, the lesser digits bear abnormal forces. Increased mechanical load in the lesser digits causes failure of key structures and compensatory muscular contraction offsets dynamic balancing of alignment of these digits. These two mechanisms have been proposed as the basis for the pathomechanics of hammertoes and other digital deformities. Understanding digital deformities requires a review of the static and dynamic structures which affect digital function and stability.

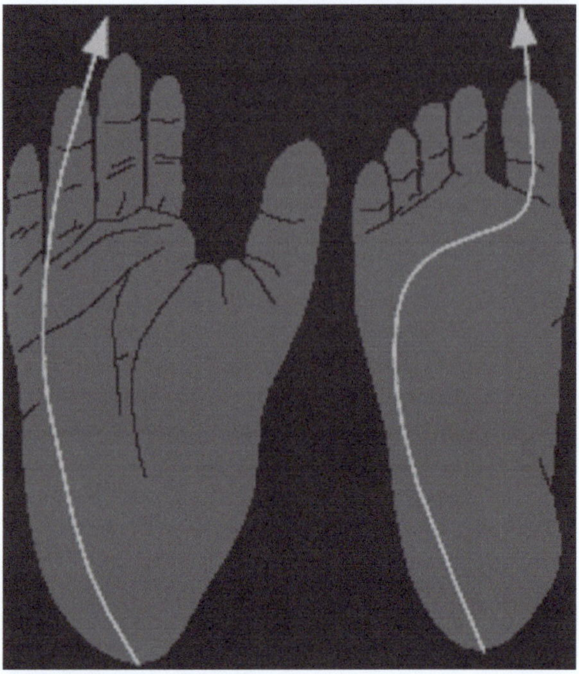

Fig. 6.7 Comparison of push of thru two different axes: oblique axis push off in ape foot (left) and transverse axis push off in human foot (right)

Static Support of the Digits

Two distinct ligaments originate from the tubercle on the medial and lateral sides of the head of the lesser metatarsal and collectively form the collateral ligaments of the MTPJ [60] (Fig. 6.8). The medial and lateral *proper metatarsophalangeal ligaments* connect the tubercles located on the dorsal-lateral and dorsal-medial metatarsal head to the plantar aspect of the proximal phalanx. The medial and lateral *accessory collateral ligaments* connect the metatarsal from the same tubercles to the plantar plate and are also called suspensory ligaments.

The *plantar plate* is also known by other terms such as the plantar metatarsophalangeal ligament, the plantar pad, or the glenoid ligament. The plantar plates are fibrocartilaginous thickenings of the plantar ligament of the MTPJ which average 2 cm in length, 1 cm in width, and between 2 and 5 mm in thickness [61] (Fig. 6.9). Each plantar plate is firmly attached to the articular cartilage at the plantar surface of the base of the proximal phalanx, forming a socket for the head of the metatarsal (Fig. 6.10). Proximally, the plantar plate has a strong fibrous attachment to the plantar fascia and the collateral ligaments. The edges of the plantar plates are attached to the deep transverse metatarsal ligaments as well as to the accessory or suspensory wing of the collateral ligament. Directly beneath the plantar plate is an attachment to the fibrous sheath of the flexor tendons. There is a weak attachment of the plantar plate to the plantar aspect of the neck of the metatarsal. The stronger attachment of the plantar plate is to the proximal phalanx.

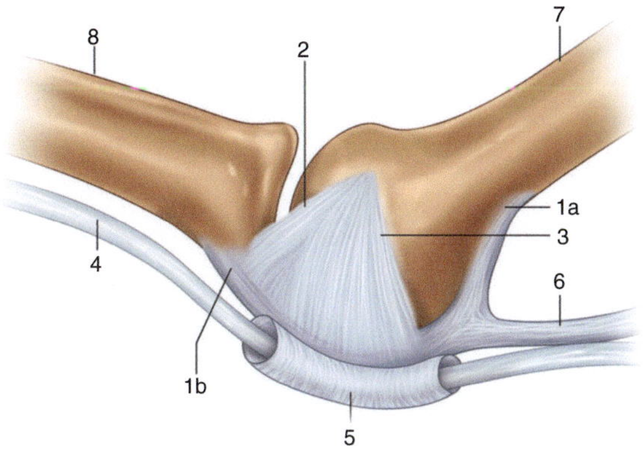

Fig. 6.8 The plantar plate and collateral ligaments. (Reprinted by permission from Springer Nature: Springer, *Hammertoes* by Emily Cook and Jeremy Cook (Eds.) © 2019)
1A = plantar plate origin, 1B = plantar plate insertion, 2 = proper collateral ligament, 3 = accessory collateral ligament, 4 = flexor tendons, 5 = flexor tendon sheath, 6 = plantar aponeurosis attaching onto plantar plate, 7 = metatarsal, 8 = proximal phalanx

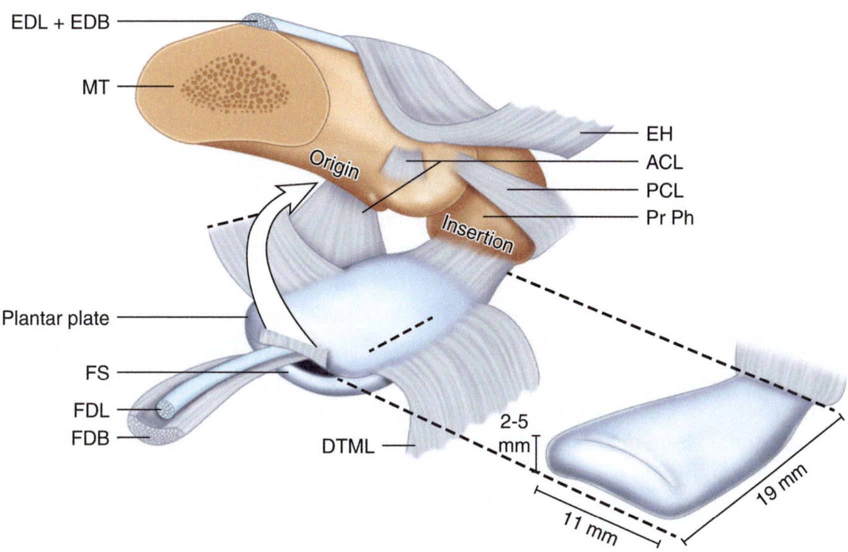

Fig. 6.9 The plantar plate: dorsal view. (From: Maas et al. [151]. Figure 2 Descriptive anatomical & histological data visualized in a schematic 3D drawing of the PP. Legend: *EDL+ EDS,* extensor digitorum longus/brevis; *MT,* metatarsal; *EH,* extensor hood; *ACL,* accessory collateral ligament; *PCL,* proper collateral ligament; *Pr Ph,* proximal phalanx; *FS,* flexor sheath; *FDL + FDB,* flexor digitorum longus/brevis; *DTML,* deep transverse metatarsal ligament, page 5. Open Access

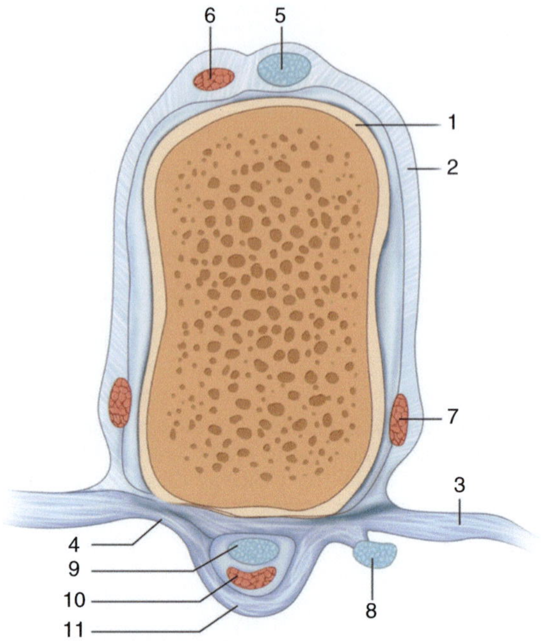

Fig. 6.10 The plantar plate: coronal view. (Reprinted by permission from Springer Nature: Springer, *Hammertoes* by Emily Cook and Jeremy Cook (Eds.) © 2019) 1 = joint capsule, 2 = extensor expansion, 3 = deep transverse metatarsal ligament, 4 = plantar plate, 5 = EDL, 6 = EDB, 7 = interossei within extensor sling, 8 = lumbrical with connection to the plantar plate, 9 = FDL, 10 = FDB, 11 = flexor tendon sheath

The plantar plates are composed 75% of type I collagen and 21% of type II collagen [62]. The dense fibrocartilaginous network of collagen is similar to that of the meniscus of the knee, designed to cushion and resist compressive forces [62, 63]. The dorsal surface of the plantar plate is smooth, providing a gliding surface for the metatarsal head. The inferior surface of the plantar plate has a groove for the FDL and FDB tendons. The plantar fascia also attaches to the plantar plate at its inferior surface while providing a sheath for the flexor tendons (Fig. 6.10).

The plantar aponeurosis attaches to the proximal phalanx via the plantar plate and then continues thru vertical fibers to attach to the plantar integument (Fig. 6.21). The plantar fascia divides near the heads of the metatarsal heads into five processes, one for each toe. Each of these processes divides into superficial and deep strata. The superficial strata insert into the dermis of the plantar integument and have a key function in stabilizing the skin and fat pad during push off. The deep stratum has multiple insertions which stabilize the digits: the plantar plate, the deep transverse metatarsal ligament, the periosteum of the metatarsals, and the fibrous sheath of the flexor tendons at the base of the proximal phalanx.

The sheath of the EDL expands to form an aponeurosis known as the extensor expansion which has two components. The *extensor hood* inserts plantarly into base of the proximal phalanx. The *extensor sling* inserts directly into the plantar plate (Fig. 6.11). The extensor expansion exerts dorsiflexion moment across the MTPJ, enabling the proximal phalanx to move into extension despite having no insertion from the EDL or the EDB.

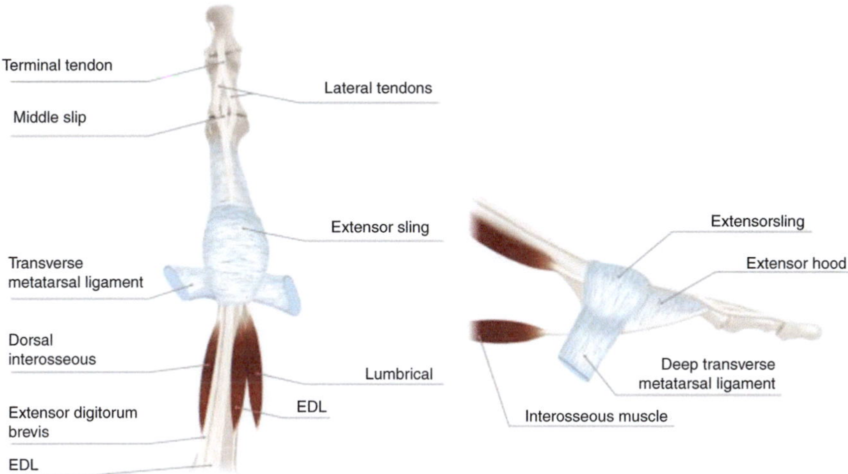

Fig. 6.11 The extensor expansion

Dynamic Support of the Digits

Saraffian describes the "motors" or functional units of the toes providing dynamic support and stabilization for push off [64] (Fig. 6.12).

1. Extensor Digitorum Longus(EDL) and Extensor Digitorum Brevis(EDB) for Digits 2–4

The extensor digitorum longus tendon divides into three slips over the proximal phalanx. The middle slips inserts onto the base of the middle phalanx, and the two lateral slips pass around and converge to form the terminal tendon inserting on the base of the distal phalanx (Fig. 6.12). The EDB inserts into the lateral aspect of the EDL tendon on digits 2, 3, and 4 just distal to the MTPJ.

2. Flexor Digitorum Longus (FDL) and Flexor Digitorum Brevis (FDB) for Digits 2–5

The flexor digitorum longus (FDL) tendon inserts onto the distal phalanx, flexing the distal interphalangeal joint (DIPJ). The FDB tendon splits at the level of the PIPJ to allow passage of the FDL and then inserts on both the medial and lateral aspects of the base of the middle phalanx (Fig. 6.13). Thus, the FDB flexes the PIP joint. There is no direct flexor insertion on the proximal phalanx; thus, with the MP joint in the extended position, the long and short flexors can only flex the PIPJ and DIPJ.

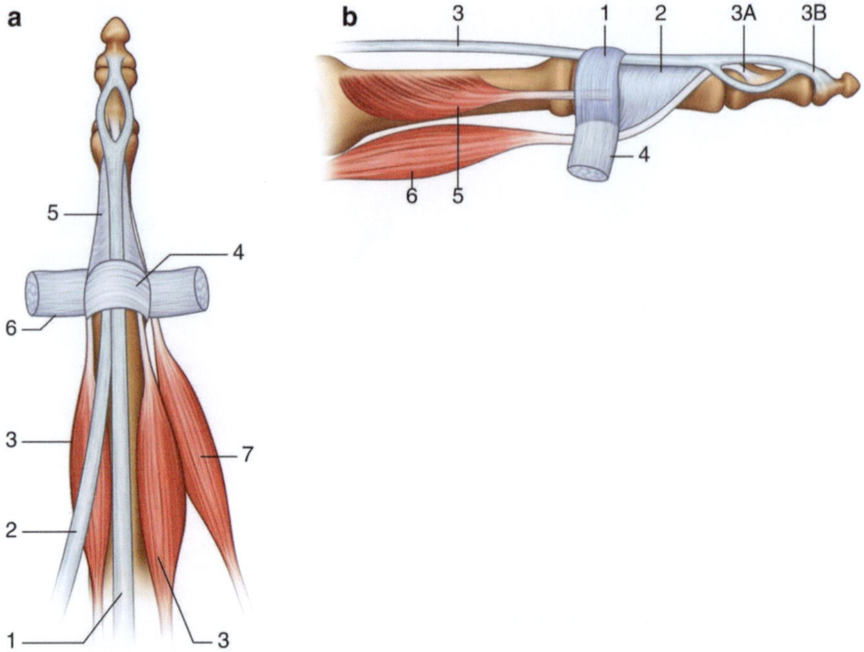

Fig. 6.12 Muscles of the digits. (Reprinted by permission from Springer Nature: Springer, *Hammertoes* by Emily Cook and Jeremy Cook (Eds.) © 2019). (**a**) Top view: 1 = EDL, 2 = EDB, 3 = interossei, 4 = extensor sling, 5 = extensor hood or wing, 6 = DTML, 7 = lumbrical. (**b**) Side view: 1 = extensor sling, 2 = extensor wing or hood, 3 = EDL, 3A = central slip middle phalanx insertion of the EDL, 3B = terminal slip distal phalanx insertion of the EDL, 4 = DTML, 5 = interossei, 6 = lumbrical

3. Plantar Interossei for Toes 3, 4, and 5

Three in number, these muscles insert on the medial sides of the bases of the proximal phalanx of digits 3–5, the MTP joint capsule, and the extensor expansion (Figs. 6.12 and 6.18). All the plantar interossei pass dorsal to the deep transverse metatarsal ligament. They adduct the third, fourth, and fifth toes toward the second digit.

4. Dorsal Interossei for Toes 2, 3, and 4

Four in number, these muscles insert on the base of the proximal phalanx (Fig. 6.12). The first dorsal interossei inserts medially on the second toe and all others laterally on the proximal phalanx of the respected digit as well as the extensor expansion of respected digit (Fig. 6.18). The three lateral dorsal interossei abduct the toes away from midline of foot (second metatarsal). The second digit does not receive an insertion of any plantar interossei, whereas the first dorsal interossei acts to pull the digit in a medial direction toward the midline of the body.

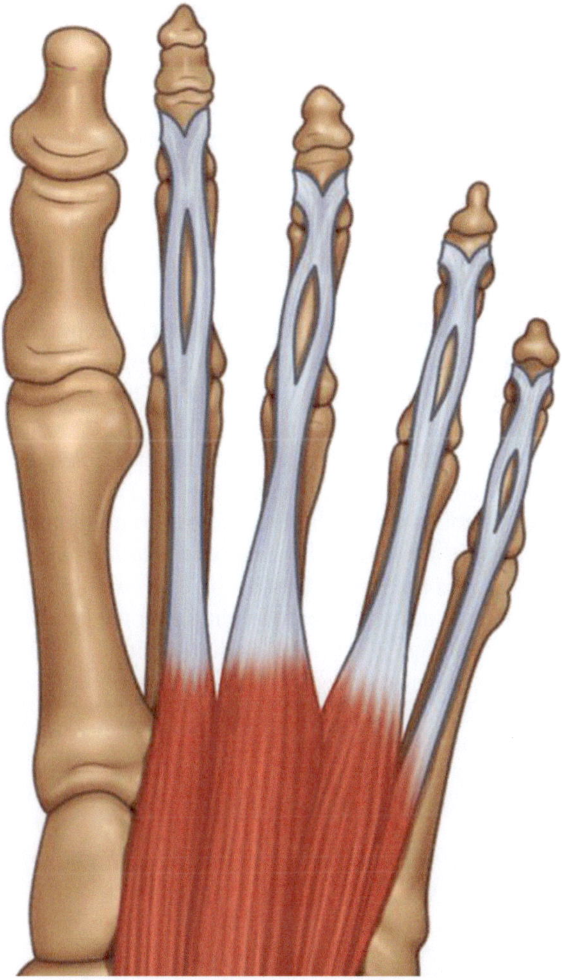

Fig. 6.13 The flexor digitorum brevis cannot plantarflex the MTPJ; its insertion acts only at the PIPJ

In summary, both sets of interossei insert on:

- The deep transverse intermetatarsal ligaments
- The suspensory portion of the collateral ligament
- The plantar plate
- The sling portion of the extensor expansion
- The base of the proximal phalanx

5. Lumbricals for Toes 2–5

Four in number, the lumbricals arise from the FDL tendon and course distally along the medial border of this tendon to cross the MTPJ medially (Figs. 6.12 and 6.17). The lumbricals pass plantar to the deep transverse metatarsal ligament and insert on the medial aspect of the extensor expansion, contributing to the wing portion of the extensor hood of the respective digit.

6. Abductor and Flexor Digiti Quinti

These muscles insert together on the lateral-plantar aspect of the base of the proximal phalanx of the fifth digit. They also insert into the plantar plate of the fifth MTPJ. Together these two muscles plantarflex the fifth digit at the MPJ and abduct the toe.

Function of the Dynamic Structures of the Digits

The EDL and EDB actually have stronger attachment plantarly than dorsally on the proximal phalanx via the extensor sling apparatus. There is only a loose connection to the dorsum of the proximal phalanx. The pull of the EDL results in hyperextension of the proximal phalanx at the MTPJ thru the dorsal pull exerted by sling portion of the extensor expansion. Pull of this tendon has minimal extension effect on the PIPJ and DIPJ, particularly when the MTPJ is in dorsiflexion. These joints rely on extension provided by the interossei and lumbricals. *The straight lesser digit requires full function of the interossei and the lumbricales in order to fully extend the PIPJ and DIPJ* (Fig. 6.14).

There are no attachments for the flexor tendons on the proximal phalanx, except for the fifth digit via the flexor digiti quinti. Because there is no insertion on the proximal phalanx, the FDB and FDL flex the interphalangeal joints but do not directly plantarflex the MTPJ. FDL acts at both the PIPJ and DIPJ, while the FDB acts only at the PIPJ.

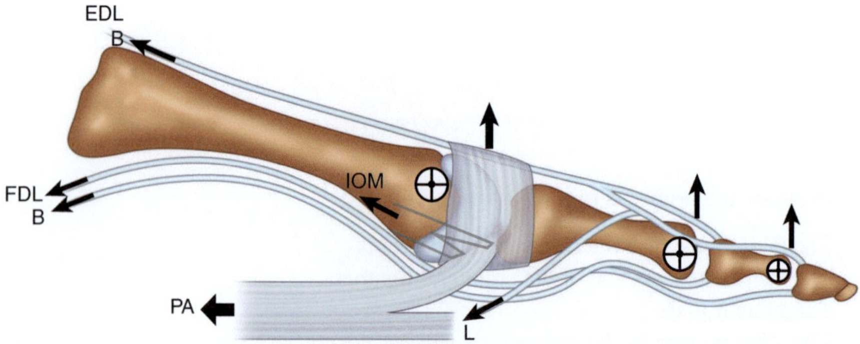

Fig. 6.14 The straight lesser digit. The interossei, the lumbrical, and the plantar aponeurosis are critical to preventing hyperextension of the metatarsophalangeal joint. (From: Sarrafian and Kelikian [64]. Figure 10.137 The straight lesser toe. The hyperextension at the metatarsophalangeal (MP) joint through the pull of the extensor digitorum longus *(EDL)* and brevis *(EDB)* is prevented by the flexion forces exerted at the same joint by the interossei muscles *(IOM),* the lumbrical *(L),* the plantar aponeurosis (PA), and the flexor digitorum longus *(FDL)* and brevis *(FDB).* The extension force exerted by the *EDL,* the *EDB,* and the *L* at the interphalangeal joints is counterbalanced by the *FDL,* with the *FDB* acting only at the proximal interphalangeal joint)

Saraffian shows that simultaneous pull on both the FDB and FDL on the plantar side and simultaneous pull of the EDB and EDL on the dorsal side will result in a claw toe [64]. This is characterized by hyperextension of the MTPJ and marked flexion of the PIPJ and the DIPJ. This un-opposed contraction of four muscles, the EDL, EDB, FDL, and FDB, is what is seen in the "intrinsic minus" condition of the neuropathic diabetic foot. *In the healthy foot, the combined action of these four muscles is counteracted by the interossei and lumbricales providing flexion stability of the MTPJ and extension of the PIPJ and DIPJ. Even more important is the static and dynamic contribution from the plantar aponeurosis* via *the plantar plate to prevent hyperextension of the MTPJ, facilitating the extensors to act distally on the toes.*

Pathomechanics of Digital Deformity

By definition, a hammertoe is a flexion deformity of the proximal interphalangeal joint (PIPJ), with or without involvement of the distal interphalangeal joint (DIPJ) [65]. Claw toes include flexion of one or both interphalangeal joints along with extension of the metatarsophalangeal joint (Fig. 6.15). Herein is a point of contention as it would be impossible during weight bearing for the PIPJ to plantarflex, moving the middle phalanx into the ground without some type of associated hyperextension of the proximal MTP or distal DIPJ (Fig. 6.16). Thus most digital deformities commonly described as hammertoes are actually claw toes.

Hammertoes and claw toes are thought to develop from imbalance between the extensors and flexors of the toes and/or loss of function of the intrinsic muscles of the foot [66–68]. Saraffian showed that even when the MTP is in a neutral or rectus position, the combined simultaneous pull of the long and short flexors and extensors will result in a claw toe with flexion of the DIPJ and PIPJ along with extension of the MTPJ [64]. This illustrates the importance of the two intrinsic muscle groups,

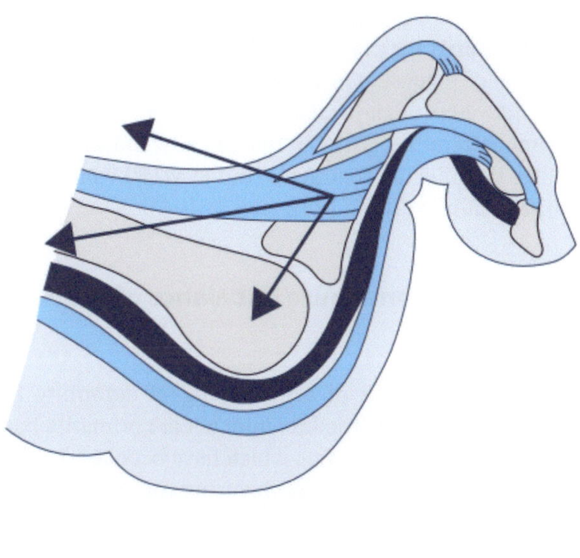

Fig. 6.15 The claw toe. Reverse buckling occurs when interossei cannot stabilize proximal phalanx at the MTPJ. The interossei and lumbrical cannot extend the PIPJ. FDL contraction causes plantarflexion of the PIPJ, dorsiflexion of the MTPJ, with retrograde plantarflexion force on head of metatarsal

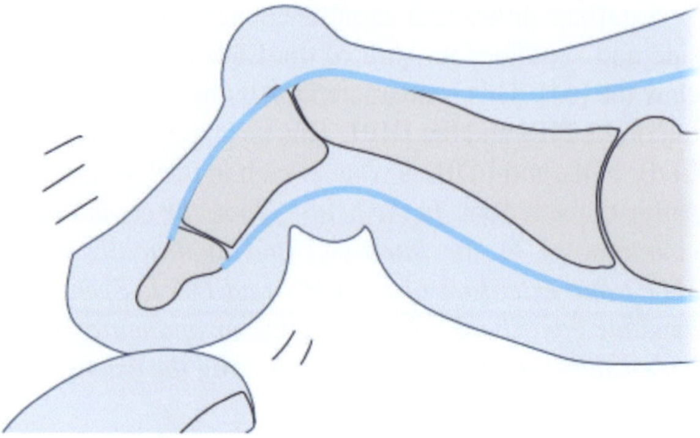

Fig. 6.16 This digit meets the definition of a hammertoe with flexion contracture of the PIPJ and DIPJ. The tip of the digit is below the weight-bearing surface of the foot. Thus, with weight bearing, the entire digit will move dorsal at the MTPJ, thereby creating a claw toe deformity

the interossei and the lumbricales, to stabilize the lesser digits. While the tendons of these two muscles pass plantar to the axis of rotation of the MTPJ, any slight dorsiflexion of the proximal phalanx will compromise the ability of these muscles to dorsiflex this joint. There is no muscle or tendon structure which inserts on the proximal phalanx to directly resist dorsiflexion of the MTPJ.

Yet, dorsiflexion is the primary motion of the MTPJs during gait, starting at terminal stance and extending thru pre-swing and all phases of swing. The digits never actually plantarflex from a neutral position during gait.

Muscle activity also favors dorsiflexion [69]. The FDL is active from 12% thru 55%, totaling 43% of the gait cycle. The EDL is active from 55% until 112%, totaling 57% of the entire gait cycle. Plantarflexion of the digits is opposed by ground reaction forces, while dorsiflexion has no opposition including from the antagonist FDL which fires at a different time. No wonder digital deformities are so common!

The solitary structure which has a primary role in resisting dorsiflexion of the lesser MTPJ is the plantar aponeurosis, via the plantar plate [70]. The collateral ligaments of the MTPJ play a secondary role in resisting hyperextension [71]. Many authorities believe that plantar plate attrition and rupture are the key events to the development of digital deformities [72–75].

Attrition of the plantar plate begins with persistent positioning of the MTPJ in a position of extension or dorsiflexion, which occurs from extrinsic or intrinsic factors. Elevated, high-heeled shoes are the most common extrinsic factor causing digital deformities [76, 77]. Intrinsically, muscle imbalance has been speculated to lead to dorsiflexion contracture of the MTPJ which starts the events leading to plantar plate failure [70, 71, 78].

What Conditions Cause Imbalance of the Flexors and Extensors of the Digits?

The pathomechanics of acquired digital deformity have been well described in orthopedic and podiatric literature but are primarily based upon observational studies. There are several terms which have been used almost exclusively in the podiatric literature to describe mechanisms for acquired digital deformity. Most of the articles using these terms came from the faculty of the Podiatry Institute, primarily Don Green DPM [67, 79–81]. Although written some time ago, the various causes of hammertoe deformity described by Green et al. are still valid today and are worthy of discussion.

Reverse buckling is a phenomenon which is attributed to contraction of the FDL and FDB tendons on the lesser digits, un-opposed by any other muscle activity. During the stance phase of gait, particularly during terminal stance and pre-swing, the FDL and FDB provide essential plantar purchase of the respective digits against the supportive surface. As long as the toe maintains a "rigid beam" function by stabilization from the plantar fascia (via the plantar plate), extensors, and plantar intrinsics, activity of the toe flexors will facilitate plantar purchase and stability for push off (Fig. 6.14). If the toe is not stabilized by the plantar fascia, the interossei, the lumbricales, and the extensors, contraction of the FDL and FDB will fail to produce adequate purchase the toe into the ground. Instead, a flexion motion will occur at the PIPJ with a retrograde extension motion at the metatarsophalangeal joint. In this "reverse buckling" phenomenon, the toe flexors have actually become extensors of the digit at the metatarsophalangeal joint (Fig. 6.15).

Reverse buckling is the end result of a process known as "flexor stabilization." Green attributes this phenomenon to a process whereby the FDL fires earlier and longer during the stance phase of gait, overpowering the interossei. This is most commonly seen in flexible flatfoot deformity where two abnormal mechanisms occur: the FDL works excessively to substitute for a weak or dysfunctional tibialis posterior, and/or the abducted position of the forefoot alters the mechanical advantage of the FDL on the digits. Green attributes much of the efficiency of the FDL tendons on the function of the quadratus plantae muscle (Fig. 6.17). This intrinsic muscle inserts on the lateral aspect of the tendons of the FDL at the level of the midfoot and functions to "straighten" or realign the FDL which approaches the midfoot from a medial direction. The FDL is thought to lose the stabilizing influence of the quadratus plantae in flatfoot deformity. Flatfoot deformity everts the calcaneus, altering the alignment of the origin of the quadratus plantae. Flatfoot deformity also abducts the forefoot relative to the rearfoot. Both scenarios result in the FDL pulling the toes from a more medial direction, losing mechanical advantage and causing a varus rotation of the digits, particularly the lateral toes.

Abnormal alignment of the FDL in flatfoot deformity would not directly explain a claw toe deformity which is primarily due to over contraction of the toe flexors. That situation would occur when the FDL is attempting to substitute for a weak or ruptured tibialis posterior muscle. The FDL has identical inversion moment arm as the tibialis posterior [82]. However, the native FDL produces 177 Newtons of force, considerably less than the larger tibialis posterior which produces 727 Newtons of force [83]. Hypertrophy of the FDL muscle has been observed with MR imaging of patients with posterior tibial tendon insufficiency [84]. Ultrasound study has revealed increased cross-sectional area of the FDL and FHL muscles in patients with pes planus compared to patients with normal arch feet [85]. This verifies the role of these muscles as arch supporters which are recruited for increased activity in pes planus.

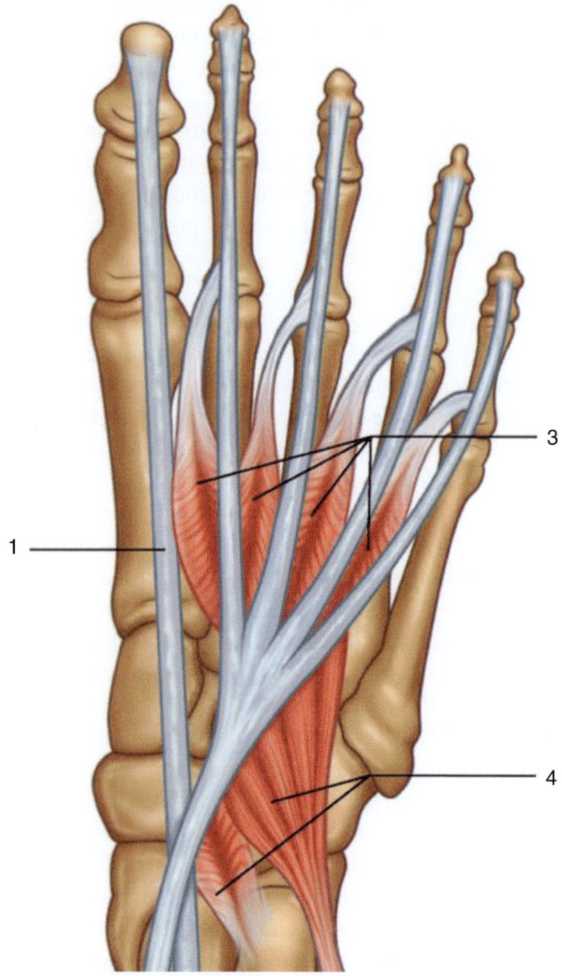

Fig. 6.17 The quadratus plantae depends on calcaneal alignment eversion of calcaneus directs a lateral vector of pull on FDL and lumbricales. (Reprinted by permission from Springer Nature: Springer, *Hammertoes* by Emily Cook and Jeremy Cook (Eds.) © 2019) 1 = FHL, 2 = FDL, 3 = lumbricales, all four lie plantar to the DTML, 4 = quadratus plantae

Thus, in adult acquired flatfoot deformity, the compensatory activity of the FDL might be better called "flexor substitution." Flexor substitution is also seen in neurologic conditions of weakness of the calf musculature or injury to the Achilles tendon. Green suggests that the flexor substitution seen in calf weakness will usually be accompanied by over supination of the foot during terminal stance. This occurs from overactivity of not only the FDL but also the FHL and tibialis posterior which attempt to provide heel rise in the absence of the gastrocnemius. These three muscles all have supination moment arm at the rearfoot complex. Conversely, when flexor substitution occurs in adult acquired flatfoot, the tibialis posterior is compromised so excessive supination does not occur.

The term "extensor substitution" refers to overactivity of the EDL during the swing phase of gait causing excessive extension of the digits at the MTPJs. Green speculates that the lumbricales must function during the swing phase of gait to stabilize the MTPJs from excessive extension due to the pull of the EDL. However, there are no EMG studies of lumbrical activity due to the difficult nature of isolating and measuring activity of this small muscle group deeply imbedded in the foot.

The EDL may be overactive in situations where there is weakness of the other ankle dorsiflexors or when there is equinus contracture of the calf musculature. Excessive positioning of the lesser MTPJs in the attitude of dorsiflexion during swing may inhibit the static and dynamic plantarflexion action of structures acting on the MTPJs during stance. Of these, the interossei are most important as they are the only muscle which inserts directly to the proximal phalanx (Fig. 6.18).

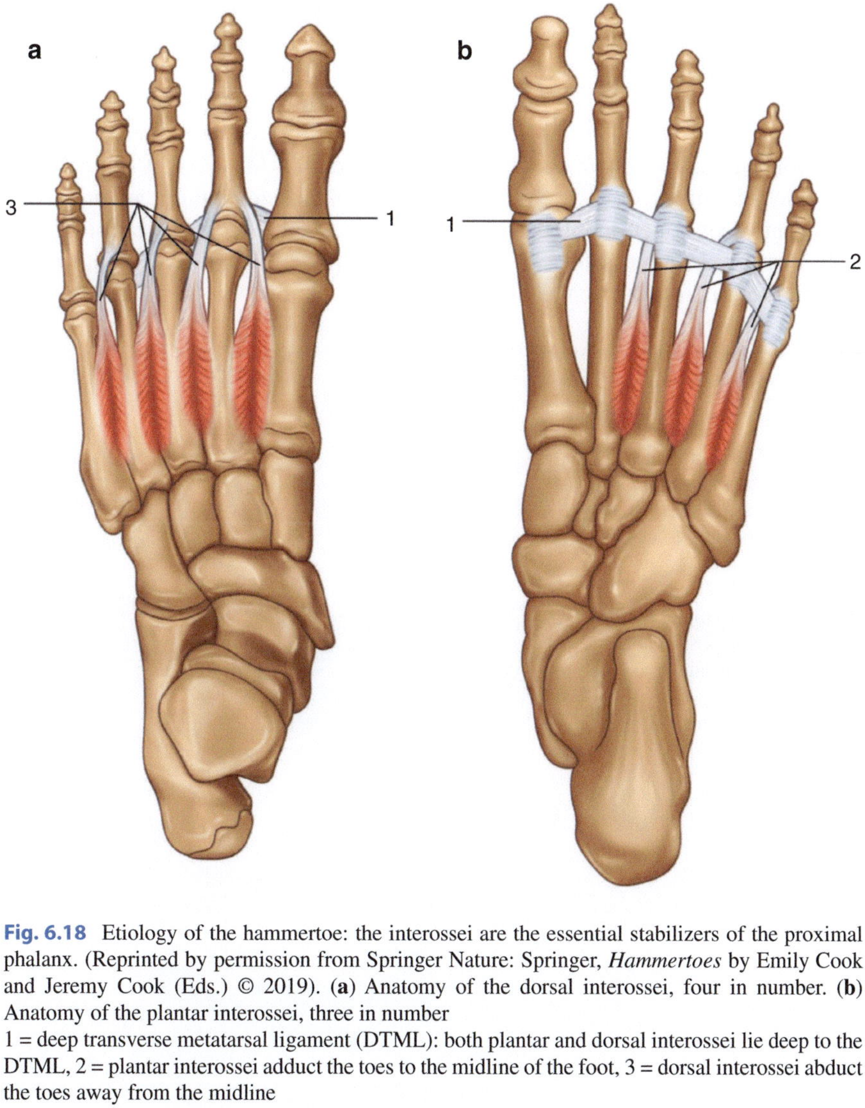

Fig. 6.18 Etiology of the hammertoe: the interossei are the essential stabilizers of the proximal phalanx. (Reprinted by permission from Springer Nature: Springer, *Hammertoes* by Emily Cook and Jeremy Cook (Eds.) © 2019). (**a**) Anatomy of the dorsal interossei, four in number. (**b**) Anatomy of the plantar interossei, three in number
1 = deep transverse metatarsal ligament (DTML): both plantar and dorsal interossei lie deep to the DTML, 2 = plantar interossei adduct the toes to the midline of the foot, 3 = dorsal interossei abduct the toes away from the midline

Extensor substitution is also seen in cavus deformity. In swing phase, a cavus foot has an anterior equinus of the forefoot requiring greater activity of all the ankle joint dorsiflexors in order for the foot to clear the ground. In stance phase, the plantar-flexed alignment of the metatarsals automatically places the digits in a dorsiflexed alignment at the MTPJ during weight bearing. The lumbricales and the interossei now lie close to or even dorsal to the axis of rotation of the MTPJ and thus lose their plantarflexion moment arm at the joint (Fig. 6.19). However, the alignment of the MTPJs in cavus feet would most likely be from a reverse buckling effect of the flexor tendons on the digits rather than from contraction of the extensors. The extensor activity sets the stage for reverse buckling by nullifying the action of the lumbricales and interossei. Finally, the cavus foot deformity is often the result of a muscle imbalance – usually from weakness of the anterior and lateral muscle groups of the legs. In these cases, the EDL remains strong and becomes the primary compensation for mild dropfoot and equinus seen in patients with cavus feet.

Flexor stabilization and flexor and extensor substitution can all create a state of chronic hyperextension of the lesser digits at the MTPJ. Not only do the dynamic stabilizers of the MTPJs loose mechanical advantage, an important compromise of function of the plantar plate occurs. The chronic dorsiflexed alignment of the digit at the MTPJ initiates the process leading to attrition and eventual rupture of the plantar plate. Once this event occurs, gross instability of the digit ensues.

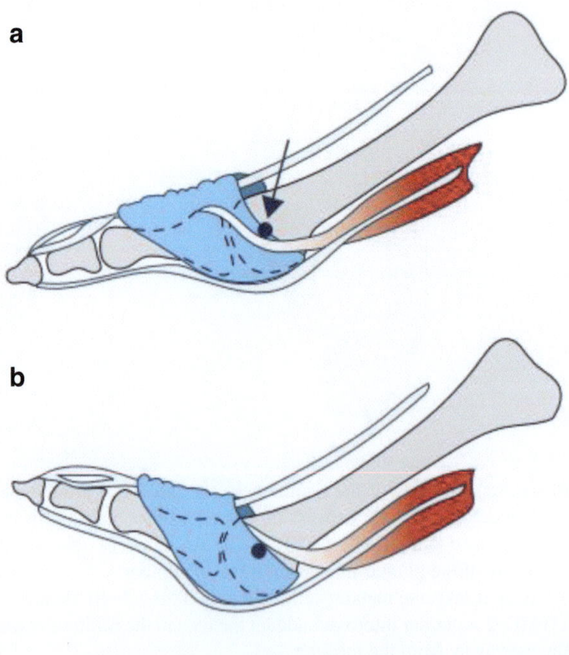

Fig. 6.19 The beginning of the claw toe deformity. (**a**) In rectus alignment, the interossei lie plantar to the axis of rotation (arrow) of the lesser MTPJ. (**b**) Chronic dorsiflexion of the MTPJ shifts the position of the interossei dorsal to the axis of rotation. Plantarflexion moment arm of the interossei is compromised

Pathomechanics of the Plantar Plate Rupture

The plantar plate is recognized as the most important stabilizing structure of the lesser toes [72, 73]. Rupture of the plantar plate compromises the action of all other dynamic structures which normally provide stability across the DIPJ, PIPJ, and MTPJ. Cadaveric and surgical studies of lesser digit instability consistently show that attenuation and rupture of the plantar plate is the primary underlying etiology of deformity [75, 76, 86–88].

When a patient is in relaxed stance, the plantar fascia is loaded, and the lesser digits should rest on the supportive surface. With contracture of the EDL, a claw toe may be evident off weight bearing which reduces to normal alignment with weight bearing (Fig. 6.20). This is due to the powerful influence of the plantar fascia providing plantarflexion moment to the lesser MTP via its attachment to the plantar plate. Loading of the plantar fascia exerts a reverse windlass mechanism across the MTPJ by restricting extension of the proximal phalanx.

Rupture or failure of the plantar plate is the result of abnormal mechanical loads placed on the structure over an extended period of time. Histologic study enables insight into the specific loads borne by the plantar plate during standing and ambulation. According to Ker, the cellular composition of a load bearing tissue will respond to specific stress and adapt accordingly [89]. Specifically, cells will display predominantly fibrocartilage when the tissue is subjected to compressive forces. Tensile forces will produce changes seen in ligament tissue, showing predominantly

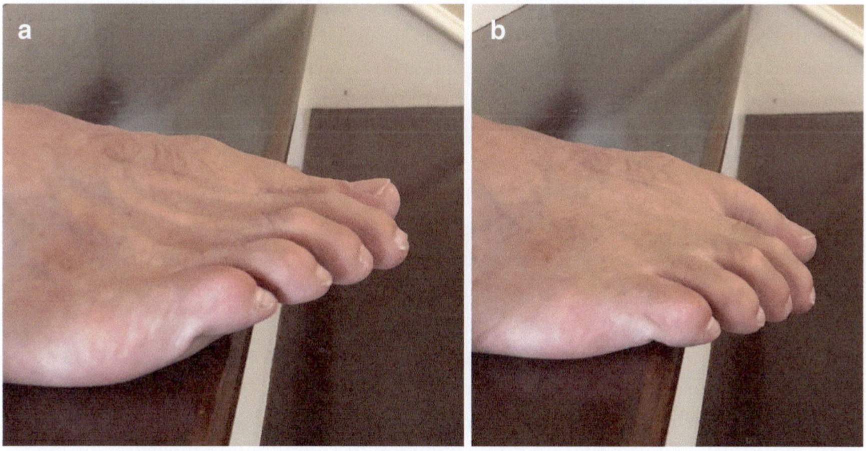

Fig. 6.20 Loading of the plantar fascia and plantar plate. (**a**) Dorsiflexion contracture of the MTPJs. (**b**) Activating the reverse windlass with loading forefoot, dorsiflexing the metatarsals, plantarflexing the digits

fibroblasts with a higher proportion of collagen. Gregg et al. performed histologic study of 16 plantar plates from 3 cadaver specimens [90]. The morphology of the plantar plate in this study suggested that the plantar portion of the plate experiences primarily tensile forces while the dorsal section experiences both tensile and compressive forces.

Maximal compressive and tensile forces are exerted on the plantar plate at the exact same phase of the walking gait cycle. Terminal stance, initiated by heel rise, causes plantarflexion of the metatarsals across the digits which are fixed on the supportive surface. Closed kinetic chain dorsiflexion of the MTPJs engages a windlass mechanism as the plantar fascia winds around the head of the metatarsal. Tensile strain is greatest in the plantar fascia at the beginning of terminal stance [91]. At the same time, heel rise also corresponds with peak plantar pressure across all of the metatarsal heads, maximizing compression force at the plantar plate [33, 35].

In a classic paper describing the effects of foot position on loading of the plantar fascia, Saraffian illustrates that internal rotation of the tibia, pronation of the rearfoot, and supination of the forefoot cause elongation strain in the plantar aponeurosis [91]. He credits MacConnaill for originally describing this position of the foot as an "untwisted plate" configuration of the forefoot relative to the rearfoot [58]. This rotation of skeletal segments between the leg and the toes also engages a "reverse windlass" mechanism originally described by Hicks [92]. Here, loading of the foot causes lowering of the arch resulting in increased tension in the plantar aponeurosis. This increased tension in the plantar aponeurosis plantar-flexes the toes at the metatarsophalangeal joints.

While many have speculated that a pes planus foot posture pre-disposes to digital deformities, there is little objective proof that such a cause-effect relationship actually exists. In the largest study (169 patients) of crossover second toe deformity (considered synonymous with plantar plate rupture), Kaz and Coughlin found that the lateral talometatarsal angle (Meary's angle) had a slight correlation with second digital deformity [14]. However, other radiographic measures of flatfoot including lateral talocalcaneal angle and A/P talonavicular coverage had no correlation with digital deformity. In that study, the most significant finding was that 86% of patients with crossover deformity of the second digit were women. This suggested that higher-heeled footwear was a significant etiology, but this factor was not directly evaluated in the investigation.

In terms of compressive loads, some investigations have suggested that an elongated metatarsal is a risk factor for plantar plate rupture [10, 13, 15]. However, as demonstrated earlier in this chapter, deviation of alignment of the first metatarsal in HAV deformity can greatly affect the accuracy of radiographic measures of length of the second metatarsal. Using the Hardy Clapham arc measure of the metatarsal parabola, Kaz and Coughlin found no significant role of an elongated second metatarsal in causing crossover toe deformity.

Stainsby speculated that the chronically dorsiflexed position of the proximal phalanx at the MTPJ, as seen in claw toe deformity, will eventually lead to rupture of the plantar plate [93]. He describes a "plunger effect" of the head of the metatarsal pushing thru the plantar plate as the digit is forced into extension at the MTPJ in claw toe deformity. Here, the combined effect of tensile strain from extension of the digit along with compression of the metatarsal head against the plantar plate could explain eventual failure of this structure.

A final etiology for plantar plate rupture is shear force, which is accentuated under the central metatarsals in patients with HAV deformity [31]. Unfortunately, there are no studies of shear stress in patients with metatarsalgia, plantar plate injuries, or digital deformity, but one must recognize the role of this deforming force when studying the pathomechanics of all these conditions. This becomes especially important when studying the detailed anatomy of the plantar plate and the fat pad of the forefoot as described by Bojsen-Moller [94].

Bojsen-Moller and Flagstad were the first to describe a mechanism whereby extension of the metatarsal-phalangeal joint, seen in hammertoe deformity, will cause the protective fat pad cushion to displace distally in the foot and correlated this with increased shear stress in the forefoot [94]. They elegantly described and illustrated the key relationships between the plantar fat pad and the distal insertions of the plantar aponeurosis (Fig. 6.21). The authors divide the "ball of the foot" into three transverse areas: (1) a proximal segment with subcutaneous transverse bands of the plantar fat pad interwoven with the deep sagittal septa of the plantar aponeurosis, (2) the area directly beneath the heads of the metatarsals which contain the submetatarsal fat cushions, and (3) the distal area which contains the insertions of

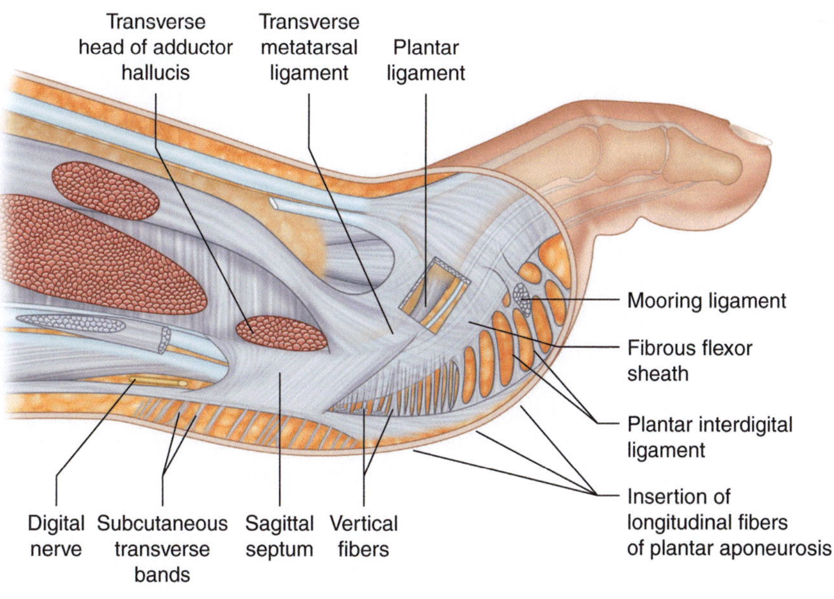

Fig. 6.21 From Bojsen-Moller and Flagstad [94]. Drawing of a sagittal section through the second interspace showing the internal architecture of the three areas of the ball of the foot. The sagittal septum is attached to the proximal phalanx through the transverse metatarsal ligament and the plantar ligament of the joint. The vertical fibers and the lamellae of the plantar interdigital ligament are attached to the proximal phalanx through the fibrous flexor sheath. The plantar longitudinal fibers of the plantar aponeurosis inserts into the plantar integument. The plantar interdigital ligament retains the plantar fat pad

the plantar aponeurosis into the skin. Just deep is the transverse interdigital ligament, also known as the superficial transverse metatarsal ligament. Bojsen-Moller describes the critical function of the plantar aponeurosis to restrain anterior migration of the integument during the third rocker or "push off" phase of gait. The integument and fat pad are speculated to undergo significant shear stress from the underlying metatarsals during push off. He suggests that failure of the plantar fascia stability is critical to displacement of the metatarsal fat pad as well as overall digital stability.

We do know that claw toes are associated with significant anterior displacement of the plantar fat pad in the forefoot as shown by Bus et al. [95]. This MRI study of diabetic patients with hammertoe or claw toe deformity revealed a significant anterior displacement of the plantar fat pad away from the weight-bearing surface of the metatarsal head. This resulted in significant thinning of the submetatarsal fat pad which would explain increased plantar pressure in this location. In a follow-up study, Bus and co-workers found that claw toe deformity caused decreased fat pad thickness under the affected metatarsal and this increased peak plantar pressure by almost twofold at that location [96]. These authors propose a "fat pad-plantar pressure exchange principle" whereby the fat pad migrates distally and plantar pressure increases proximally.

The question is do patients with elevated shear force under the lesser metatarsals eventually develop claw toes and plantar plate tears? Or is the anterior displacement of the plantar fat pad in the forefoot a result, not a cause of digital deformity? The anatomy of the distal insertion of the plantar fascia in the forefoot suggests that this structure is important in tethering the skin and fat pad and stabilizing these structures from shear forces which develop during push off in walking gait. When the plantar plate ruptures, the stabilizing influence of the distal extensions of this structure into the integument and fat pad are lost, thus explaining the common finding of distal fat pad migration in patients with plantar plate ruptures. Whether abnormal shear stress is the underlying mechanical force causing plantar plate failure is yet to be proven but should definitely be considered as an important underlying etiology. This would become important when evaluating footwear and lifestyle as risk factors for plantar plate rupture and digital deformity.

Evaluating the anatomy of the tear may reveal the specific forces which cause rupture of the plantar plate. Coughlin and co-workers evaluated patients with second MTPJ instability and evaluated the plantar plate surgically [76]. They proposed a classification and staging system based upon clinical findings as well as the anatomic type of tear in the plantar plate. They propose that attenuation of the plantar plate (Grade 0) is followed by partial (Grade 1) or complete (Grade 2) tear thru the distal aspect of the plantar plate, just proximal to its insertion on the proximal phalanx. Longitudinal tears are combined with a transverse tear (Grade 3), and finally extensive degeneration is associated with a button hole tear (Grade 4). Later, Nery and co-workers prospectively evaluated 68 patients with lesser MTP joint instability and found that this classification system accurately predicted severity of injury and prognosis for surgical repair [76]. The finding that transverse tears appeared most common in the distal portion of the plantar plate was confounding given that this area is thicker and stronger than the more proximal section. The transverse tear pattern suggests that tensile strain may be the most damaging force acting on the plantar plate. If compressive forces were most responsible, the tear pattern would be expected to show diffuse attrition, primarily directly beneath the metatarsal head. However, this is not seen until stage 4 of the Coughlin classification system.

There are many static and dynamic structures which insert on the plantar plate and which are affected by rupture (Fig. 6.3). The four lumbrical muscles arise from the medial border of the flexor digitorum longus tendons and insert to the medial side of the proximal phalanx on the plantar surface and insert into the distal plantar aspect of the plantar plate (Figs. 6.10 and 6.17). The tendons of the plantar interossei muscles lie just dorsal to the DTML and run a course adjacent to the plantar plate, with some fibers of the interossei inserting directly into the plantar plate (Figs. 6.12 and 6.18). The sheath of the FDL and the FDB insert on the plantar portion of the plantar plate. The interossei and the extensor sling also insert on the plantar plate. The accessory (suspensory) collateral ligament attaches directly to the plantar plate along with the deep transverse metatarsal ligament (Figs. 6.8 and 6.10).

Cadaveric studies by Suero [97], Bhatia [98], and Chalayon [99] have shown that solitary disruption of the plantar plate will cause significant sagittal plane instability of the MTPJ, while combined rupture with the collateral ligaments will cause gross instability in multiple planes. There is some debate whether collateral ligament rupture precedes or follows plantar plate rupture.

Deland and Sung reported on a crossover toe deformity which occurred from rupture of the lateral collateral ligament with the plantar plate still intact [17]. However, the notion that collateral ligament failure precedes plantar plate attenuation and rupture has been disproven by the largest anatomic study of 2nd MTPJ deformity published to date. Coughlin and co-workers performed anatomic dissection of 16 cadaver specimens with deformity of the 2nd MTPJ [100]. Five of the specimens had only dorsal deformity, while the other eleven had combined dorsal and medial deviation of the 2nd MTPJ. All 16 specimens had transverse tears of the plantar plate adjacent to the insertion at the base of the proximal phalanx. Six specimens had more extensive midsubstance tears as well as rupture of the lateral collateral ligament. The authors concluded that plantar plate insufficiency precedes collateral ligament tear. Once the lateral collateral ligament fails, gross instability and dislocation are likely.

When studying the pathomechanics of acquired instability of the 2nd MTPJ, one cannot ignore the strong association with this disorder and hallux abductovalgus (HAV) deformity [71, 72, 101]. Kaz and Coughlin found that 49% of patients with crossover second digit deformity also had HAV deformity [14]. Coughlin et al. in a cadaveric study found HAV deformity along with crossover second digit in 88% of the specimens. How HAV and acquired instability of the adjacent 2nd MTPJ are related remains open to speculation. Laterally directed pressure of the hallux against the second digit in HAV deformity may somehow compromise the lever arm for the interossei or lumbricales, but no study has verified this concept. Certainly lateral deviation of the 2nd MTPJ would place load on the medial collateral ligament which is less likely involved in crossover toe deformity than lateral collateral ligament failure [100]. As shown in Chap. 5, instability of the first ray does not always shift plantar pressure to the second MTPJ. In their study of 169 patients with crossover deformity of the second toe, Kaz and Coughlin found plantar callus under the second metatarsal in only 7% of the patients [14].

While rearfoot pronation and flatfoot deformity are associated with HAV, there appears to be no direct association of rearfoot pronation with plantar plate tears and digital deformity. Finally, the role of footwear contributing to HAV as well as second MTP pathology has been recognized by many authorities, and this extrinsic factor may be the common link to the two conditions [14, 102, 103].

How Do Digital Deformities Cause Metatarsalgia?

Chronic extension of the lesser MTPJ causes attenuation and eventual rupture of the plantar plate. Rupture of the plantar plate causes the proximal phalanx to assume a dorsally subluxed position at the head of the metatarsal which is different from its position in hyperextension [17] (Fig. 6.22). With hyperextension, the MTPJ is still a congruent joint. With plantar plate rupture, the proximal phalanx migrates dorsally in the sagittal plane, without moving in the normal arc motion at the MTPJ, thus creating a subluxed, incongruent joint.

Therefore, claw toes and digits with a plantar plate tear are two different pathologies, but they may co-exist. Both conditions result in instability of the metatarsophalangeal joint. Claw toes demonstrate excessive dorsiflexion of the lesser MTPJ, but the joint surfaces remain congruent. Excessive MTPJ dorsiflexion in claw toes is caused by reverse buckling from tendon imbalance and will reduce with weight bearing as long as the plantar plate is intact. Plantar plate tears cause dorsal translation of the proximal phalanx with weight bearing. The dorsal motion does not follow the normal arc of dorsiflexion of the MTPJ. Rather, plantar plate tears cause a sagittal plane subluxation as the proximal phalanx will float and gap away from the dorsal surface of the head of the metatarsal (Fig. 6.23). Claw toes will have

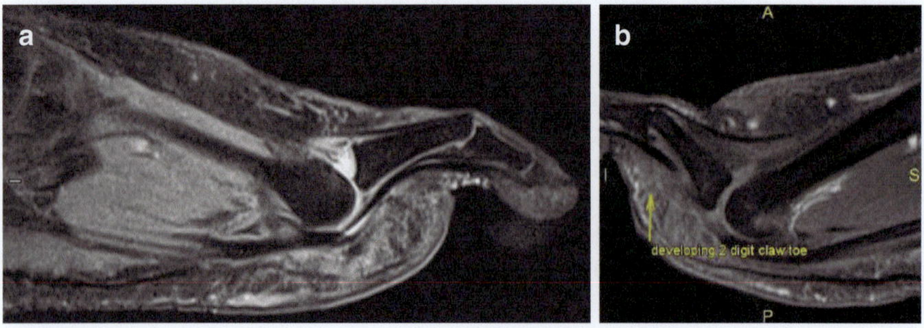

Fig. 6.22 (**a**) Plantar plate tear: subluxation with incongruent MTPJ. (**b**) Congruent joint with claw toe

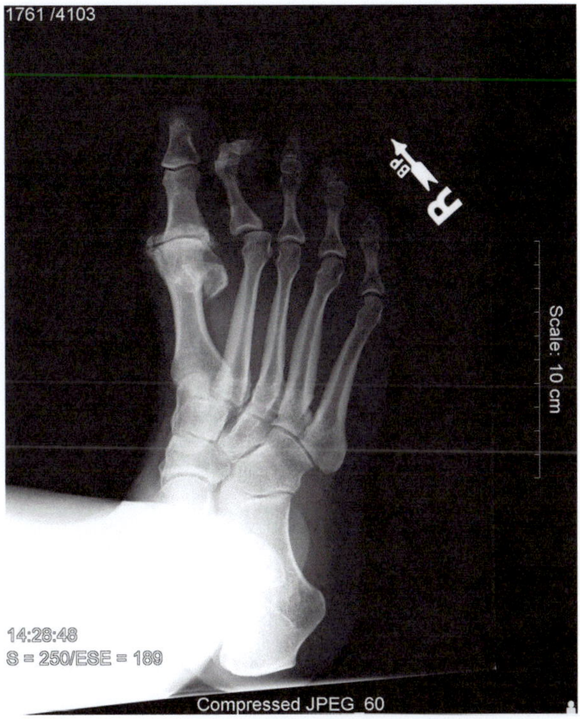

Fig. 6.23 Dorsal subluxation/gapping with plantar plate tear

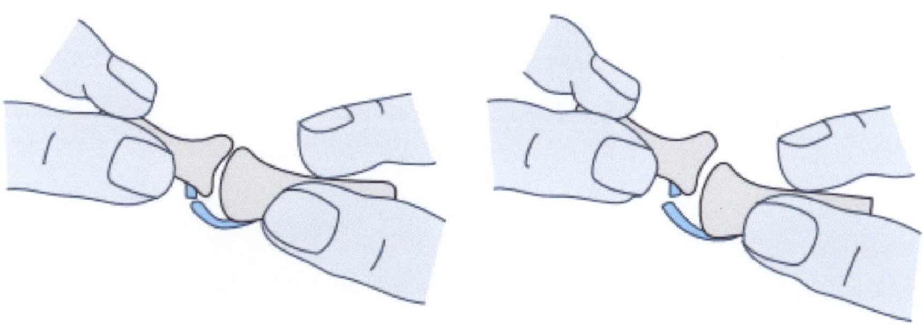

Fig. 6.24 The drawer test for plantar plate tear

excessive dorsiflexion, and the base of the proximal phalanx will follow the curvature of the head of the metatarsal, keeping the joint congruent.

The difference between the two types of instability of the lesser MTPJ can be accurately detected using the "drawer test" for sagittal plane instability (Fig. 6.24). This will be positive in plantar plate tears but negative with claw toes. Clinically, plantar plate tears will show a "floating toe" above the plane of the lesser digits with mild to moderate contracture of the PIPJ (Fig. 6.25). Claw toes demonstrate dorsiflexion, not floating, of the proximal phalanx with moderate to severe plantarflexion

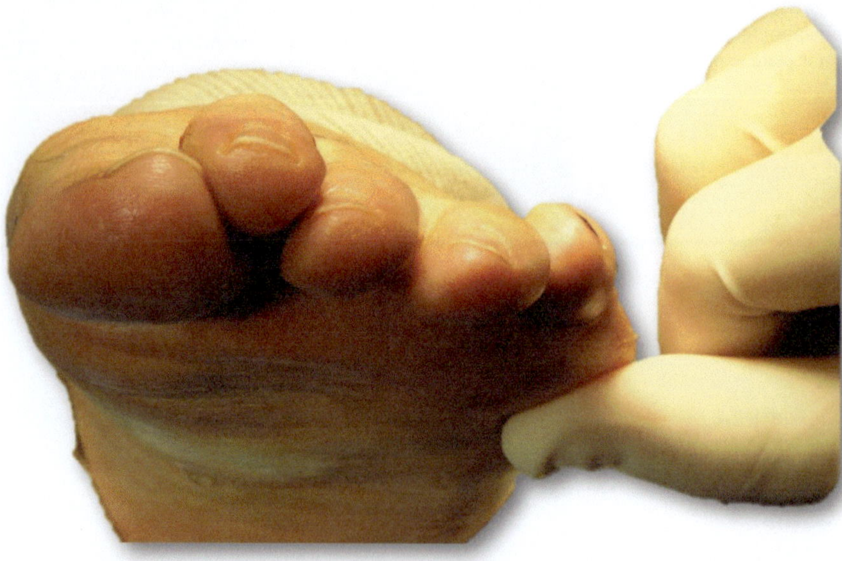

Fig. 6.25 The floating toe with mild interphalangeal contracture

Fig. 6.26 The claw toe with severe interphalangeal joint contractures at digits 3–5. Floating toe with plantar plate tear at second digit

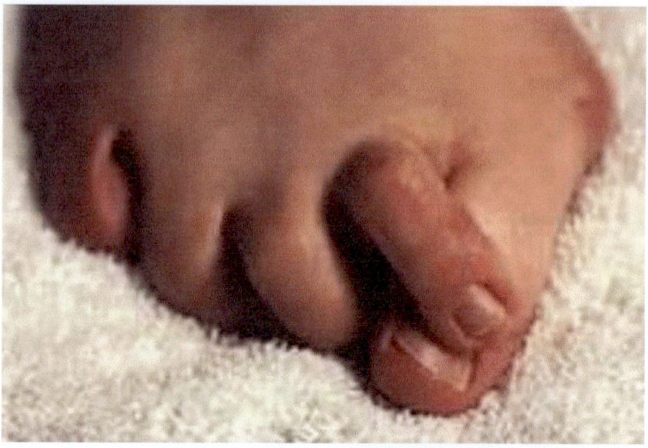

contracture of the PIPJ (Fig. 6.26). Also, plantar plate tears will progress to rupture of the lateral collateral ligament with medial subluxation combined with dorsal subluxation (Fig. 6.27). This transverse plane instability of the lesser MTPJ is rarely seen in claw toe deformity unless there has been amputation or plantar plate tear of the adjacent digit as shown in Fig. 6.27.

Whether subluxed or hyperextended, the dorsal position of the MTPJ weakens the lever arm of the EDL and EDB which can no longer effectively extend the PIPJ and DIPJ. The dorsal subluxed MTPJ also weakens the plantarflexion moment arm of the lumbricales and the interossei which are now unable to plantarflex the proximal phalanx against the supportive surface. Finally, the plantarflexion effect of the

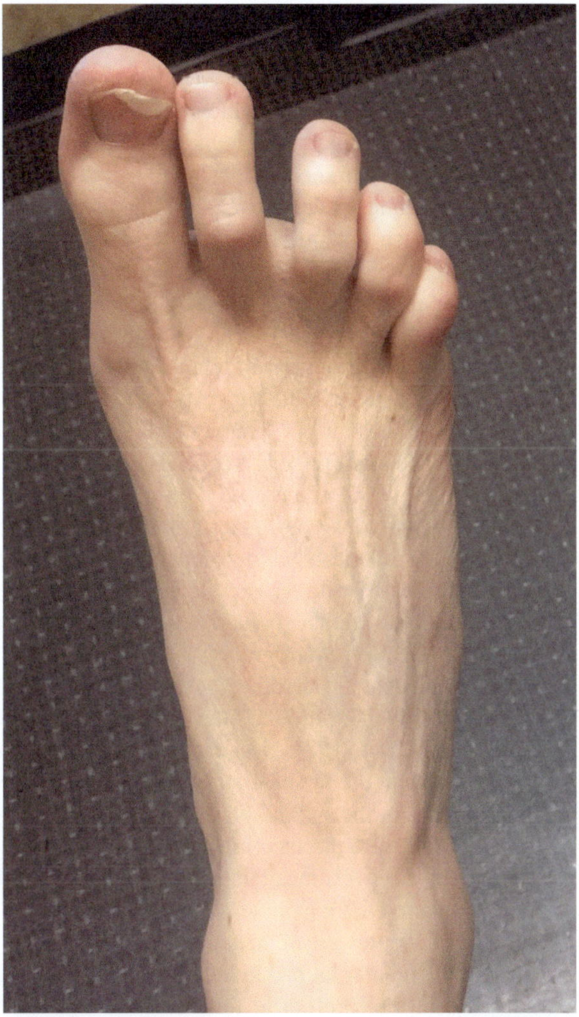

Fig. 6.27 Plantar plate tears will progress with medial subluxation of the MTPJ

FDL and FDB on the middle and distal phalanges is compromised due to reverse buckling at the MTPJ. Now the two essential structures delivering plantarflexion moment to the entire digit are lost: the plantar fascia and the flexor tendons. With both plantar plate tears and with claw toes, there is significant loss of plantar purchase of the toe.

Hamel et al. describe a windlass effect of the plantar aponeurosis at the lesser digits which applies flexion moment to the proximal phalanx [104]. This flexion moment develops during passive dorsiflexion of the digit during heel rise. Thus the plantar aponeurosis plays a critical role in stabilizing the toe and resisting dorsiflexion during toe off. Hamel et al. propose that a claw toe deformity can occur if the long flexor tendons are overloaded when the plantar fascia stabilizing mechanism is lost. Hamel et al. compromised the plantar fascia by performing a complete fasciotomy in cadaver specimens. They observed that loading shifted laterally to the lesser metatarsals during toe off and as they describe “a shift from the normal high gear push off to the less efficient low gear push off” [104]. This may explain the correlation of HAV deformity which has compromised windlass function at the 1st MTPJ, causing a lateral shift of load in the forefoot and a low-gear push off.

Sharkey and co-workers demonstrated that the plantar fascia can provide passive tension during static stance, as well as dynamic tension via the windlass mechanism during late stance and toe-off which will reduce the plantar-to-dorsal bending moments across the metatarsals [105]. In another study, these investigators demonstrated a similar offloading or shielding of bending moments of the first metatarsal by the flexor hallucis longus (FHL) [106]. Similar shielding of bending moments is provided by the flexor digitorum longus at the lesser metatarsals [107]. Proper purchase of the digits against the supportive surface will reduce plantar pressure under the heads of the metatarsals. This will offload or shield the metatarsal from dorsal displacement. Reduction of dorsal bending moment is also counteracted by retrograde plantarflexion moment placed on the metatarsal head as the digit dorsiflexes and engages the windlass during terminal stance. The net result of both mechanisms is a reduction of dorsal bending moment or dorsal displacement of the metatarsal during standing and walking.

Extension of the metatarsophalangeal joint could increase plantar pressure under the metatarsal head via retrograde force from the proximal phalanx causing plantarflexion moment at the tarsometatarsal joint (Fig. 6.15). Multisegment foot modeling has allowed insight into the motion of key skeletal segments of the human foot during dynamic gait [108]. Nester and co-workers found a significant amount of motion available at the tarsometatarsal joints [109]. Metatarsals 1–3 move an average of 6 degrees in the sagittal plane, while metatarsals 4–5 move 11 degrees. This dispels the myth that the tarsometatarsal joints are relatively rigid. It also allows understanding of how a metatarsal can plantarflex in response to extension of the proximal phalanx with a hammertoe deformity.

Digital deformities create metatarsalgia via another mechanism involving anterior shift of the plantar fat pad in the forefoot. The plantar fascia extends beyond the plantar plate in the forefoot to form a network of structures which anchor to the plantar integument [94] (Fig. 6.21). Thus, the calcaneus is actually attached to the plantar integument and fat pad in the ball of the foot via the plantar fascia. The tracts of the plantar fascia fan out beyond the MTPJs in vertical and transverse orientation anchoring the integument to the fibrous tunnel of the flexor tendons. Distally, the plantar digital ligament also known as the interdigital ligament runs transversely, sending vertical fibers to the dermis. The adipose tissue is retained within transverse bands of the plantar digital ligament.

The skin and fat pad in the ball of the foot is thus anchored longitudinally, transversely, and vertically by extensions of the plantar fascia. As the digits move into extension during terminal stance, tension develops in the plantar fascia which stabilizes the fad pad and integument to the underlying flexor tendons as well as the plantar plate. Bojsen-Moller and Lamoreux speculate that this mechanism dissipates shear force against the skin during push off [110]. They measured displacement of the skin at the ball of foot relative to dorsiflexion of the toes. They found that full extension of the digits would reduce mobility of the integument and fat pad by 50% [110].

Patients with claw toe deformity have been shown with MRI study to have significant anterior displacement of the plantar fat pad in the forefoot [95]. Reduction of this displacement of the fat pad with tape will reduce extension of the 1st MTP in claw toe deformity (Fig. 6.28).

It is plausible that attenuation and rupture of the plantar plate will compromise the distal extensions of the plantar fascia beyond the plantar plate which anchor to the plantar integument. Pressure studies by Bus et al. of patients with claw toe deformity show increased plantar pressure in the forefoot by over 50% compared to

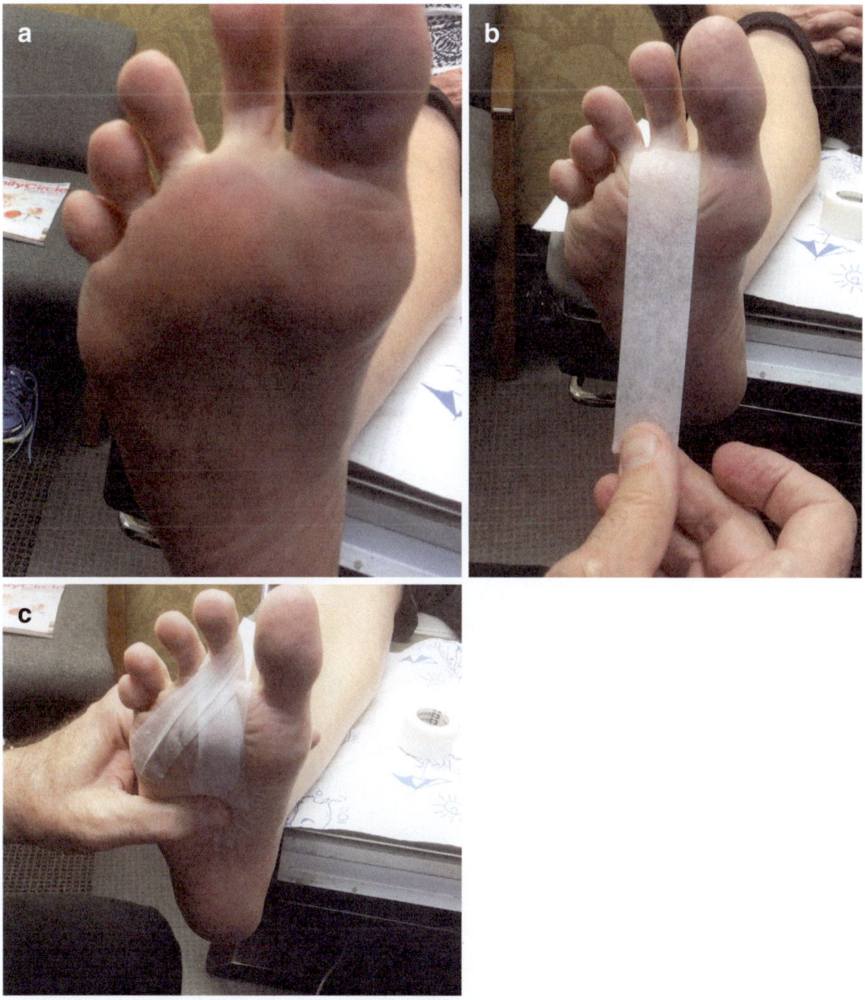

Fig. 6.28 (**a**) Crossover 2nd digit deformity with plantar plate tear. (**b**) Plantar fad pad displacement can be reduced with retrograde taping. (**c**) Correcting crossover toe with plantarflexion-abduction of digit along with proximal pull on plantar fad pad

healthy subjects [100]. The authors attributed this finding primarily to displacement of the fat pad rather than loss of mechanical stability of the digits.

Pressure Studies on Hammertoes

To verify the connection between digital deformities and metatarsalgia, we can look at pressure studies which show a pattern of similar findings. It is interesting to see that compared to HAV deformity, digital deformity can be expected to more predictably cause elevated plantar pressures under the central metatarsals [52] (Fig. 6.29).

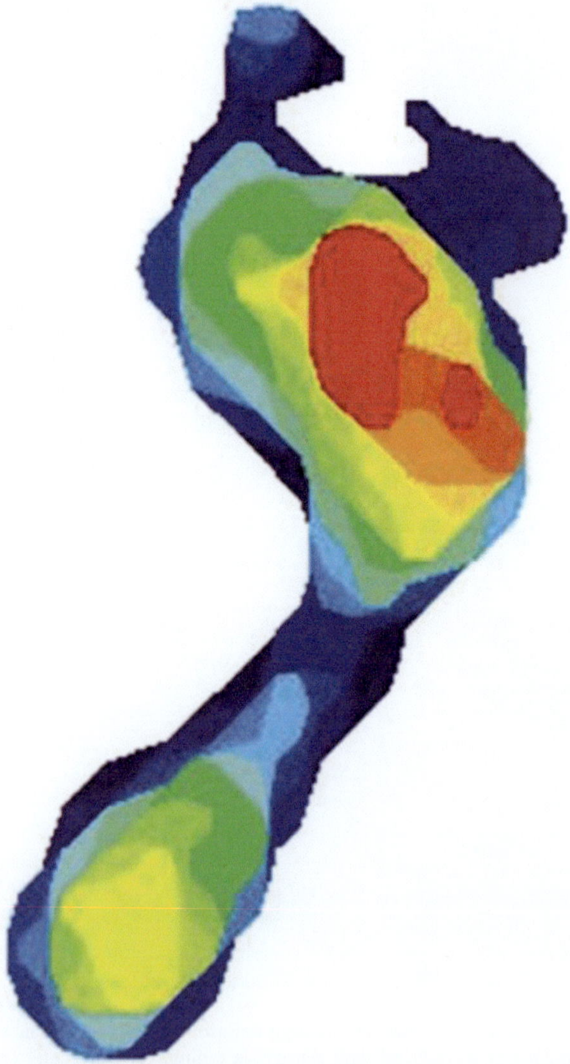

Fig. 6.29 Claw toes predictably increase plantar pressure under the lesser metatarsals

However, there are very few plantar pressure studies of hammertoe deformities, and these studies mainly focus on patients with diabetes. Two of these studies showed that claw toes or hammertoes will cause lower hallux pressures but higher metatarsal pressures [53, 111].

High plantar pressures in diabetic patients are greatly exacerbated by digital deformities. Bus and co-workers showed that both peak plantar pressure and pressure-time integrals were significantly increased in diabetic patients with claw toe deformities compared to age- and gender-matched diabetic patients without toe deformities [96]. Digital deformities are a significant risk factor for diabetic foot ulcerations [112]. In a study of 92 patients with diabetes, Holewski found that hammertoe deformities were present in 82% of patients with a history of ulceration and amputation, while these deformities were only found in only 23% patients who had no history of ulceration [113].

Finally a pressure study was conducted by Menz and co-workers with several findings which verified previous studies of patients with metatarsalgia [114]. They studied 43 older patients with forefoot pain, measuring plantar pressures as well as metatarsal length pattern on standing A/P radiographs, and then compared results to 75 patients with no forefoot pain. As seen with previous studies, there was no correlation between metatarsal length and prevalence of forefoot pain, and metatarsal length did not predict increased plantar pressure. However, the group with metatarsalgia demonstrated significant increased pressure under metatarsal heads 3–5, but not the second metatarsal head. There was a higher prevalence of HAV deformity in the metatarsalgia group, but not a higher incidence of digital deformity compared to the asymptomatic group. The authors speculated that increased stiffness of the digits and plantar fat pad, without digital deformity, might cause elevated plantar pressures, but most likely the lateral shift of pressure would be influenced by the larger number of HAV patients. Stiffness of the more proximal TMT joints might also generate higher plantar pressures, but this was not discussed by Menz et al.

The Interdigital Neuroma

A final cause of metatarsalgia is a condition which is mis-named and which has more long-standing myths explaining its etiology than any other forefoot condition. Commonly referred to as Morton's neuroma or Morton's metatarsalgia, this condition of forefoot pain continues to be elusive in terms of understanding its underlying cause. As a result, no significant advances have been made in the treatment of this pathology over the past 50 years.

A symptomatic enlargement of a plantar nerve of the foot was first described by Pasero in 1835 [115]. The painful condition resulting from a plantar nerve injury was named metatarsalgia by Thomas Morton (no relation to Dudley Morton) in 1876 [116]. There is universal agreement that a neuroma is more common in the 3rd intermetatarsal space, followed by the 2nd intermetatarsal space [117, 118]. Traditionally, the term Morton's neuroma refers to a compression neuropathy in the third intermetatarsal space. [119, 120] (Fig. 6.30). Alternative descriptions for a

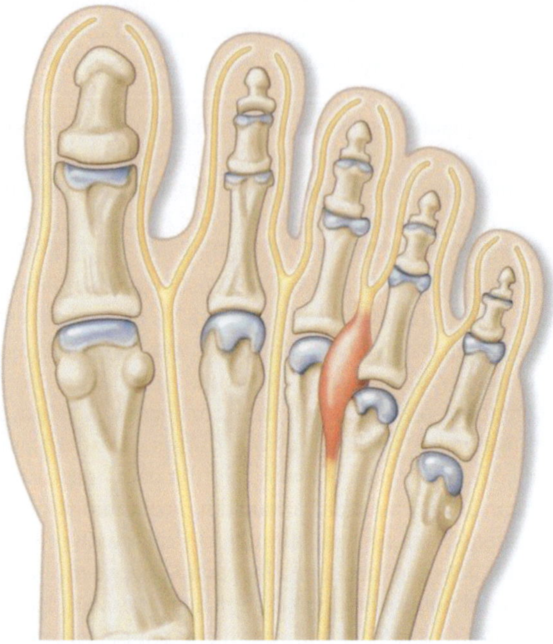

Fig. 6.30 The classic Morton's neuroma

neuroma of the foot include intermetatarsal neuroma, plantar interdigital neuroma, and interdigital neuroma [119–121].

A neuroma by definition is a neoplasm or benign tumor derived from nerve tissue, with growth of specific structures of the nerve itself [122]. In the foot, the term neuroma refers to a chronic inflammatory change in one of the plantar common digital nerves of the forefoot [123]. The histopathologic changes seen in the neuroma of the foot suggest a non-neoplastic reactive process in response to repeated minor trauma and are characterized by an abundant reactive hyperplasia of elastic fibers [124, 125]. In addition, Gianni found that specimens of neuromas removed from the third innerspace showed perineural fibrosis with local vascular proliferation and intraneural sclerohyalinosis of the common digital nerve, resulting in neuropathic pain [126]. None of these changes are consistent with a true neuroma which would involve proliferation of nerve tissue only. Therefore, in the human foot, the term neuroma refers to the non-neoplastic enlargement of a common digital nerve. Hence, the term interdigital neuroma appears more appropriate when describing this common cause of metatarsalgia.

Neuromas are common condition in the human foot. A survey in the United Kingdom found a diagnosis of neuroma in 88 of 100,000 women and 50 in every 100,000 men presenting for primary care in the United Kingdom, causing the most common compressive neuropathy after carpal tunnel syndrome [127]. In another study, it was shown that women age 50–55 were threefold more likely to be admitted to the hospital to be treated for neuroma compared to men [128].

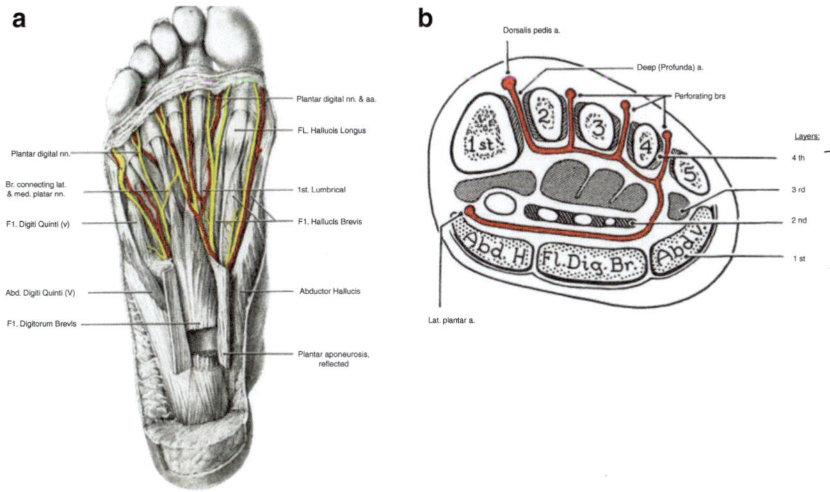

Fig. 6.31 Third plantar common digital nerve wraps around the FDB muscle

Neuromas are commonly found in patients with HAV deformity, digital deformities, and flatfoot [129, 130]. This observation along with the histopathologic findings of neuroma suggests a mechanical etiology. Numerous theories have been proposed about the pathomechanics of the interdigital neuroma, particularly specific to the anatomy of the third intermetatarsal space. Entrapment of the third plantar common digital nerve has been theorized to be the result of compression or traction or both mechanisms.

It has been previously proposed that the third plantar common digital nerve branches from an anastomosis of both the medial and lateral plantar nerves, thus occupying more space and also wrapping around the flexor digitorum brevis muscle belly [131, 132] (Figs. 6.31 and 6.32). Compression from increased thickness of the third digital nerve branch as well as traction placed upon it from contraction of the flexor digitorum muscle, according to past theories, make this nerve uniquely vulnerable to entrapment [131, 132]. However, in a cadaver study of 71 feet, Levitsky and co-workers found that the third plantar common digital nerve, originating as a communicating branch from the medial and lateral plantar nerves, was found in only 27% of the specimens and, when present, was not thicker than any of the other common digital nerves [123] (Fig. 6.33).

An alternative explanation for the mechanical cause of the interdigital neuroma was proposed by Bojsen-Moller and Flagstad [133]. Their elegant dissection of the ball of the foot in 20 cadaver specimens revealed several unique characteristics of the third intermetatarsal space (Fig. 6.34). The third plantar digital nerve was found to originate from both the medial and lateral plantar nerve and was postulated to be less mobile and vulnerable to traction when the toes move into extension. This traction was proposed to increase compression of the third plantar common digital nerve against the deep transverse metatarsal ligament at the dorsal surface of the

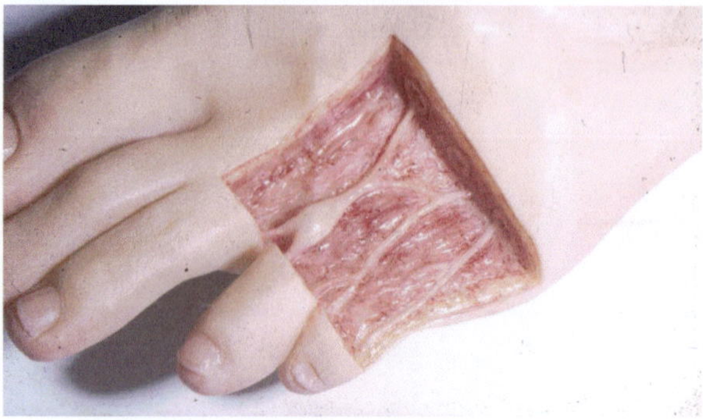

Fig. 6.32 Communicating branches from the lateral and medial plantar nerves form the third plantar common digital nerve which has developed a neuroma

nerve. Also, the deep sagittal septa extensions of the plantar fascia revealed differences between the second and the third intermetatarsal spaces. The plantar common digital nerves course along the posterior margin of the septa to enter the intermetatarsal spaces. The third plantar common digital nerve makes a more oblique turn around this posterior margin of the sagittal septa compared to the other common digital nerves and is pulled upon by the flexor digitorum muscle belly lying adjacent. This places traction upon the third plantar common digital nerve. Interestingly, Bojsen-Moller and Flagstad describe a unique role of footwear as an etiology of entrapment of the digital nerves. They speculate that stiffer shoes inhibit dorsiflexion of the digits. Normally, dorsiflexion motion of the digits pulls the deep sagittal septa and superficial transverse metatarsal ligament in an anterior direction which will decompress the digital nerves. Reduced mobility of the digits leaves the septa and distal extensions of the plantar fascia directly on top of the nerve during push off. This may verify why patients suffering from a painful Morton's neuroma commonly take off their shoes and feel relief walking barefoot. It is possible that increased extension motion of the lesser digits may actually facilitate decompression of the digital nerves.

The exact site of entrapment of a plantar common digital nerve leading to neuroma formation has been subject to debate. Gauthier embraced a previous held notion that the entrapment occurred at the deep transverse metatarsal ligament and described a surgical remedy which sectioned the anterior margin of the ligament without nerve resection [134]. Release of the deep transverse metatarsal ligament via endoscopic approach was later promoted as an alternative to neurectomy for treatment of neuroma [135].

However, the true location of the Morton's neuroma relative to the deep transverse metatarsal ligament was elucidated in a cadaveric and clinical study conducted by Kim and co-workers [136]. Kim et al. studied 17 cadaver specimens and found that the bifurcation of the common digital nerve in the second and third innerspace was located distal to the anterior margin of the deep transverse metatarsal ligament by 16.7 mm and 15.1 mm, respectively. This position changed minimally when looking at foot position at midstance and at heel off. In their clinical study of surgical evaluation of 32 feet, Kim et al. noted that the neuroma started at the bifurcation and extended an average 7.5 mm proximal, still located distal to the deep transverse

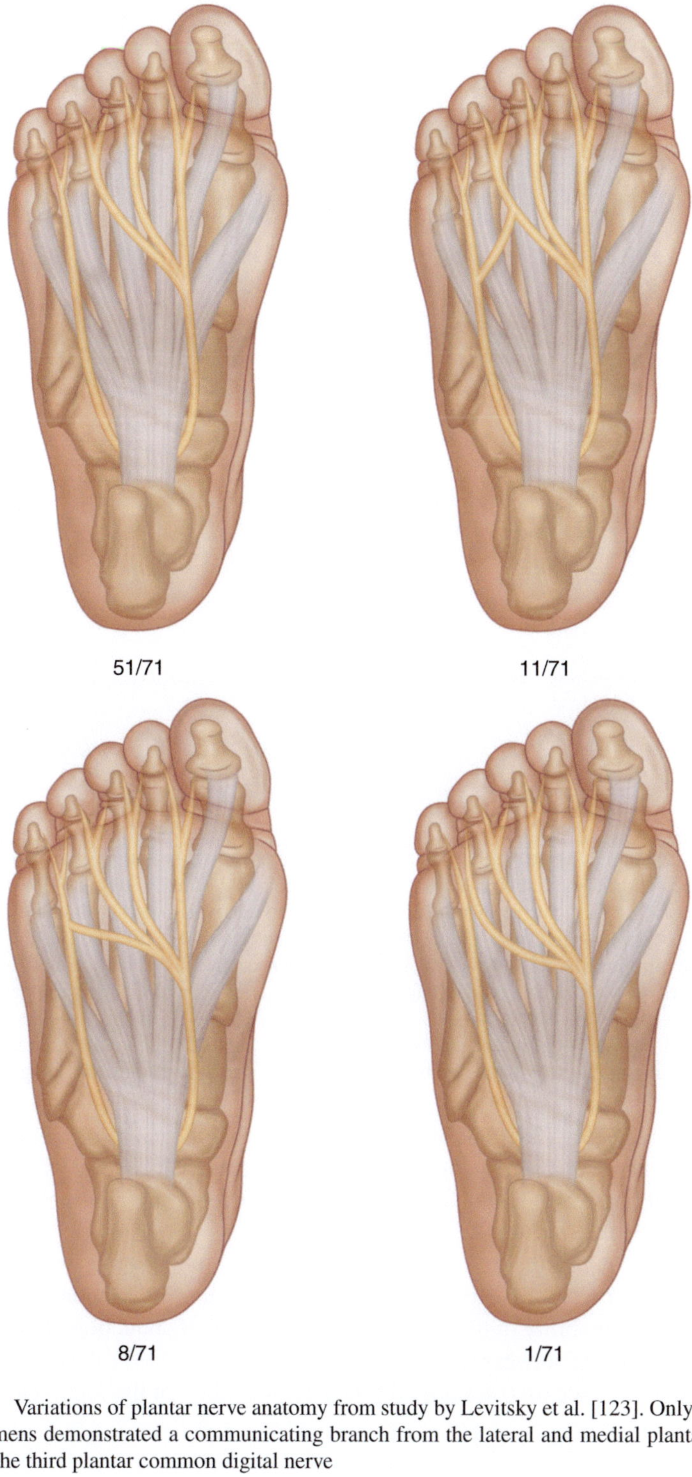

Fig. 6.33 Variations of plantar nerve anatomy from study by Levitsky et al. [123]. Only 11 out of 71 specimens demonstrated a communicating branch from the lateral and medial plantar nerves, forming the third plantar common digital nerve

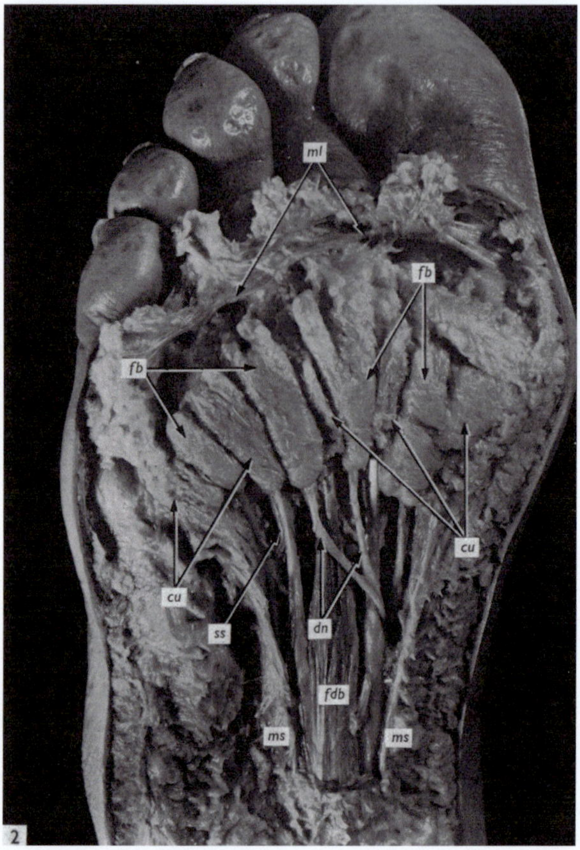

Fig. 6.34 The digital nerve (dn) to the third innerspace passes makes an oblique turn below the belly of the flexor digitorum brevis (fdb). (From; Bojsen-Moller and Flagstad (1976). Figure 2 Dissection of right foot. The plantar aponeurosis has been removed to demonstrate the superficial layer of the central compartment and the five submetatarsal cushions (cu). The cushion under the head of the third metatarsal bone is very narrow in this specimen. Four fat bodies (fb) cover the digital nerves (dn), while they pass through the weight-bearing area of the ball. The nerve for the third interstice takes an oblique course below the belly of the short flexor (fdb) to pass behind the sagittal septa. One of the septa of the fourth toe is shown (ss). Note the mooring ligament (ml) which in the interdigital web arches from one fibrous flexor sheath to the next. *ms* marginal septum

metatarsal ligament [136]. The authors concluded that the cause of the interdigital neuroma in the foot was pinching of the nerve by the metatarsal heads or the base of the proximal phalanx at the joint level, distal to the deep transverse metatarsal ligament (Fig. 6.35).

Pinching or compression of the third plantar common digital nerve has been proposed to be caused by excessive pronation of the foot. Pronation is thought to cause excessive mobility of the forefoot in the direction of abduction possibly causing compression of footwear against the lateral metatarsals [137]. Also, pronation is thought to allow excessive mobility off the lateral column of the foot relative to the

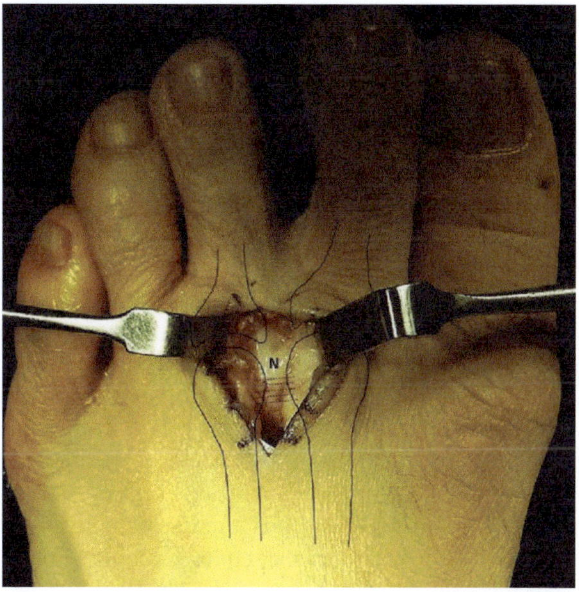

Fig. 6.35 The neuroma is located distal to the deep transverse metatarsal ligament

more stable medial column, setting up a movement interface thru the third intermetatarsal space [138, 139].

Conversely, cavus or supinated foot types have also been associated with formation of interdigital neuroma [140]. In particular, neuroma formation in the second intermetatarsal space have been identified in cavus feet and have been attributed to close metatarsal impingement and/or increased plantar declination of the metatarsals [141, 142].

Using the Foot Posture Index, the Framingham Foot Study found no association between pronated and supinated foot types and the presence of neuroma [143]. Naraghi and co-workers studied the Foot Posture Index of 68 patients with a diagnosis of interdigital neuroma and found no correlation between foot posture and incidence or location of neuroma [144]. Patients with an interdigital neuroma demonstrated foot posture falling within normative range. However this study did show that patients with interdigital neuroma had significant reduced ankle joint dorsiflexion compared to healthy controls.

Naraghi et al. followed up with another study measuring plantar pressure in 52 patients with interdigital neuroma compared to 31 healthy control subjects. There were no differences in peak plantar pressure or in contact time when comparing the two groups [145]. It is important to note that this study selected neuroma patients who did not have other forefoot pathologies such as HAV or digital deformities which could affect plantar pressure distribution. In this study, there was no difference in forefoot width and length measurements when comparing neuroma patients with healthy subjects. This is in agreement with a previous study showing no relationship between forefoot width, as well as metatarsal spacing and risk of neuroma [146].

It is interesting that Naraghi et al. found a strong association between restriction of ankle joint dorsiflexion and presence of neuroma *yet did not find increased plantar pressure in the forefoot of patients with an interdigital neuroma* [144, 145]. Naraghi et al. acknowledge that their method of measurement of ankle joint dorsiflexion using a hand held goniometer had previously shown to be unreliable [147]. Gatt and Chockalingam have shown that the traditional technique of measuring ankle joint dorsiflexion actually measures a combination of both foot and ankle dorsiflexion [148]. They demonstrated that with simple goniometer measure of ankle joint dorsiflexion taken off weight bearing and with full weight-bearing measurement such during a lunge test, the measurement is really a total of both *ankle and foot* sagittal plane motion [148]. In fact, the motion across the midtarsal joint and tarsometatarsal joints in the sagittal plane during gait totals 50% greater than that of the ankle joint alone. This contribution of the midfoot joints to overall dorsiflexion of the foot and ankle during gait has been verified in other studies [149, 150]. Therefore, what Naraghi et al. observed in patients with neuroma was a relative decreased motion of dorsiflexion *of the combined foot and ankle joints* which did not result in increased plantar pressure in the forefoot during walking. This lack of motion may have caused compression or traction of the common digital nerves via a mechanism which has yet to be identified.

In conclusion, despite its common occurrence, the interdigital neuroma of the human foot remains an elusive entity. Despite a wide range of theories of etiology, no underlying mechanism or intrinsic risk factor for the interdigital neuroma has been validated and confirmed with laboratory or clinical study. Treatments for the interdigital neuroma are not based upon addressing the etiology. Instead, interventions which ablate or excise the damaged nerve continue to be the preferred treatment [138, 139].

References

1. Dunn JE, Link CL, Felson DT, et al. Prevalence of foot and ankle conditions in a multiethnic community sample of older adults. Am J Epidemiol. 2004;159(5):491–8.
2. O'Kane C, Kilmartin TE. The surgical management of central metatarsalgia. Foot Ankle Int. 2002;23:415–9.
3. Coughlin MJ. Common causes of pain in the forefoot in adults. J Bone Joint Surg Br. 2000;82:781–90.
4. Scranton PE Jr. Metatarsalgia: diagnosis and treatment. J Bone Joint Surg Am. 1980;62(5):723–32.
5. Espinosa N, Maceira E, Myerson MS. Current concept review. Metatarsalgia. Foot Ankle Int. 2008;29:871–9.
6. Morton D. Metatarsus atavicus: the identification of a distinctive type of foot disorder. J Bone Joint Surg. 1927;9:531–44.
7. Barouk LS. Weil's metatarsal osteotomy in the treatment of metatarsalgia. Orthopade. 1996;25(4):338–44.
8. Hardy R, Clapham J. Observations on hallux valgus. J Bone Joint Surg. 1951;33-B:376–91.
9. Nilsonne H. Hallux rigidus and its treatment. Acta Orthop Scand. 1930;1:295–303.
10. Coughlin MJ. Crossover second toe deformity. Foot Ankle. 1987;8:29–39.
11. Meschan I. Radiology of the normal foot. Semin Roentgenol. 1970;5(4):327–40.

12. Maestro M, Besse JL, Ragusa M, Berthonnaud E. Forefoot morphotype study and planning method for forefoot osteotomy. Foot Ankle Clin. 2003;8:695–710.
13. Coughlin MJ. Second metatarsophalangeal joint instability in the athlete. Foot Ankle. 1993;14:309–19.
14. Kaz AJ, Coughlin MJ. Crossover second toe: demographics, etiology, and radio- graphic assessment. Foot Ankle Int. 2007;28:1223–37.
15. Fleischer AE, Ahmad M, Shah S, Catena F, Weil LS Sr, Weil L Jr. Association of abnormal metatarsal parabola with second metatarsophalangeal joint plantar plate pathology. Foot Ankle Int. 2017;38(3):289–97.
16. Slullitel G, Lopez V, Calvi JP, Seletti M, Bartolucci C, Pinton G. Effect of first ray insufficiency and metatarsal index on metatarsalgia in hallux valgus. Foot Ankle Int. 2016;37(3):300–6.
17. Deland J. Sung Il-Hoon: the medial crossover toe: a cadaveric dissection. Foot Ankle Int. 2000;21:375–8.
18. Thompson FM, Hamilton WG. Problems of the second metatarsophalangeal joint. Orthopedics. 1987;10:83–9.
19. Bhutta MA, Chauhan D, Zubairy AI, Barrie J. Second meta- tarsophalangeal joint instability and second metatarsal length association depends on the method of measurement. Foot Ankle Int. 2010;31(6):486–91.
20. Maceira E, Monteagudo M. Transfer metatarsalgia post hallux valgus surgery. Foot Ankle Clin. 2014;19:285–307.
21. Albrecht GH. The pathology and treatment of hallux valgus. RussVrach. 1911;10:14.
22. Scioli MW. Complications of hallux valgus surgery and subsequent treatment options. Foot Ankle Clin. 1997;2:719–39.
23. Lehman DE. Salvage of complications of hallux valgus surgery. Foot Ankle Clin. 2003;8:15–35.
24. David-West KS. Complications associated with Mitchell's osteotomy for hallux valgus correction: a retrospective hospital review. FAOJ. 2011;4(3):1.
25. Kuo CH, Huang PJ, Cheng YM, et al. Modified Mitchell osteotomy for hallux valgus. Foot Ankle Int. 1998;19(9):585–9.
26. Roukis TS. Clinical outcomes after isolated periarticular osteotomies of the first metatarsal for hallux rigidus: a systematic review. J Foot Ankle Surg. 2010;49:553–60.
27. Postema K, Burm PE, Zande ME, Limbeek J. Primary metatarsalgia: the influence of a custom moulded insole and a rocker bar on plantar pressure. Prosthetics Orthot Int. 1998;22:35–44.
28. Hsi WL, Kang JH, Lee XX. Optimum position of metatarsal pad in metatarsalgia for pressure relief. Am J Phys Med Rehabil. 2005;84:514–20.
29. Kang JH, Chen MD, Chen SC, Hsi WL. Correlations between subjective treatment responses and plantar pressure parameters of metatarsal pad treatment in metatarsalgia patients: a prospective study. BMC Musculoskelet Disord. 2006;7:95.
30. Yavuz M, Botek G, Davis BL. Plantar shear stress distributions: comparing actual and predicted frictional forces at the foot–ground interface. J Biomech. 2007;40(13):3045–9.
31. Yavuz M, Hetherington VJ, Botek G, Hirschman GB, et al. Forefoot plantar shear stress distribution in hallux valgus patients. Gait Posture. 2009;30:257–9.
32. Dockery GL. Evaluation and treatment of metatarsalgia and keratotic disorders. In: Myerson M, editor. Foot and ankle disorders. Philadelphia: Saunders; 2000. p. 359–77.
33. Dreeben SM, Noble PC, Hammerman S, Bishop JO, Tullos HS. Metatarsal osteotomy for primary metatarsalgia: radiographic and pedobarographic study. Foot Ankle. 1989;9:214–8.
34. Jung HG, Zaret DI, Parks BG, Schon LC. Effect of first meta- tarsal shortening and dorsiflexion osteotomies on forefoot plantar pressure in a cadaver model. Foot Ankle Int. 2005;26:748–53.
35. Kaipel M, Krapf D, Wyss C. Metatarsal length does not correlate with maximal peak pressure and maximal force. Clin Orthop Relat Res. 2011;469:1161–6.
36. Coughlin M, Jones C. Hallux Valgus; demographics, etiology and radiographic assessment. Foot Ankle Int. 2007;28:759–77.

37. Meyer JM, Tomeno B, Burdet A. Metatarsalgia due to insufficient support by the first ray. Int Orthop. 1981;5:193–201.
38. Grebing BR, Coughlin MJ. Evaluation of morton's theory of second metatarsal hypertrophy. J Bone Joint Surg Am. 2004;86-A(7):1375–84.
39. Cooper AJ, Clifford PD, Parikh VK, Steinmetz ND, Mizel MS. Instability of the first metatarsal-cuneiform joint: diagnosis and discussion of an independent pain generator in the foot. Foot Ankle Int. 2009;30(10):928–32.
40. Glasoe WM, Allen MK, Kepros T, Stonewall L, Ludewig PM. Dorsal first ray mobility in women athletes with a history of stress fracture of the second or third metatarsal. J Orthop Sports Phys Ther. 2002;32(ll):560–5.
41. Shibuya N, Roukis T, Jupiter DC. Mobility of the first ray in patients with or without hallux valgus deformity: systematic review and meta-analysis. J Foot Ankle Surg. 2017;56:1070–5.
42. Doty J, Coughlin M. Hallux valgus and hypermobility of the first ray: facts and fiction. Int Orthop. 2013;37:1655–60.
43. Greisberg J, Sperber L, Prince DE. Mobility of the first ray in various foot disorders. Foot Ankle Int. 2012;33:44–9.
44. Coughlin MJ, Jones CP. Hallux valgus and first ray mobility: a prospective study. J Bone Joint Surg Am. 2007;89:1887–98.
45. Mueller MJ, Strube MJ. Generalizability of in-shoe peak pressure measures using the f-scan system. Clin Biomech. 1996;11:159–64.
46. Abouaesha F, van Schie C, Griffths G, Young R, Boulton A. Plantar tissue thickness is related to peak plantar pressure in the high-risk diabetic foot. Diabetes Care. 2001;24:1270–4.
47. Dietze A, Bahlke U, Martin H, Mittlmeier T. First ray instability in hallux valgus deformity: a radiokinematic and pedobarographic analysis. Foot Ankle Int. 2013;34:124–30.
48. Waldecker U. Pedographicanalysis of hallux valgus deformity. Foot Ankle Surg. 2004;10:121–4.
49. Nyska M, Liberson A, McCabe C, Linge K, Klenerman L. Plantar foot pressure distribution in patients with hallux valgus treated by distal soft tissue procedure and proximal metatarsal osteotomy. Foot Ankle Surg. 1998;4:35–41.
50. Kernozek TW, Elfessi A, Sterriker S. Clinical and biomechanical risk factors of patients diagnosed with hallux valgus. J Am Podiatr Med Assoc. 2003;93:97–103.
51. Bryant A, Tinley P, Singer K. Plantar pressure distribution in normal, hallux valgus and hallux limitus feet. Foot. 1999;9:115–9.
52. Mickle KJ, Munro BJ, Lord SR, Menz HB, Steele JR. Gait, balance and plantar pressures in older people with toe deformities. Gait & Posture. 2011;34:347–51.
53. Mueller MJ, et al. Forefoot structural predictors of plantar pressures during walking in people with diabetes and peripheral neuropathy. J Biomech. 2003;36:1009–17.
54. Kadono K, Tanaka Y, Sakamoto T, Akiyama K, Komeda T, Taniguchi A, Takakura Y. Plantar pressure distribution under the forefeet with hallux valgus during walking. J Nara Med Assoc. 2003;54:273–81.
55. Hoffman UK, Gotze M, Wiesenreiter K, Muller O, et al. Transfer of plantar pressure from the medial to the central forefoot in patients with hallux valgus. BMC Musculoskelet Disord. 2019;20:149.
56. Galica AM, Hagedorn TJ, Dufour AB, Riskowski JL, Hillstrom HJ, Casey VA, Hannan MT. Hallux valgus and plantar pressure loading: the Framingham foot study. J Foot Ankle Res. 2013;6(1):42.
57. Bojsen-Moller F. Calcaneocuboid joint and stability of the longitudinal arch of the foot at high and low gear push off. J Anat. 1979;129:165.
58. MacConaill MA. The postural mechanism of the human foot. Proc R Ir Acad. 1944–1945;50B:265–78.
59. Griffin NL, D'Aout K, Richmond B, Gordon F, Aerts P. Comparative in vivo forefoot kinematics of Homo sapiens and Pan paniscus. J Hum Evol. 2010;59:608e619.

60. Barg A, Courville XF, Nickisch F, et al. Role of collateral ligaments in metatarsophalangeal stability: a cadaver study. Foot Ankle Int. 2012;33:877–82.
61. Johnston RB 3rd, Smith J, Daniels T. The plantar plate of the lesser toes: an anatomical study in human cadavers. Foot Ankle Int. 1994;15:276–82.
62. Deland JT, Lee KT, Sobel M, et al. Anatomy of the plantar plate and its attachments in the lesser metatarsal phalangeal joint. Foot Ankle Int. 1995;16:480–6.
63. Deland JT, Sung IH. The medial crossover toe: a cadaveric dissection. Foot Ankle Int. 2000;21:375–8.
64. Sarrafian SK, Kelikian AS. Functional anatomy of the foot and ankle. In: Kelikian AS, editor. Sarrafian's anatomy of the foot and ankle. 3rd ed. Philadelphia: Wolters Kluwer; 2011. p. 586–93.
65. Schrier JC, Verheyen CC, Louwerens JW. Definitions of hammer toe and claw toe. An evaluation of the literature. J Am Podiatr Med Assoc. 2009;99(3):194–7.
66. Coughlin MJ. Lesser toe deformities, surgery of the foot and ankle. In: Coughlin MJ, Mann CL, Saltzman CL, editors. Surgery of the foot and ankle. 9th ed. Philadelphia: Mosby Elsevier Inc; 2007. p. 366–400.
67. Green DR, Brekke M. Anatomy, biomechanics, and pathomechanics of lesser digital deformities. Clin Podiatr Med Surg. 1996;13:179.
68. Harmonson JK, Harkless LB. Operative procedures for the correction of hammertoe, claw toe, and mallet toe: a literature review. Clin Podiatr Med Surg. 1996;13:211.
69. Perry J. Gait analysis: Normal and pathological function. 2nd ed. Thorofare: SLACK Inc.; 1995, 2010 Figure 4-5, page 57-59.
70. Coughlin MJ. Crossover second toe deformity. Foot Ankle Int. 1987;8:29–39.
71. Coughlin MJ. Subluxation and dislocation of the second metatarsophalangeal joint. Orthop Clinics of North Am. 1989;20:539–51.
72. Yu GV, Judge MS, Hudson JR, et al. Predislocation syndrome. Progressive subluxation/dislocation of the lesser metatarsophalangeal joint. J Am Podiatr Med Assoc. 2002;92:182–99.
73. Sarrafian SK, Topouzian LK. Anatomy and physiology of the extensor apparatus of the toes. J Bone Joint Surg Am. 1969;51:669–79.
74. Cooper MT, Coughlin MJ. Sequential dissection for exposure of the second metatarsophalangeal joint. Foot Ankle Int. 2011;32:294–9.
75. Coughlin MJ, Baumfeld DS, Nery C. Second MTP joint instability: grading of the deformity and description of surgical repair of capsular insufficiency. Phys Sportsmed. 2011;39:132–41.
76. Nery C, Coughlin MJ, Baumfeld D, et al. Lesser metatarsophalangeal joint instability: prospective evaluation and repair of plantar plate and capsular insufficiency. Foot Ankle Int. 2012;33:301–11.
77. Frey C, Thompson F, Smith J. Update on women's footwear. Foot Ankle Int. 1995;16(6):328–31.
78. Doty JF, Coughlin MJ. Metatarsophalangeal joint instability of the lesser toes. J Foot Ankle Surg. 2014;53:440–5.
79. Green DR, Ruch JA, McGlamry ED. Correction of equinus related forefoot deformities. J Am Podiatr Assoc. 1976;66:768–79.
80. Jarrett BA, Manzi JA, Green DR. Interossei and lumbrical muscles of the foot-an anatormical and functional study. J Am Pod Assoc. 1980;70:1–13.
81. McGlamry ED. Floating toe syndrome. J Am Podiatric Med Assoc. 1982;72:561–8.
82. McCullough MB, Ringleb SI, Arai K, Kitaoka HB, Kaufman KR. Moment arms of the ankle throughout the range of motion in three planes. Foot Ankle Int. 2011;32(3):300–6.
83. Hui HJ, Beals TC, Brown NT. Influence of tendon transfer site on moment arms of the flexor Digitorum Longus muscle. Foot Ankle Int. 2007;28(4):441–7.
84. Wacker J, Calder JD, Engstrom CM, Saxby TS. MR morphometry of posterior tibialis muscle in adult acquired flat foot. Foot Ankle Int. 2003;24(4):354–7.
85. Angin S, Crofts G, Mickle KJ, Nester CJ. Ultrasound evaluation of foot muscles and plantar fascia in pes planus. Gait Posture. 2014;40:48–52.

86. Bouche RT, Heit EJ. Combined plantar plate and hammertoe repair with flexor digitorum longus tendon transfer for chronic, severe sagittal plane instability of the lesser metatarsophalangeal joints: preliminary observations. J Foot Ankle Surg. 2008;47:125–37.
87. Powless SH, Elze ME. Metatarsophalangeal joint capsule tears: an analysis by arthrography, a new classification system and surgical management. J Foot Ankle Surg. 2001;40:374–89.
88. Barg A, Courville XF, Nickisch F, Bachus KN, Saltzman CL. Role of collateral ligaments in metatarsophalangeal stability: a cadaver study. Foot Ankle Int. 2012;33:877–82.
89. Ker RF. The design of soft collagenous load-bearing tissues. J Exp Biol. 1999;202:3315–24.
90. Gregg J, Marks P, Silberstein M, Schneider T, Kerr J. Histologic anatomy of the lesser metatarsophalangeal joint plantar plate. Surg Radiol Anat. 2007;29(2):141–7.
91. Sarrafian SK. Functional characteristics of the foot and plantar aponeurosis under tibiotalar loading. Foot Ankle. 1987;8(1):10–9.
92. Hicks JH. The three weight-bearing mechanisms of the foot. In: Evans FG, editor. Biomechanical studies of the musculoskeletal system. Springfield: Charles C Thomas; 1961. p. 161–91.
93. Stainsby GD. Pathological anatomy and dynamic effect of the displaced plantar plate and the importance of the integrity of the plantar plate- deep transverse metatarsal ligament tie-bar. Ann R Coll Surg Engl. 1997;79(1):58–68.
94. Bojsen-Moller F, Flagstad KE. Plantar aponeurosis and internal architecture of the ball of the foot. J Anatolia. 1976;121:599–611.
95. Bus SA, Maas M, Cavanagh PR, Michels RP, Levi M. Plantar fat-pad displacement in neuropathic diabetic patients with toe deformity: a magnetic resonance imaging study. Diabetes Care. 2004;27(10):2376–81.
96. Bus S, Maas M, de Lange A, Michels R, Levi M. Elevated plantar pressures in neuropathic diabetic patients with claw/hammer toe deformity. J Biomech. 2005;38:1918–25.
97. Suero EM, Meyers KN, Bohne WHO. Stability of the metatarsophalangeal joint of the lesser toes: a cadaveric study. J Orthop Res. 2012;30(12):1995–8.
98. Bhatia D, Myerson MS, Curtis MJ, Cunningham BW, Jinnah RH. Anatomical restraints to dislocation of the second metatarsophalangeal joint and assessment of a repair technique. J Bone Joint Surg Am. 1994;76:1371–5.
99. Chalayon O, Chertman C, Guss AD, Saltzman CL, Nickisch F, Bachus KN. Role of plantar plate and surgical reconstruction techniques on static stability of lesser metatarsophalangeal joints: a biomechanical study. Foot Ankle Int. 2013;34(10):1436–42.
100. Coughlin MJ, Schutt SA, Hirose CB, Kennedy MJ, Grebing BR, Smith BW, Cooper MT, Golano P, Viladot R. Metatarsophalangeal joint pathology in crossover second toe deformity: a cadaveric study. Foot Ankle Int. 2012;33:133–40.
101. Branch HE. Pathologic dislocation of the second toe. J Bone Joint Surg. 1937;19:978–84.
102. Nix SE, Vicenzino BT, Collins NJ, Smith MD. Characteristics of foot structure and footwear associated with hallux valgus: a systematic review. Osteoarthr Cartil. 2012;20:1059–74.
103. Thompson FM, Coughlin MJ. The high price of high-fashion footwear. J Bone Joint Surg Am. 1994;76-A:1586–93.
104. Hamel AJ, Donahue SW, Sharkey NA. Contributions of active and passive toe flexion to forefoot loading. Clin Orthop Relat Res. 2001;393:326–34.
105. Sharkey NA, Donahue SW, Ferris L. Biomechanical consequences of plantar fascial release or rupture during gait – Part II: Alterations in forefoot loading. Foot Ankle Int. 1999;20(2):86–96.
106. Kirane YM, Michelson JD, Sharkey NA. Contribution of the flexor hallucis longus to loading of the first metatarsal and first metatarsophalangeal joint. Foot Ankle Int. 2008;29(4):123–34.
107. Sharkey NA, Ferris L, Smith TS, et al. Strain and loading of the second metatarsal during heel-lift. J Bone Joint Surg Am. 1995;77:1050–7.
108. Leardini A, Benedetti MG, Catani F, Simoncini L, Giannini S. An an atomically based protocol for the description of foot segment kinematics during gait. Clin Biomech. 1999;14:528–36.
109. Nester CJ, Liu AM, Ward E, Howard D, Cocheba J, Derrick T, Patterson P. In vitro study of foot kinematics using **a** dynamic walking cadaver model. J Biomech. 2007;40:1927–37.

110. Bojsen-Moller F, Lamoreux L. Significance of free dorsiflexion of the toes in walking. Acta Orthop Scand. 1979;50(4):471–89.
111. Ahroni JH, Boyko EJ, Forsberg RC. Clinical correlates of plantar pressure among diabetic veterans. Diabetes Care. 1999;22:965–72.
112. Lavery LA, Armstrong DG, Vela SA, Quebedeaux TL, Fleischli JG. Practical criteria for screening patients at high riskfor diabetic foot ulceration. Arch Intern Med. 1998;158:157–62.
113. Holewski JJ, Moss KM, Stess RM, Graf PM, Grunfeld C. Prevalence of foot pathology and lower extremity complications in a diabetic outpatient clinic. J Rehabil Res Dev. 1989;26:35–44.
114. Menz HB, Fotoohabadi MR, Munteanu SE, Zammit GV, Gilheany MF. Plantar pressures and relative lesser metatarsal lengths in older people with and without forefoot pain. J Orthop Res. 2013;31(3):427–33.
115. Pasero G, Marson P. Filippo Civinini (1805–1844) and the discovery of plantar neuroma. Reumatismo. 2006;58:319–22.
116. Morton TG. A peculiar and painful affliction of the fourth metatarsophalangeal joint articulation. Am J Med Sci. 1876;71:37–45.
117. Cashley DG, Cochrane L. Manipulation in the treatment of plantar digital neuralgia: a retrospective study of 38 cases. J Chiropr Med. 2015;14:90–8.
118. Saygi B, Yildirim Y, Saygi EK, Kara H, Esemenli T. Morton neuroma: comparative results of two conservative methods. Foot Ankle Int. 2005;26:556–9.
119. Bennett GL, Graham CE, Mauldin DM. Morton's interdigital neuroma: a comprehensive treatment protocol. Foot Ankle Int. 1995;16:760–3.
120. Morscher E, Ulrich J, Dick W. Morton's intermetatarsal neuroma: morphology and histological substrate. Foot Ankle Int. 2000;21:558–62.
121. Dockery GL. The treatment of intermetatarsal neuromas with 4% alcohol sclerosing injections. J Foot Ankle Surg. 1999;38:403–8.
122. https://medical-dictionary.thefreedictionary.com/neuroma
123. Levitsky KA, Alman BA, Jevsevar DS, Morehead J. Digital nerves of the foot: anatomic variations and implications regarding the pathogenesis of interdigital neuroma. Foot Ankle 1993;14(4):208–214.
124. Shereff MJ, Grande DA. Electron microscopic analysis of the interdigital neuroma. Clin Orthop Relat Res. 1991;271:296–9.
125. Graham CE, Graham DM. Morton's neuroma: a microscopic evaluation. Foot Ankle. 1984;5:150–3.
126. Giannini S, Bacchini P, Ceccarelli F, Vannini F. Interdigital neuroma: clinical examination and histopathologic results in 63 cases treated with excision. Foot Ankle Int. 2004;25:79–84.
127. Latinovic R, Gulliford MC, Hughes RA. Incidence of common compressive neuropathies in primary care. J Neurol Neurosurg Psychiatry. 2006;77:263–5.
128. Naraghi R, Bryant A, Slack-Smith L. Description of total population hospital admissions for Morton's metatarsalgia in Australia. J Am Podiatr Med Assoc. 2014;104(5):451–4.
129. Guiloff RJ, Scadding JW, Klenerman L. Morton's metatarsalgia: clinical, electrophysiological and histological observations. J Bone Joint Surg. 1984;66-B(4):586–91.
130. Morscher E, Urlich J, Dick W. Morton's intermetatarsal neuroma: morphology and histological substrate. Foot Ankle. 2000;21(7):558–62.
131. Jones JR, Klenerman L. A study of the communicating branch between the medial and lateral planter nerves. Foot Ankle. 1984;4(6):313–7.
132. Mann RA, Reynolds JC. Interdigital neuroma: a critical clinical analysis. Foot Ankle. 1983;3(4):238–43.
133. Bojsen-Moller F, Flagstad KE. Plantar aponeurosis and internal architecture of the ball of the foot. J Anat. 1976;121:599–611.
134. Gauthier G. Thomas Morton's disease : a nerve entrapment syndrome. A new surgical technique. Clin Orthop. 1979;142:90–2.

135. Barrett SL, Pignetti TT. Endoscopic decompression for intermetatarsal nerve entrapment-the EDIN technique : preliminary study with cadaveric specimens ; early clinical results. J Foot Ankle Surg. 1994;33:503–8.
136. Kim J, Choi J, Park J, Wang J. An anatomical study of Morton's interdigital neuroma: the relationship between the occurring site and the deep transverse metatarsal ligament (DTML). Foot Ankle Int. 2007;28(9):1007–10.
137. Root ML, Orien WP, Weed JH, Hughes RJ. Clinical biomechanics: Normal and abnormal function of the foot. 1st ed. Los Angeles: Clinical Biomechanics Corporation; 1977.
138. Wu K. Morton's interdigital neuroma: a clinical review of its etiology, treatment, and results. J Foot Ankle Surg. 1996;35(2):112–9.
139. Jain S, Mannan K. The diagnosis and management of Morton's neuroma: a literature review. Foot Ankle Spec. 2013;6(4):307–17.
140. Pazzaglia U, Moalli S, Leutner M, Gera R. Morton's neuroma: an immunohistochemical study. Foot. 1996;6(2):63–5.
141. Abassian A, Pomeroy G. The idiopathic cavus foot-Not so subtle after all. Foot Ankle Clin N Am. 2013;18:629–42.
142. Bygrave CJ, Jones S, Smith TW, Flowers MJ, Betts RP. Ultrasonic diagnosis of morton's neuroma: a guide to problems, pointers, pitfalls and prognosis. Foot. 2003;13(2):92–9.
143. Hagedorn TJ, Dufour AB, Riskowski JL, Hillstrom HJ, Menz HB, Casey VA, et al. Foot disorders, foot posture, and foot function: the Framingham Foot Study. PLoS One. 2013:1–7.
144. Naraghi R, Bremner A, Slack-Smith L, Bryant A. The relationship between foot posture index, ankle equinus, body mass index and intermetatarsal neuroma. J Foot Ankle Res. 2016;9(1):46.
145. Naraghi R, Slack-Smith L, Bryant A. Plantar pressure measurements and geometric analysis of patients with and without Morton's neuroma. Foot Ankle Int. 2018;39(7):829–35.
146. Park YH, Jeong SM, Choi GW, Kim HJ. The role of the width of the forefoot in the development of morton's neuroma. Bone Joint J. 2017;99-B(3):365.
147. Gatt A, Chockalingam N. Clinical assessment of ankle joint dorsiflexion: a review of measurement techniques. J Am Podiatr Med Assoc. 2011;101(1):59–69.
148. Gatt A, DeGiorgio S, Chockalingam N. Formosa. A pilot investigation into the relationship between static diagnosis of ankle equinus and dynamic ankle and foot dorsiflexion during stance phase of gait: time to revisit theory? Foot. 2017;30:47–52.
149. Lundgren P, Nester C, Liu A, Arndt A, Jones R, Stacoff A, Wolf P, Lundberg A. Invasive in vivo measurement of rear-, mid- and forefoot motion during walking. Gait Posture. 2008;28(1):93–100.
150. Nester CJ, Jarvis HL, Jones RK, Bowden PD, Anmin L. Movement of the human foot in 100 pain free individuals aged 18–45: implications for understanding normal foot function. J Foot Ankle Res. 2014;7:51.
151. Maas NMG, van der Grinten M, Bramer WM, Gert-Jan Kleinrensink. Metatarsophalangeal joint stability: a systematic review on the plantar plate of the lesser toes. J Foot Ankle Res (2016) 9:32

The Adult Acquired Flatfoot

7

"Doc, my feet feel like suction cups on the floor."

Anonymous patient of Dr. Richie with bilateral Adult Acquired Flatfoot deformity

Key Points

- Flatfoot deformity is common in the newborn and usually spontaneously resolves by age 10
- The adult acquired flatfoot (AAF) occurs in middle-aged adults who have pre-existing flatfoot deformity.
- The shape of the talus in patients with AAF shows valgus angulation on both the superior and inferior articular surfaces
- AAF begins with a gradual degeneration of the posterior tibial tendon which results from abnormal loads and increased friction which occurs in flatfoot deformity
- AAF is characterized by progressive valgus deformity of the hindfoot, collapse of the medial longitudinal arch, and abduction of the forefoot
- The progressive structural changes in AAF are not solely due to rupture of the posterior tibial tendon alone but more likely due to rupture of the spring ligament and the medial collateral ligaments of the ankle.
- Gait studies of AAF patients show delayed heel rise, excessive sagittal plane motion in the midfoot, and a lack of pronation of the forefoot during terminal stance.
- Coupling between the forefoot and rearfoot as well as between the rearfoot and the leg is disrupted in AAF deformity.

D. H. Richie Jr, *Pathomechanics of Common Foot Disorders*,
https://doi.org/10.1007/978-3-030-54201-6_7

Introduction

We are all born with flat feet. Most people develop an arch by age 10, and among those who do not, the vast majority remain relatively asymptomatic for the rest of their life. However, a percentage of people with flatfeet will develop chronic symptoms, and some will suffer progressive collapse of foot alignment later in life. That condition has become known as the adult acquired flatfoot (AAF) deformity.

The adult acquired flatfoot is the result of failure of multiple static and dynamic supportive structures of the longitudinal arch of the foot. Ironically, these same structures are unique to the human foot and provide specialized function not seen in other primates. With failure of these unique support elements, the end stage adult acquired flatfoot closely resembles the ape foot in both in structure and function.

Before studying the adult acquired flatfoot, it is appropriate to begin with a review of the pediatric flatfoot. It is important to clarify that the pediatric flatfoot and the adult acquired flatfoot are two entirely different conditions involving different anatomic structures and posing vastly different indications and prognosis for treatment.

The Pediatric Flatfoot

According to Tachdjian, a "flatfoot" is a generic term to describe any condition where the longitudinal arch of the human foot is "abnormally low or absent" [1]. Thirty years after that definition appeared in an authoritative pediatric textbook, there remains no agreed criteria for establishing the diagnosis of "flatfoot." Furthermore, there remains considerable debate about the long-term sequelae of the pediatric flatfoot and the need to treat the condition from the outset. Regarding the lack of agreement for the criteria in diagnosing flatfoot, Evans and Rome state in their 2011 Cochrane review, *"In the 21st century this situation remains and stymies clinical practice and research globally. There is a need for a standardized framework from which to evaluate the pediatric flatfoot"* [2].

The condition known as flatfoot has been also described with various names including pes planus, pes valgus, flexible pes valgus, flexible pes planus, calcaneal valgus, and pronated foot. Despite the fact that the deformity presents in all three body planes, most authorities have focused on simple alignment of the arch (flat) and position of the hindfoot (valgus) [3–5]. Studies of pediatric patients which diagnosed flatfoot have used various measures including ink footprint, arch index, photography, pedobarograph, standing radiograph, and rearfoot angle. The reliability of these measures and the validity of using them to diagnose a flatfoot deformity have not been proven. However studies of the natural resolution of flatfoot in the growing child have shown consistent findings.

All newborn children have flatfeet, and subsequently the arch begins to develop over the first 6 years of life [6, 7] (Fig. 7.1). Staheli and co-workers studied 441 asymptomatic subjects over a wide age range and concluded that flatfeet are common in infants, are less common in children, and when seen in adults, are usually

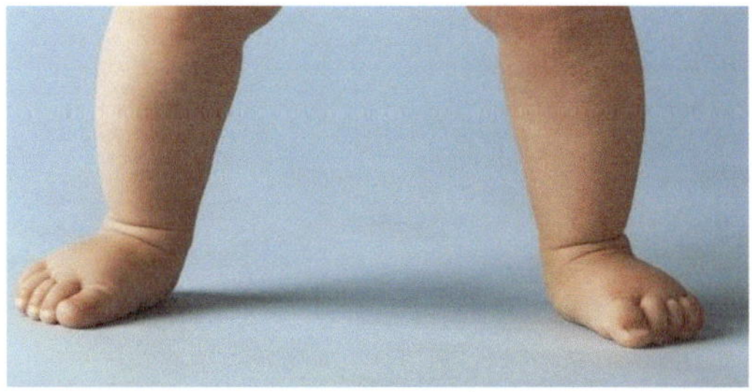

Fig. 7.1 All newborn and early walking infants have the appearance of flatfeet

asymptomatic [8]. Staheli et al. recommended observation only, without treatment of the pediatric flatfoot.

Pfeiffer and co-workers studied 835 children between the ages of 3 and 6 years old, and based upon visual evaluation of a flat arch and valgus heel alignment, diagnosed flatfoot deformity in 44% of the subject population [5]. The flatfoot deformity reduced with age starting with the 3 year olds who had a 54% incidence of flatfoot while the 6 year olds had a 24% incidence. Also, there was an increased risk of flatfoot in males and overweight subjects. This study also differentiated between "physiologic" and "pathologic" flatfoot deformity whereby a heel valgus position greater than 20 degrees was deemed pathologic and warranted treatment. Therefore, 90% of the patients with flatfoot did not require treatment according to the authors.

El and co-workers evaluated 579 school children aged 10 years and measured the arch index using footprints [4]. Flatfoot deformity was found in 17.2% of the children of whom 25% had evidence of ligament laxity. In their summary of 15 studies of the prevalence of flatfoot deformity in the pediatric population, Evans and Rome found that between ages 2 and 6, 46.3% of children have a flatfoot deformity [2]. This prevalence decreases to 14.2% by 10 years of age. The authors concluded that flatfoot deformity was a relatively normal finding in children under age 6 and, when present, will spontaneously improve by age 10.

A review of these studies shows a consensus of opinion that the vast majority children with flatfeet are asymptomatic. In particular, the majority of children with flatfeet are noted to have a *flexible* deformity. By definition, a flexible flatfoot demonstrates some arch structure off weight bearing which disappears upon weight bearing [9]. Conversely, a *rigid* flatfoot deformity shows no arch, whether weight bearing or not. The congenital rigid flatfoot deformity is usually associated with a tarsal coalition or a vertical talus, and both conditions are almost always symptomatic and worthy of treatment [2, 6, 7] (Figs. 7.2 and 7.3).

The American College of Foot and Ankle Surgeons classified the pediatric flatfoot as being either flexible or rigid [10]. Flexible deformities have two subdivisions: symptomatic and asymptomatic. Symptoms are more likely in patients with increased rearfoot eversion and in patients with tight heel cords. This opinion echoes the findings from the classic article published by Harris and Beath in 1948 where 3619 Canadian soldiers were studied [11]. Flexible flatfeet were found in 14% of the adult subjects, and 25% of these people reported regular pain in their feet. Harris and Beath observed that a tight heel cord was the most common finding in

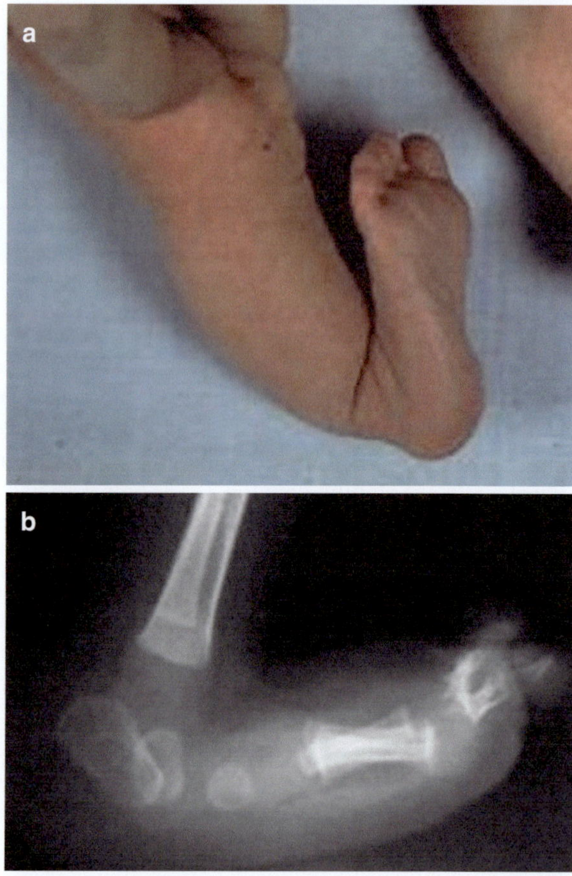

Fig. 7.2 (**a**) Vertical talus. (**b**) Radiograph of vertical talus

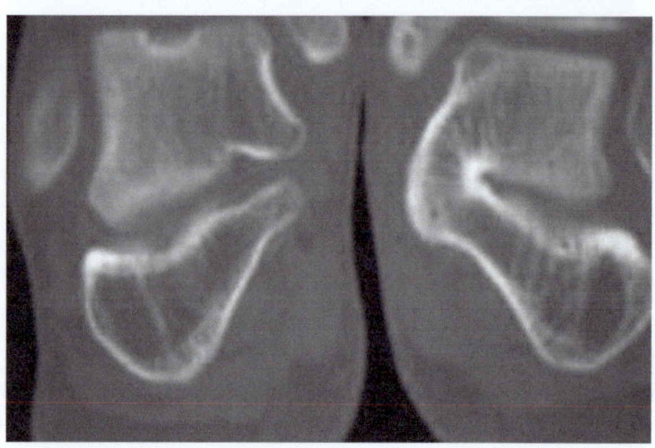

Fig. 7.3 Tarsal coalition in coronal readiograph: note loss of middle facet of subtalar joint in right image

symptomatic adults with flexible flatfoot and concluded that this most likely contributed to worsening of deformity with age.

When evaluating systematic reviews of the pediatric flatfoot deformity, several findings have consensus of opinion [2, 9, 12]. Flatfoot deformity in the newborn will spontaneously resolve in most cases by age 10. The majority of children with flatfeet have a flexible deformity which is asymptomatic. Therefore, treatment of flatfoot in asymptomatic children is not warranted. However, most authorities agree that a subgroup of asymptomatic children with flatfoot will develop symptoms later in life. However extrinsic and intrinsic risk factors for these young, asymptomatic patients to develop future disability have not been universally agreed upon.

Flexible pes planus or flatfoot deformity has been associated with a myriad of lower extremity pathologies proximal to the feet themselves. Studies have shown that adults with pes planus are more likely to suffer from low-back or diffuse lower limb pain [13, 14]. Pes planus is associated with generalized foot pain, hallux abductovalgus, metatarsalgia, and hammertoes [15–20]. These findings continue to challenge clinicians who evaluate young patients presenting with asymptomatic flatfoot deformity. Which of these children will eventually progress to develop significant symptoms later in life? And, is there any evidence that early intervention will minimize the risk of developing symptoms later in life?

Of all the adult pathologies associated with pediatric flatfoot deformity, the adult acquired flatfoot (AAF) certainly has the greatest potential for significant and permanent disability. As will be shown, the pediatric flatfoot may be a precursor to the adult acquired flatfoot, but the two conditions are entirely different in their pathoanatomy and pathomechanics.

Definition of the Adult Acquired Flatfoot

The adult acquired flatfoot is a symptomatic, progressive deformity of the foot caused by a loss of dynamic and static supportive structures of medial longitudinal arch [21]. Previously described as posterior tibial tendon dysfunction (PTTD), the term adult acquired flatfoot has become a more accurate description of the global nature of the condition involving multiple anatomic structures.

History and Terminology

Symptomatic, progressive flatfoot deformity in the adult has traditionally been attributed to degeneration or rupture of the posterior tibial tendon [21]. While tenosynovitis of the posterior tibial tendon associated with flatfoot deformity was described in the medical literature over 75 years ago, these reports did not mention actual rupture of the tendon or progressive deterioration of foot alignment in the patients studied [22–24]. Cozen was the first to describe a progressive deformity of the foot resulting from attenuation of the posterior tibial tendon [25]. He proposed that a planovalgus foot deformity caused “bowstringing” of the posterior tibial tendon, leading to mechanical irritation and synovitis.

Actual rupture of the posterior tibial tendon with subsequent development of acquired flatfoot deformity was first reported in 1969 by Kettlekamp and Alexander [26]. The first description of acquired flatfoot in an adult secondary to both tendon and ligament failure was published by Goldner et al. in 1974 [27]. They described a sequence of events starting with progressive failure of the posterior tibial tendon and eventual rupture of the medial calcaneonavicular (spring) ligament leading to flatfoot deformity. In this article, Goldner et al. were the first to describe an operative procedure to treat this condition involving a transfer of the flexor digitorum longus tendon to replace the ruptured posterior tibial tendon.

In the 1980s a sudden awareness took place in the orthopedic and podiatric communities that patients could undergo spontaneous rupture of the posterior tibial tendon and subsequently develop a flatfoot deformity [27–32]. Why this cause-effect relationship had never been reported before remains an enigma, particularly when recognizing the monumental volume of research which subsequently was published focusing on the adult acquired flatfoot after the initial reports in the 1980s.

After initial reports of complete rupture of the posterior tibial tendon with acquired flatfoot deformity, other surgeons began documenting the existence of an attenuated or "deficient" posterior tibial tendon which could also lead to progressive deformity [33–35]. This observation led Johnson and Strom to develop and publish their landmark classification system describing a progressive flatfoot deformity based upon the integrity of the posterior tibial tendon [36]. This published article also propagated the popular use of the term "posterior tibial tendon dysfunction" (PTTD) which is still used today. Notwithstanding, other authors began documenting associated ligament failure with the development of progressive flatfoot in adult patients and began using the term adult acquired flatfoot (AAF) instead of PTTD [37, 38].

Recognizing the shortcoming of attributing the condition solely to rupture of the posterior tibial tendon, many authorities began using the term adult acquired flatfoot to describe a condition which involved failure of multiple anatomic structures in the foot and ankle [38–41]. In 2017, Ross and co-workers evaluated the evolution of the terms "posterior tibial tendon dysfunction" and "adult acquired flatfoot deformity" [42]. They point out the shortcomings of using the term PTTD which focuses only on tendon integrity while overlooking the role of ligament failure in progressive adult flatfoot deformity. Ross et al. also clarify that a progressive flatfoot deformity can occur in adult patients for reasons other than tendon failure such as sequela from Charcot arthropathy, rheumatoid arthritis, or certain neurologic disorders [43–45].

Notwithstanding, the terms AAF and PTTD are used interchangeably today both in peer-reviewed publications and in academic symposia. In this chapter, both terms will be used depending upon the particular study which is referenced, using the terms chosen by the specific authors.

Epidemiology

The true prevalence of AAF or PTTD is not known because patients often do not seek treatment and when they do, the condition is often misdiagnosed [46, 47]. A survey of elderly patients (mean age 80) revealed a 10% incidence of PTTD which had been misdiagnosed by their treating doctors [27]. A subsequent study by the same researchers, surveying a younger population, found that among women age 40 and older, the prevalence of PTTD was 3.3%, and all of the patients were previously undiagnosed despite characteristic deformity and long-standing symptoms [48].

The prevalence of AAF in women is startling and undeniable. A higher rate of rupture of the posterior tibial tendon in women vs men was first documented in some of the early reports of the condition [30, 49]. Subsequent to those reports, many articles were published showing a consistent higher prevalence of AAF in women compared to men. A review of 5 published studies of conservative treatment of PTTD revealed that among the 170 subjects, 128 were women (75%) [31]. A recent systematic review of ten studies of patients with PTTD revealed that 78% of the subjects were women [51]. Further discussion about this finding will be found in the pathomechanics section of this chapter.

Risk Factors

It is well recognized that patients with AAF have a history of flatfeet all their life [34, 35, 52]. Around the fifth decade, some of these people, usually women with flatfeet, become markedly symptomatic. The symptoms begin usually on one foot only, with swelling and pain along the distal course of the posterior tibial tendon [30, 34–36, 41]. Why certain people with congenital flatfoot deformity suffer this significant, potentially catastrophic breakdown of foot alignment with loss of general mobility has remained an unanswered question for many years.

Potential groundbreaking research recently published from the Hospital for Special Surgery in New York gives insight into this issue. Using weight-bearing, multiplanar CT imaging, a group of researchers identified a specific deformity in the shape of the talus in patients with stage II adult acquired flatfoot deformity [52, 53]. Specifically, the inferior surface of the talus (part of the posterior subtalar joint) and the superior surface of the talus (part of the ankle joint) have a significant valgus angulation in AAF deformity (Fig. 7.4). This valgus angulation of the talus is not seen in healthy controls. Interestingly, the significant valgus alignment of the inferior surface of the talus is seen in other non-human primates, where the calcaneus is located lateral rather than directly beneath the talus. This is discussed in Chap. 2. The entire posterior facet of the subtalar joint of the chimpanzee is oriented vertical or in extreme valgus attitude relative to the weight-bearing surface as shown in Fig. 7.1, Chap. 2. The human embryo demonstrates this same alignment early in

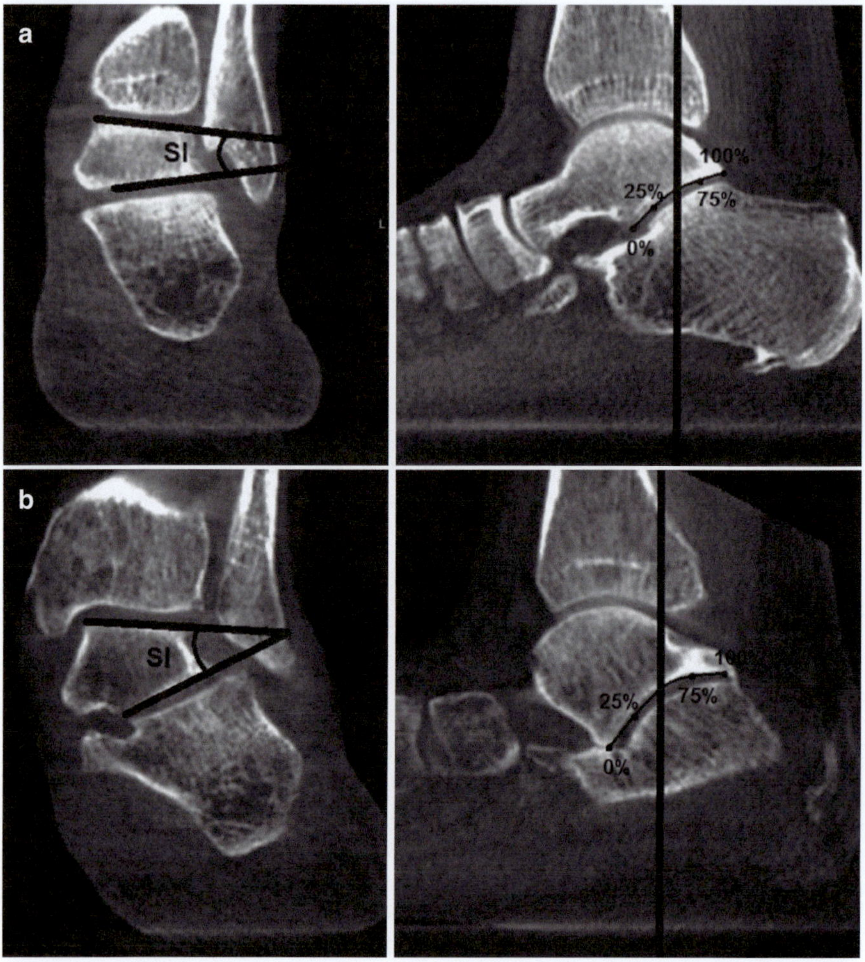

Fig. 7.4 Valgus angulation of superior and inferior articular surfaces of the talus in stage II adult acquired flatfoot. (From: Cody et al. [53]. Figure 2 MP-WB coronal plane images (left) demonstrate how to measure the superior to inferior angle at 50% of the anterior-posterior dimension of the posterior facet of the talus on sagittal images (right). (**a**) The superior to inferior angle (SI) is shown in a control patient versus (**b**) in a flatfoot patient. Increased talar valgus is evident in the flatfoot patient)

development (see Fig. 7.3 in Chap. 2). The authors who discovered the deformity of the talus in AAF patients believe that it is an "innate deformity" and not acquired [53]. Severity of deformity of the talus does not change with progression of AAF. Further research is needed to determine if valgus deformity of the articular surfaces of the talus are unique to the pathologic adult acquired flatfoot or might also be seen in other adults with non-pathologic, asymptomatic flat feet.

Thirty years ago, Holmes and Mann published an article citing certain risk factors for rupture of the posterior tibial tendon [49]. Of the 67 patients diagnosed with rupture of the posterior tibial tendon, 76% were women. Fifty-two percent of the patients had either diabetes, hypertension, or obesity. The authors proposed that hyperlipidemia could affect perfusion of the posterior tibial tendon. In reality, Holmes and Mann were the first to propose that a relationship exists between metabolic syndrome and tendinopathy. That process has been studied extensively since 1989 and clearly has relevance to the pathomechanics of AAF.

Metabolic syndrome identifies a group of risk factors which increases the risk of coronary artery disease, stroke, and diabetes. The five main risk factors are increased abdominal fat, high triglycerides, low HDL cholesterol, high blood pressure, and high fasting blood sugar [54]. Increased circulating lipids as well as fat deposition in the abdominal (belly) area are associated with elevated markers for inflammation in various tissues around the body [54, 55]. Research has shown a strong association between metabolic syndrome and certain tendinopathy conditions [56, 57]. Also, generalized musculoskeletal pain has been associated with metabolic syndrome [58, 59]. Since people with metabolic syndrome are usually overweight, the relationship between increased body mass index (BMI) causing overload of tendons and increased inflammation from hyperlipidemia can both become precursors to degeneration and eventual rupture of tendons in the human body [60, 61]. Obesity by itself a risk factor for PTTD [30, 49, 50]. When combined with metabolic syndrome, these two factors would likely increase the risk of degeneration of the posterior tibial tendon and certain ligaments in the hindfoot which leads to AAF deformity.

There is a growing body of evidence for genetic risk factors for tendinopathy. Certain genes have been identified as important regulators of fibrillogenesis and collagen cross bridging which are essential for establishing integrity to tendons and ligaments [62–64]. These genetic factors may interact with other intrinsic and extrinsic risk factors to contribute to degeneration of the posterior tibial tendon and the spring ligament which are both critical structures in the pathogenesis of AAF deformity.

Anatomy

The tibialis posterior muscle (TP) originates from the tibia, the fibula, and the interosseous membrane in the proximal portion of the lower leg. Morimoto performed cadaver studies and observed that the fibular side of the origin of the TP is the strongest which improves the lever arm for transverse and frontal plane control of the foot and ankle [65] (Fig. 7.5).

The tibialis posterior is a multipennate muscle with short fibers designed for limited excursion and more powerful contraction. Shorter muscle fiber length equates to increased strength. However, as demonstrated in Chap. 3, the tibialis posterior is relatively weak compared to the gastrocnemius and soleus musculature. According to the study by Silver, the soleus provides 30% of the strength of the lower leg musculature, while the gastrocnemius provides 19.2% of overall strength [66]. In comparison, the tibialis posterior provides only 6.4% of the strength of the lower extremity musculature, while the flexor digitorum longus (FDL) provides 1.8%.

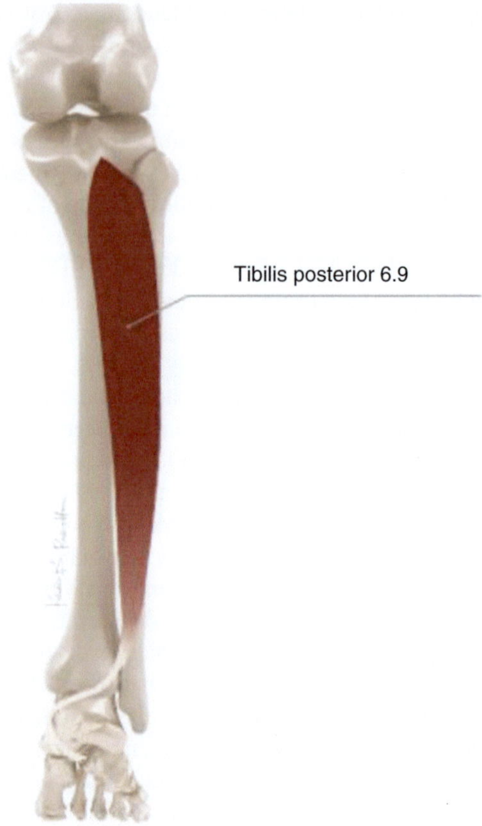

Fig. 7.5 The tibialis posterior: tendon crosses lower leg in transverse plane and pulls against the fibular origin

The tendon of the tibialis posterior, also known as the posterior tibial tendon, originates at the myotendon junction located in the distal third of the lower leg. Relative to the other deep leg flexors (FDL and FHL), the tendon of the TP is the longest and plays a critical role in storage and return of elastic energy during walking [67]. If this tendon becomes stretched or attenuated, the short fibers of the tibialis posterior muscle are unable to compensate to increase tension.

There are six sections of the posterior tibial tendon: myotendon transition, proximal retinaculum, retromalleolar, inframalleolar, distal retinaculum, and insertion [68]. The TP tendon changes direction at the distal margin of the medial malleolus and takes on a distinct histologic appearance due to the physical forces which are created by the bony pulley at this location (Fig. 7.6). In this *retromalleolar region* is a section of fibrocartilage on the anterior surface of the TP tendon providing a gliding surface which protects against compression and shear stress [69, 70] (Fig. 7.7). This gliding zone of the posterior tibial tendon is unique among the other ankle flexors and renders the TP tendon vulnerable to degeneration and rupture. This will be discussed further in the pathoanatomy and pathomechanics sections.

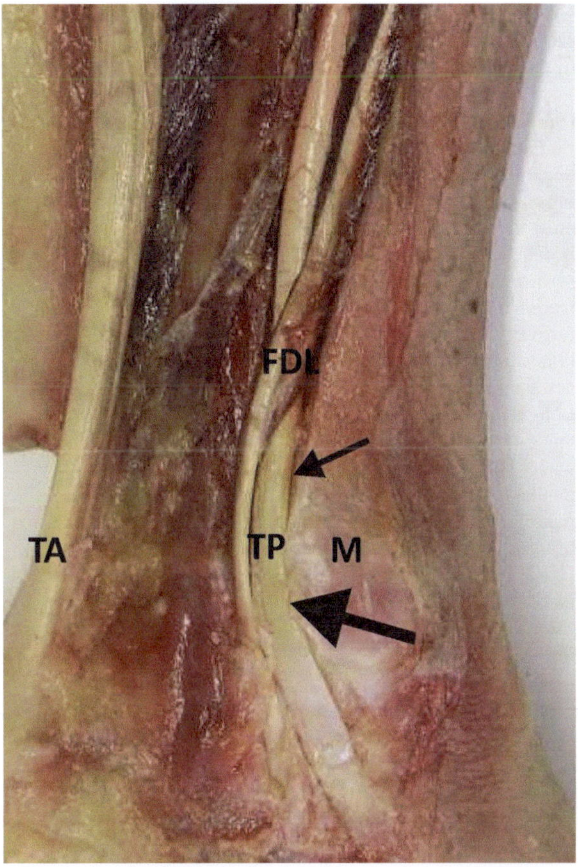

Fig. 7.6 The retromalleolar pulley sets up a "gliding zone" for the TP tendon. (From: Semple et al. [182] Figure 2 Gross anatomy of retromalleolar region. Gross anatomy of retromalleolar region indicating flexor digitorum longus tendon (FDL), tibialis posterior tendon (TP), medial malleolus (M), and tendo Achilles (TA). Small arrow indicates rounded TP tendon proximally, and large arrow indicates the flattened area of tendon in retromalleolar region)

There are three main components of the insertion of the TP tendon. Often overlooked, the multiple insertions of the TP tendon demonstrate the far-reaching influence of this muscle on foot function and stability (Fig. 7.8). The anterior component of the TP insertion is the largest and attaches to the tuberiosity of the navicular as well as the first (medial) cuneiform [68]. The middle component is most important functionally as it inserts on the second and third cuneiforms as well as the cuboid. This component appears to stabilize the transverse arch across the midfoot. Other slips of the middle component of the TP tendon insert on the base of second, third, fourth, and sometimes the fifth metatarsals [68]. The middle component also gives attachment to the medial limb of origin of the flexor hallucis brevis muscle. Finally, the third component is a "recurrent" slip of the TP tendon which inserts on the anterior aspect of the sustentaculum tali. Olewnik studied 80 cadaver specimens and found four different patterns of insertion of the TP tendon [71] (Fig. 7.9). Consistently, the TP tendon inserts on the navicular and medial cuneiform bones. The tendon did not always insert on the base of the metatarsals and was more likely to insert on the remainder of the cuneiform bones. Therefore, in some patients, the

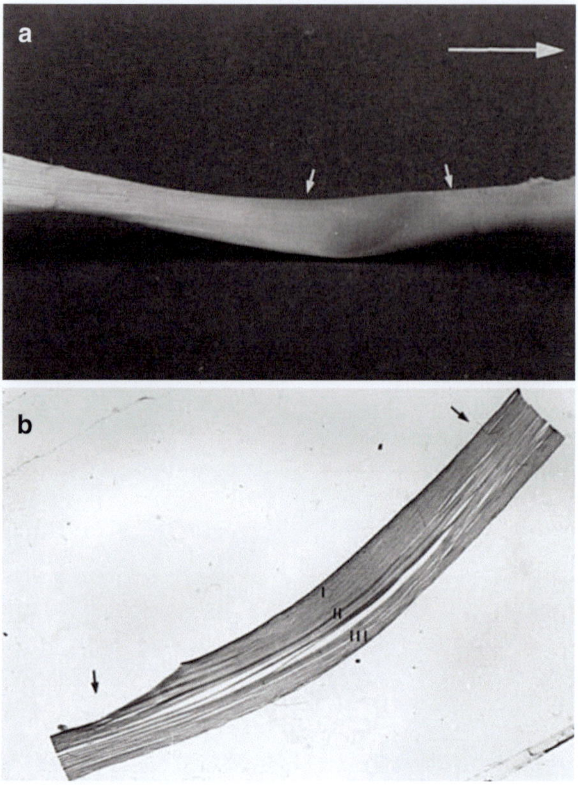

Fig. 7.7 The retromalleolar gliding zone of the TP tendon. (From: Petersen et al. [69]. (Springer Publication). Figure 1 (**a**) Posterior tibial tendon. In the retromalleolar region, the cross section of the tendon is oval, and the tendon surface directed toward the medial malleolus is smooth (arrows). The large arrow points toward the insertion at the navicular. (**b**) Longitudinal section of a tendon from the region where it wraps around the medial malleolus. In this region the tissue consists of three different zones (I, II, III). The small arrows mark the longitudinal extension of the fibrocartilaginous zone

TP tendon does extend beyond the cuneiforms with its insertion and therefore does not have a long lever arm to apply torque to the midtarsal joint. It is unknown if these patients are more prone to develop AAF deformity later in life.

The Talonavicular Joint

The talonavicular joint is the pivotal joint of the adult acquired flatfoot deformity. Radiographic studies show that as the adult acquired flatfoot progresses, the primary changes occur at the talonavicular joint in both the sagittal and transverse planes [72, 73] (Fig. 7.10). Surgical correction of the AAF deformity is most effective when, compared to all the other rearfoot joints, realignment of the talonavicular joint is performed [74].

The talonavicular joint is the most mobile of all the joints in the human foot [75]. Compared to the calcaneocuboid joint, the talonavicular joint lacks any type of intrinsic osseous features to provide anatomic "locking" for stability. Instead, the talonavicular joint relies on an intricate complex of ligaments for stability. MacConnail draws an analogy between the talonavicular joint and the hip joint. The hip joint is referred to as the "acetabulum coxae," while the talonavicular

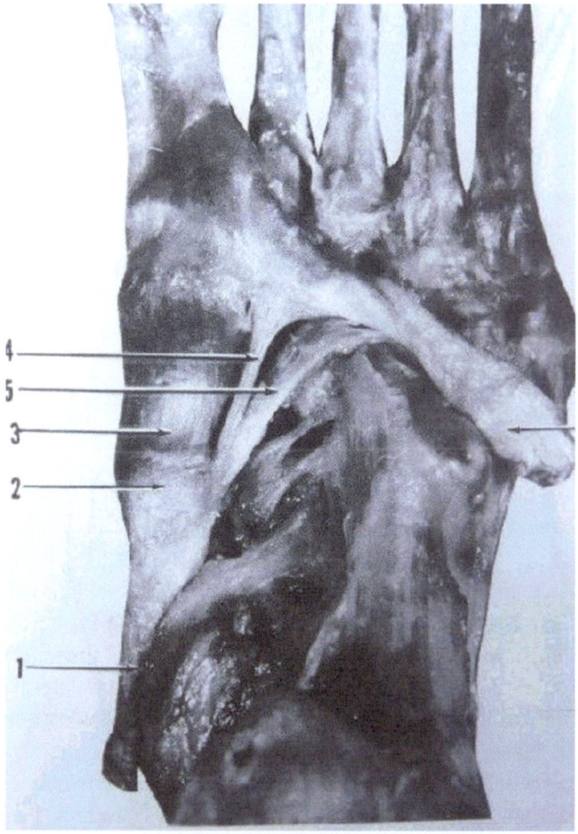

Fig. 7.8 From Kelikian and Saffafian [68]. The three zones of insertion of the TP tendon are anterior (4, 5), middle (2, 3) and recurrent (1)

joint was described as the "acetabulum pedis" [76]. Unlike the hip, however, MacConnail observed that the talonavicular joint was a hinged structure in the transverse plane, pivoting laterally at the medial sling of the bifurcate ligament or the calcaneonavicular ligament. He also supported the notion proposed by Wood-Jones that the talus is an intercalary bone, moving with the leg rather than with the bones of the foot [77]. MacConnail describes the unique anatomy and function of the talus:

> "The talus has seven articular surfaces. Three of these are within the ankle joint, one touches the navicular bone, one touches the dorsum of the calcaneus, and two articulate with the sustentaculum tali, each by its proper facet. This seven-fold articulation of the talus is of major importance. The talus has no muscles, and its ligaments come into action only as the result of external forces; it is, therefore, to be taken as a rigid body with six degrees of freedom: three axes of rotation, and three planes of translation."

The intercalary position of the talus and the envelope of articular cartilage around its body led Pisani to propose that this bone functions as a "meniscus" between the leg and the foot [78]. He coined the term "coxa pedis" to describe a similarity between the head and neck of the femur with the talus. In combination with the tibia, the body and

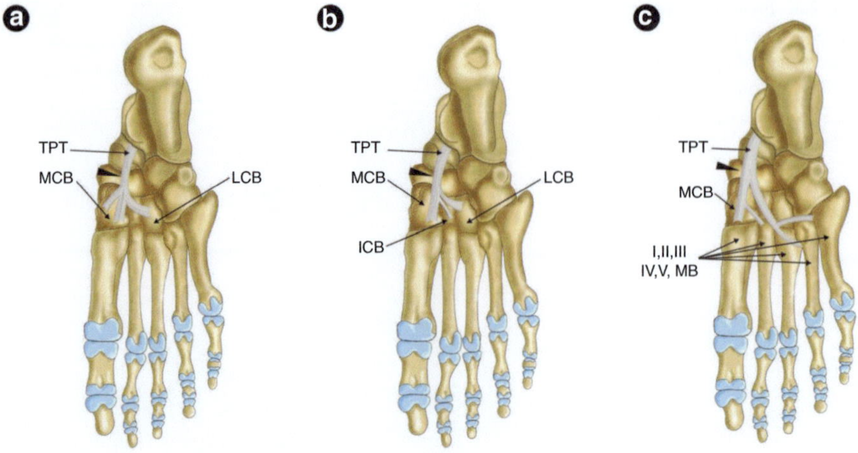

Fig. 7.9 Variation of insertion of the TP tendon. (From: Olewnik [71])
Legend: TPT tibialis posterior tendon, MCB medial cuneiform bone, LCB lateral cuneiform bone

neck and the talus resemble the femur and articulate with an acetabulum, the navicular, to form a ball and socket joint (Fig. 7.11). Pisani describes two extremes of foot deformity, both of which involve dislocation of the head of the talus from the acetabulum of the navicular. In club foot there is a dorsal-lateral dislocation from the acetabulum, while in vertical talus, there is a plantar dislocation [78].

Pisani also argues that the term "subtalar joint" should be abandoned as it suggests one single articulation. Instead, there are two distinct articulations on the inferior surface of the talus [78]. He reviews the fact that the two articulations of the talus appear at different times during embryonic development. The most important articulation involves the talus and the navicular as well as the anterior and middle facets of the calcaneus. This articulation, described by Pisani as the *talocalcaneonavicular joint*, develops as a distinct joint in the 16th embryonic week and remains anatomically separated from the posterior facet of the *talocalcaneal joint* (Fig. 7.12). Each joint has its own stabilizing ligaments and distinctly separate joint capsules [78].

Pisani also studied the critical role of the coxa pedis in the pathomechanics of the adult acquired flatfoot [79]. He describes AAF as a destabilization of the talocalcaneonavicular joint which begins with a degenerative process within the supportive ligaments of the joint itself. Specifically, Pisani describes the fibrocartilage within the superomedial calcaneal navicular (spring) ligament as a "glenoid structure" of the coxa pedis which is subject to considerable loads in the pes planus foot condition. In particular, the glenoid is subjected to compressive load applied laterally from the head of the talus and medially from the TP tendon. In pes planus, the talus protrudes medially, and the TP tendon is tensioned to resist this subluxation.

In a subsequent paper, Pisani reports on several surgical cases of "degenerative glenopathy" whereby the spring ligament was fully or partially ruptured while the

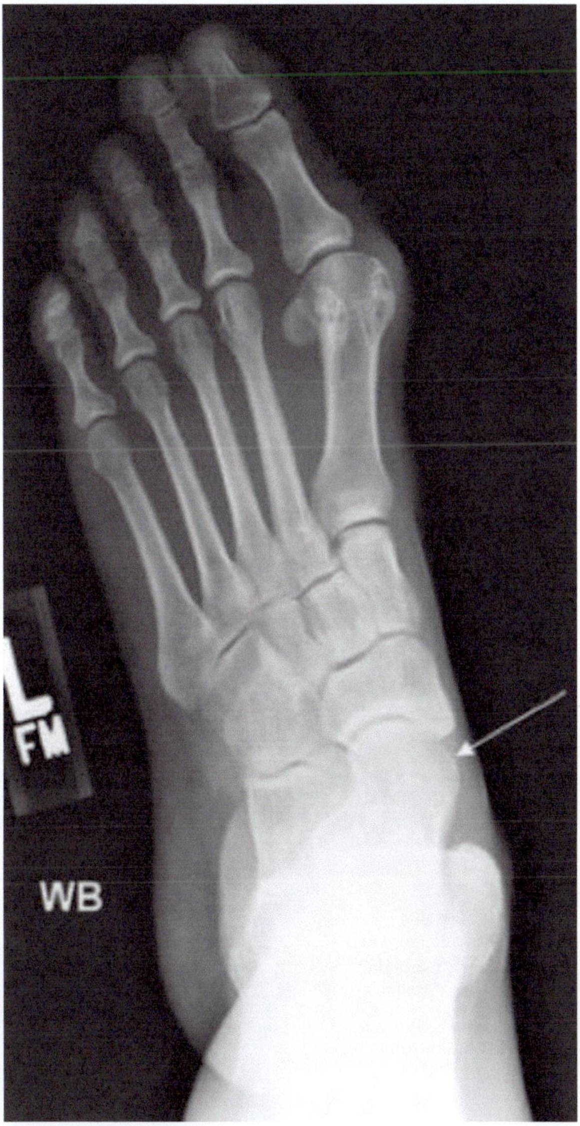

Fig. 7.10 Subluxation of the talonavicular joint in the transverse plane is the hallmark of the adult acquired flatfoot deformity

posterior tibial tendon was intact [80]. In each of these cases, there was presence of an accessory navicular bone which Pisani proposed contributes to excessive compression of the fibrocartilage of the spring ligament against the medially protruding head of the talus. Once the glenoid structure fails, the subluxation of the talocalcaneal navicular joint progresses and secondarily causes valgus rotation of the posteriorly located talocalcaneal joint. This entire process was labeled by Pisani as "peritalar destabilization syndrome" [80, 81].

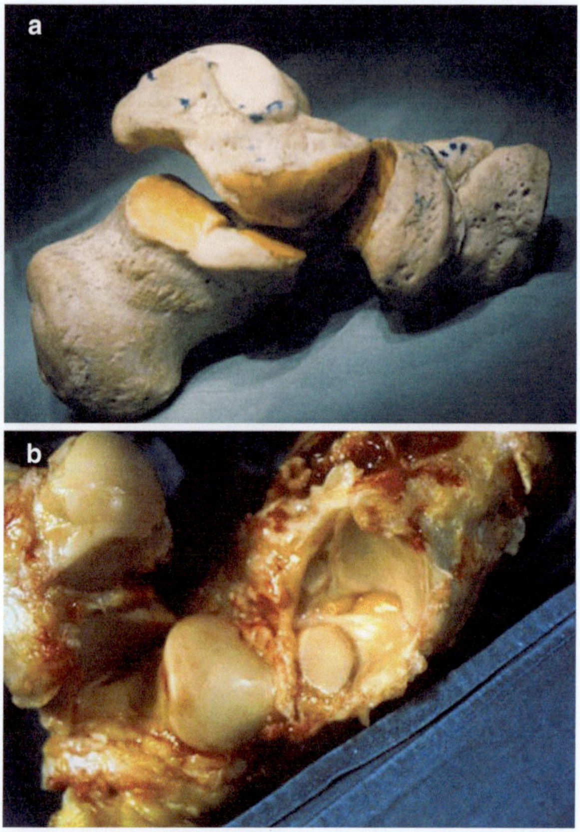

Fig. 7.11 The acetabulum of the talocalcaneonavicular joint. (From: Pisani [78]. (**a**) The talus, navicular, and middle facet of the calcaneus are one distinct articulation. The middle facet of the calcaneus is clearly part of the acetabulum pedis, while the posterior facet of the calcaneus is a separate joint) (**b**) Talus is removed to expose the surfaces of the articulations between the talus, navicular and calcaneus. The posterior facet of the calcaneus, forming the talocalcaneal joint, is clearly anatomically separated from the anterior and middle facets which are part of the talocalcaneonavicular joint

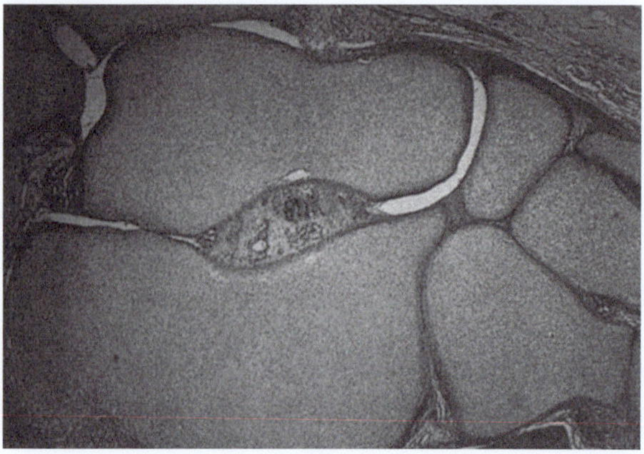

Fig. 7.12 12-week embryo: talocalcaneonavicular or anterior subtalar joint is clearly differentiated from posterior subtalar joint. (From Pisani [77])

Key Ligaments

The deep concavity of the talocalcanconavicular articulation is enveloped by a broad network of ligamentous and capsular structures to restrain the wide range of freedom provided by the anatomic configuration of this ball and socket joint (Fig. 7.13).

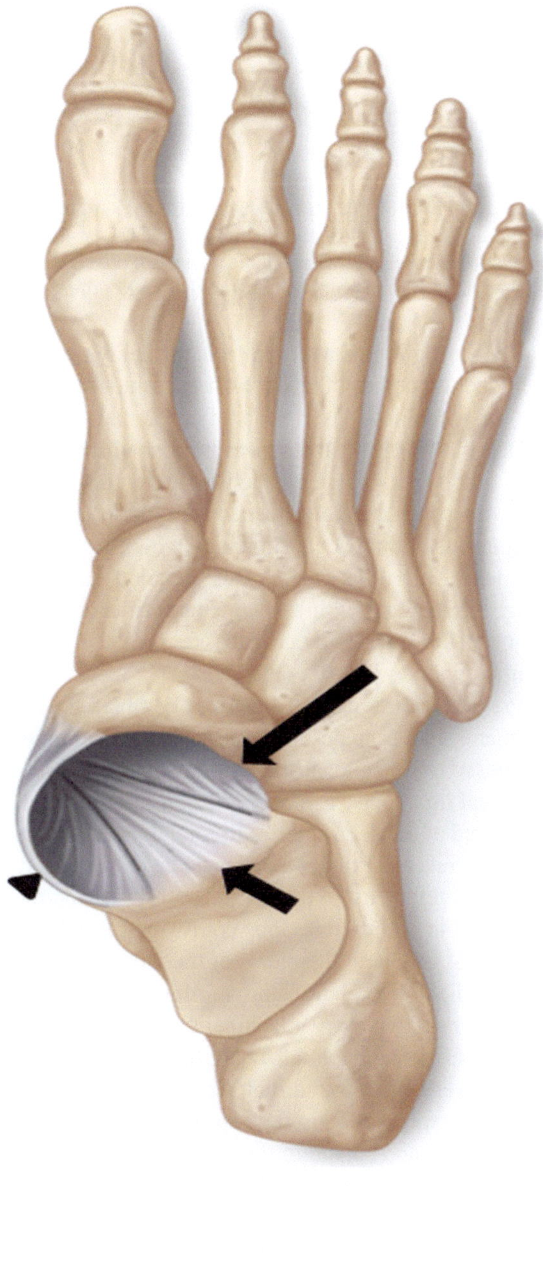

Fig. 7.13 From: Spring ligament complex: illustrated normal anatomy and spectrum of pathologies on 3T MR imaging. Omar et al. [83]. Figure 1 Illustration showing the normal anatomy of the SLC (with the talar head removed). Arrowhead = SM band, small arrow = MPO band, large arrow = IPL band)

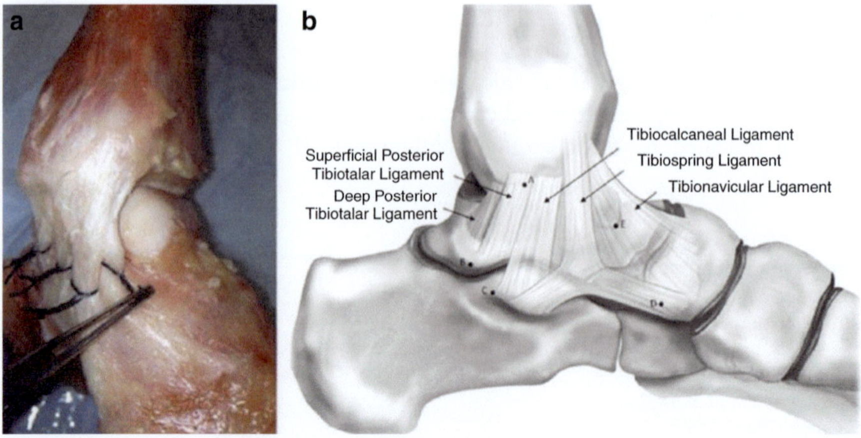

Fig. 7.14 From Figure 1 (**a**) Cadaveric dissection demonstrating an anteromedial view of the deltoid ligament complex with the tibionavicular ligament grasped. (**b**) Schematic diagram of the origins and attachments of the ligamentous components of the deltoid complex. (Source: Reprinted with permission from Campbell et al. [90]. and Reprinted in: Ormsby et al. [155])

The spring ligament complex is probably the most important anatomic structure of the adult acquired flatfoot as it is commonly ruptured in this condition [82, 83]. The first detailed description of an actual "complex" of ligaments collectively termed the spring ligament was published by Davis and co-workers in 1996 [84]. The complex described by Davis et al. included the inferior calcaneonavicular ligament, the superomedial calcaneonavicular ligament, the posterior tibial tendon, and the anterior tibionavicular ligament which is a component of the superficial deltoid ligament. Since that time, many detailed anatomic studies have been published which verify that multiple ligaments form a complex. Furthermore, one continuous joint capsule spans and encompasses both the spring ligament and the deltoid ligament complexes [85–92] (Fig. 7.14).

Previously, the deltoid ligament complex, also known as the medial collateral ligaments, had been described as having two components. The *superficial portion* is composed of the tibiospring, tibionavicular, superficial posterior tibiotalar, and talocalcaneal ligaments. The *deep portion* of the medial collateral ligament complex is composed of the deep anterior tibiotalar and deep posterior tibiotalar ligaments [88].

Cromeens et al. dissected nine cadavers and used a novel technique to isolate the individual ligaments of the deltoid and spring ligament complexes [91]. Their study attempted to address previous confusion regarding terminology and description of the medial collateral ligaments of the ankle, i.e., the deltoid ligament. Furthermore, they studied the relationship of the deltoid and the spring ligament complex. Dissection started at the attachment of each ligament, and then mapping was carried out with CT imaging and 3-D digitization (Fig. 7.15).

Cromeens et al. found that the combined medial collateral and spring ligament complexes were composed of multidirectional bands of tissue connecting four different bones and spanning three synovial joints. They designated five components to the two ligament complexes. These were named the tibiocalcaneonavicular,

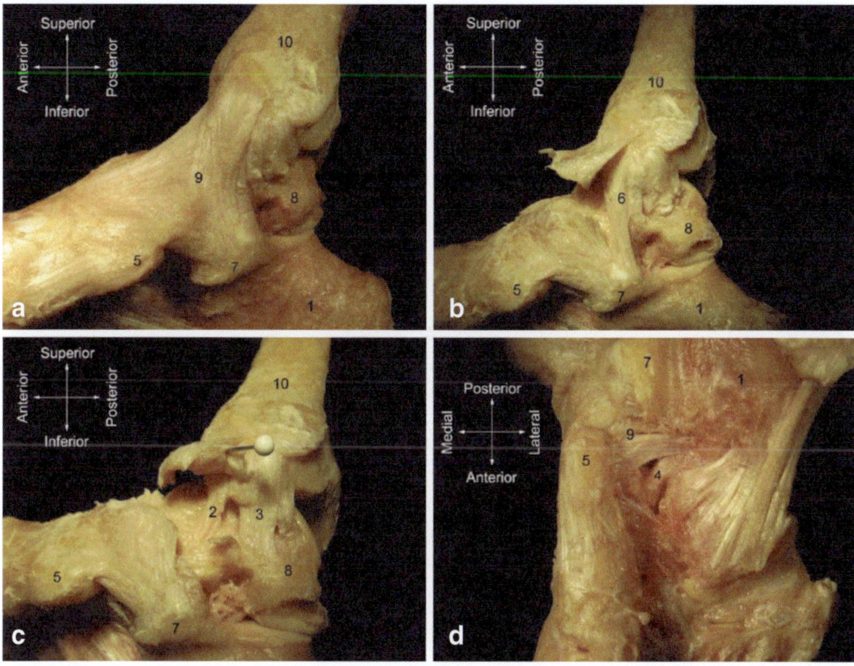

Fig. 7.15 (From: Cromeens et al. [91]. Fresh-frozen cadaveric dissections demonstrating the five components contributing to the medial collateral and spring ligament complexes. (**a**) Medial view of the hindfoot showing the tibiocalcaneonavicular ligament. (**b**) Medial view of the hindfoot with the tibiocalcaneonavicular ligament transected and reflected, showing the superficial posterior tibiotalar ligament. (**c**) Medial view of the hindfoot with the tibiocalcaneonavicular and superficial posterior tibiotalar ligaments transected and reflected, showing the deep anterior and posterior tibiotalar ligaments. (**d**) Plantar view of the hindfoot showing the tibiocalcaneonavicular and inferoplantar longitudinal ligaments. (1) Calcaneus; (2) deep anterior tibiotalar ligament; (3) deep posterior tibiotalar ligament; (4) inferoplantar longitudinal ligament; (5) navicular; (6) superficial posterior tibiotalar ligament; (7) sustentaculum tali; (8) talus; (9) tibiocalcaneonavicular ligament; (10) tibia)

superficial posterior tibiotalar, deep anterior tibiotalar, deep posterior tibiotalar, and inferoplantar longitudinal ligaments. These ligaments were interwoven into one continuous network extending from the medial malleolus to the navicular.

Of importance from the work of Cromeens et al. was the recognition that the two branches of the superficial deltoid ligament, i.e., tibiocalcaneal and tibionavicular, were intimately connected and part of the main component of the spring ligament, the superomedial calcaneonavicular ligament. They name the combined structure of all three ligaments the tibiocalcaneonavicular (TCN) ligament. The articulation of this ligament with the head of the talus has cartilage tissue forming a "glenoid" structure as observed by Pisani [80]. The TCN ligament encompasses multiple ligaments to form a continuous band of tissue which surrounds the head of the talus and maintains a proximal stabilization of the navicular against the calcaneus. The blending of ligaments around the head of the talus, the sustentaculum, and the navicular verifies the distinct anatomical and functional articulation named the

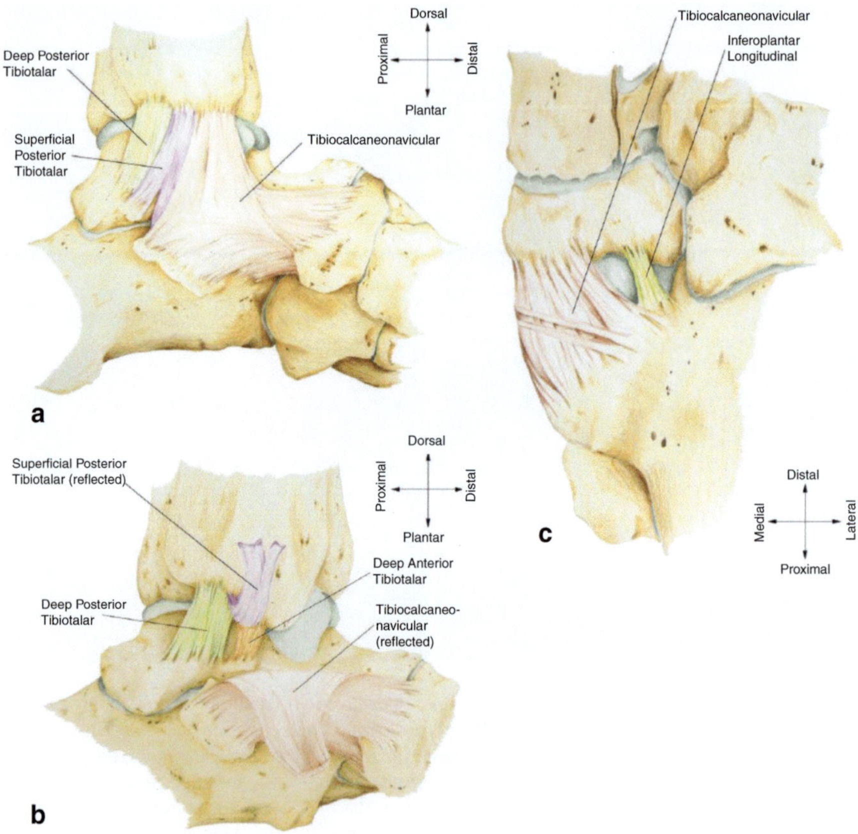

Fig. 7.16 Caption (From: Cromeens et al. [91]) The tibiocalcaneonavicular ligament combines the superficial deltoid with the spring ligaments

talocalcaneonavicular joint by Pisani [78]. It is now clear that this articulation is separate from the posterior facet of the calcaneus which makes up the talocalcaneal joint. Hence the term "subtalar joint" is misleading as it suggests one articulation, whereas the talus actually has two distinctly different articulations on its inferior surface. Comparing the motion which occurs at the talocalcaneonavicular joint versus the talocalcaneal joint as shown in Chap. 3, it is clear that these two articulations affect function of the foot in very different ways.

Amaha verified the observations of Cromeens et al. that the medial collateral ligaments of the ankle blend with and become part of the spring ligament complex [92]. Furthermore, Amaha et al. showed that the capsule of the ankle, subtalar, and talonavicular joints could be detached as one continuous sheet and the key ligaments could not be easily separated from this capsule (Figs. 7.16 and 7.17). The cartilaginous tissue within the capsule articulated with the head of the talus and provides an important connection between the navicular and the calcaneus. The authors also showed that the capsule of the combined ankle and talonavicular joints joins with the talocalcaneonavicular ligament to form the floor of the TP tendon sheath. They propose a close association between all these structures and the adult

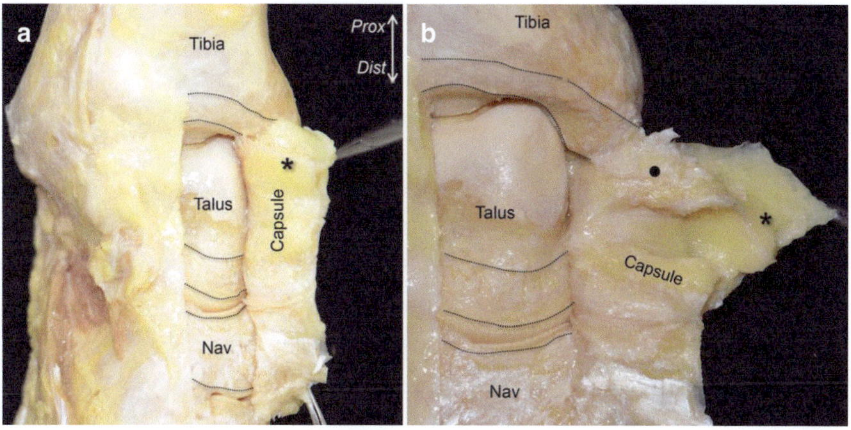

Fig. 7.17 Caption (From: Amaha et al. [92]) The capsule of the ankle, subtalar and talonavicular joints are one continuous structure

acquired flatfoot deformity. This verifies a long accepted notion that the TP tendon actually reinforces and protects the spring ligament complex from excessive loads [38, 41, 84].

Muscle Function and the Adult Acquired Flatfoot

Traditionally, the tibialis posterior has been called the "strongest invertor of the foot" [35, 51, 52, 93]. One point of confusion is the common practice of interchangeably using the terms "foot" motion and subtalar joint motion. The belief that the tibialis posterior is the strongest invertor of the foot has been based upon the location of the proximal portion of the TP tendon relative to the axis of the subtalar joint which passes thru the head and neck of the talus [94].

The location of the TP tendon in the transverse plane, relative to the axis of the subtalar joint, places it most medially located of all the tendons of the foot (Fig. 7.18). Based in this simple evaluation of location of the TP tendon in one plane, it could be concluded that it has the longest lever arm for inversion of the subtalar joint compared to all other tendons of the lower leg. This notion was supported by Hinterman et al. in an in vitro cadaver study of ankle and subtalar joint moment arms [95]. In this study, the TP tendon was given a value of 1.0 inversion moment arm, while the FDL scored 0.75 and the tibialis anterior (TA) scored 0.59. Klein showed similar dominance of inversion moment arm at the subtalar joint of the TP (19.2 mm) compared to the tibialis anterior (3.8 mm) [96]. It should be noted that the inversion lever arm of the TP and TA are shorter as the foot is moved into a pronated position [95] (Fig. 7.18).

Hinterman also showed that as the ankle moves into plantarflexion, there is greater increase in plantarflexion moment arm for the gastrocnemius and soleus [95]. As the ankle moves into dorsiflexion, the moment arm of all the invertors of the ankle decreases. This relates to the position of the insertion of the Achilles on the calcaneus and the perpendicular distance from the ankle joint axis. This study showed that as the foot moves into eversion, moment arm of the invertors,

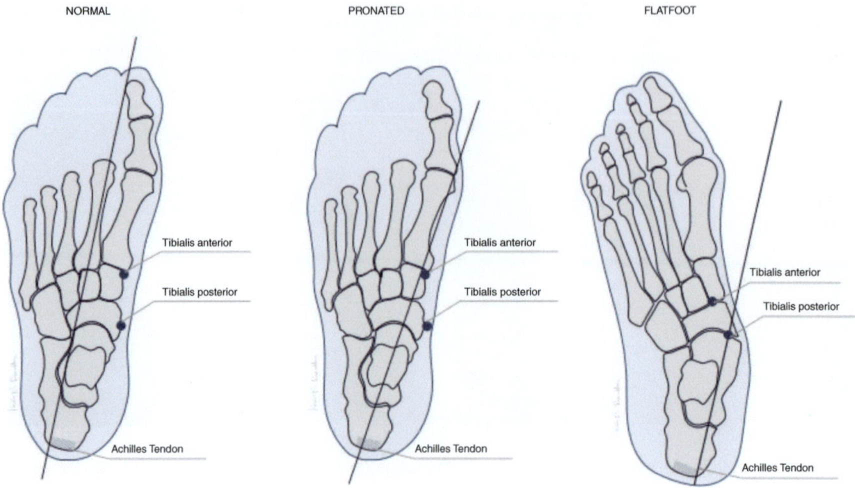

Fig. 7.18 Loss of inversion lever arm for TA, TP, and Achilles as pronation and flatfoot deformity progress. The TA and Achilles eventually become evertors rather than invertors

particularly the tibialis posterior, decreases. As Fig. 7.18 demonstrates, with progressive change of alignment of the foot from neutral to AAF deformity, the lever arm for inversion provided by the TP, TA, and Achilles becomes significantly compromised. In particular, the TA and Achilles actually become evertors of the foot as deformity progresses.

Looking at muscle moment arm isolated only to the subtalar joint is short sighted when studying adult acquired flatfoot deformity. Radiographic studies of patients with AAF demonstrate that the primary deformity occurs at the talonavicular joint [72, 73]. Gait studies of patients in stage II AAF show that excessive motion is located in the midfoot joints and not the subtalar joint [97–101]. These studies also showed that motion in the transverse plane, not the frontal plane, dominates the AAF deformity.

A more relevant study of muscle moment arms in regard to the AAF deformity was conducted by McCullough and co-workers [102]. Although not intended to simulate the AAF deformity, this cadaver study did allow unconstrained motion across the rearfoot and midfoot joints and also measured moment arm in the transverse plane. The studies conducted by Hinterman and Klein isolated motion to the subtalar joint only. Using a simulator with more modern methods of measuring tendon moment arms, McCullough et al. determined that the longest, most effective moment arm for *total foot inversion* is provided by the tibialis anterior (16 mm) which was 30% longer than the tibialis posterior (10 mm). The most effective moment arm provided by the tibialis posterior is in the transverse plane (*adduction/abduction*) where the moment arm measures 21.4 mm.

Surprisingly, McCullough et al. showed that the FDL and tibialis posterior had identical total foot *inversion* moment arms of 10 mm which were 30% lower than the tibialis anterior. The FDL tendon is closely aligned to the posterior tibial tendon as they pass along the medial aspect of the calcaneus. Pisani points out the unique position of both the FDL and FHL beneath the sustentaculum which are suited to resist frontal plane eversion of the calcaneus even though neither tendon inserts on this bone [80].

Pisani points out the futility of performing an FDL tendon transfer as part of surgical correction of the AAF deformity [80]. This notion was later verified in a study by Hui and co-workers [103]. They measured the inversion motor arm of the intact, native flexor digitorum longus (FDL) and found that it was reduced or shortened by 45% when the tendon was transferred to the navicular. Furthermore, Hui factored the anticipated loss of one grade of strength with any tendon transfer, which amounts to a loss of 20% total force, and then calculated overall loss of moment or torque. Reducing the total force of the transferred FDL by 20% and then calculating moment arm, Hui et al. show that the intact FDL inversion capacity drops from 3.5 Newton-meters to 1.5 Newton-meters when transferred to the navicular [103]. Therefore, the FDL tendon transfer, a popular surgical procedure for correcting the adult acquired flatfoot, results in a *loss of over 60% inversion moment* (torque) for the entire foot.

This study by Hui illustrates that moment arm is only one component of overall torque-producing capacity of a muscle across a joint. In Chap. 3, cross-sectional area of muscles of the lower extremity was used to make an estimate of maximal isometric force. Multiplying this force by the moment arm, estimated by McCullough et al., the total moment or torque in Newton-meters was calculated for various muscles in the lower extremity. From these calculations, it was demonstrated that for inversion of the foot, the tibialis posterior is capable of generating 7.5 Newton-meters (Nm) of torque while the tibialis anterior can generate 5.8 Nm. For adduction, the TP generates 15.7 Nm, while the FDL generates 3.8 Nm (see Fig. 7.15, Chap. 3).

Clearly, the torque-producing activity of the tibialis posterior is twice as large in the transverse plane as the frontal plane. If lost, the TP can be supported in the direction of inversion by another muscle with similar torque-producing capacity: the tibialis anterior. However, in the transverse plane, the tibialis anterior is unable to compensate. Supplemental or replacement adduction torque produced by the FDL achieves only one fourth the power of the tibialis posterior. This finding explains why the AAF deformity is most dominant in the transverse plane rather than the frontal plane [97–101].

Studies show evidence of overactivity and underactivity of certain muscle groups in patients with pes planus, with or without PTTD. In a study of 12 patients undergoing surgery for partial or complete rupture of the TP tendon, Wacker et al. used MRI to measure significant hypertrophy of the FDL muscle compared to the unaffected leg [104]. This may not be the result of PTTD as ultrasound study of patients with pes planus compared to normal arch feed also demonstrates hypertrophy of the FDL muscle as well as the FHL [87]. Conversely, this same study showed decreased cross-sectional area of the peroneal musculature in patients with pes planus [105].

EMG studies of muscular activation mirror the findings of muscle volume changes relative to foot posture. Increased EMG activity of the ankle inverters has been shown in patients with pes planus compared to control patients with normal foot posture [106, 107]. Decreased EMG activation of the peroneus longus and brevis has been measured in patients with pes planus compared to normal arched individuals [106–108]. These findings of change in magnitude of muscle activation suggest a compensation for abnormal foot posture to improve alignment and offset progression of deformity.

Pathoanatomy

The posterior tibial tendon inserts on the navicular via an enthesis or fibrocartilage attachment. An enthesis has four zones starting with the tendon itself which blends into fibrocartilage and calcified fibrocartilage and then becomes part of the underlying bone [109]. The enthesis of any tendon insertion is designed to dissipate stress at the anchorage point to bone which result from tensile, compressive, shear, and bending forces. Studies have shown that immunocytes can be activated and attach to receptors in the enthesis located at these high points of stress [110, 111]. Hence, there is evidence linking posterior tibial tendinopathy with rheumatoid arthritis as well as certain seronegative inflammatory arthropathies such as ankylosing spondylitis, ulcerative colitis, and psoriatic arthritis [43–45, 112].

The enthesis of the posterior tibial tendon has been studied in detail by Moriggl and colleagues [113]. They identified two locations of fibrocartilage: one within the enthesis itself and one more proximal and imbedded within the TP tendon itself. This second location was closely opposed to the superomedial calcaneonavicular (spring) ligament. Both locations of fibrocartilage were proposed by the authors to protect the tendon from friction or bending forces.

Moriggl et al. found that the proximal section of fibrocartilage at the insertion of the TP tendon may differentiate into bone, forming an accessory navicular [113]. The presence of an accessory navicular has been associated with AAF [114, 115]. Pisani reported on several surgical cases of AAF where an accessory navicular bone was attributed to cause degenerative changes in the adjacent spring ligament [81].

The presence of fibrocartilage in the distal portion of the posterior tibial tendon creates a hypovascular zone which appears vulnerable to degeneration and rupture. Since the early reports in the 1980s, the area behind the medial malleolus was observed to be the most common site of rupture of the posterior tibial tendon [28, 30, 35, 36]. The TP tendon changes direction at the distal margin of the medial malleolus and demonstrates a distinct histologic appearance due to the physical forces which result from the bony pulley action at this location. The fibrocartilage on the anterior surface of the TP tendon in the retromalleolar region provides a gliding surface which protects against compression and shear stress [69, 70]. This fibrocartilage has been speculated to also interrupt the normal blood flow to the PT tendon [116] (Fig. 7.19).

Vascular compromise has been speculated to be a primary cause of degeneration and non-traumatic rupture of the posterior tibial tendon [27–30]. However studies of perfusion of the posterior tibial tendon have shown conflicting results. Several investigations have identified a so-called hypovascular zone of the TP tendon

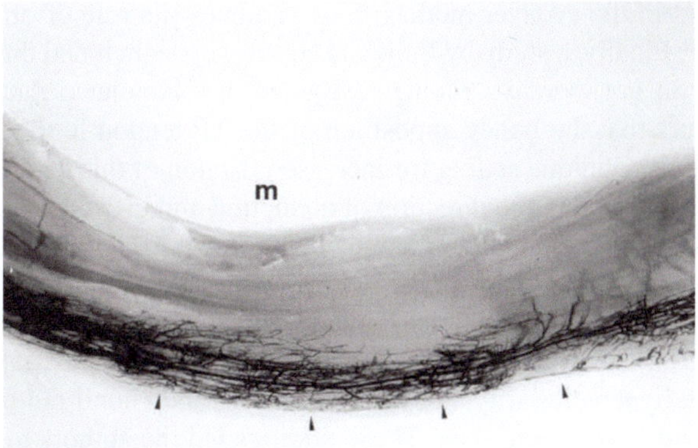

Fig. 7.19 The retromalleolar section of the TP tendon. (From: Petersen et al. [116]. Figure 2 Injection of Indian ink into a posterior tibial tendon of a 65-year-old man after clearing of the tissue shows that the terminal endings of the peroneal artery enter the tendon via a mesotenon (arrowheads) from the posterior aspect. In the region where the tendon turns around the medial malleolus (m), the well-vascularized paratenon is absent, and the distribution of vessels in the tendon is inhomogeneous. The posterior part of the tendon has a complete vascular network. The anterior part which is directed toward the medial malleolus is avascular. Detail is taken from the region where the tendon passes around the medial malleolus)

beginning in the retromalleolar region and extending 1.4 cm distal to the tip of the medial malleolus [117, 118]. This finding was verified by Peterson et al. who used immunohistochemical staining and identified an avascular region in the anterior portion of the posterior tibial tendon where it passes behind the medial malleolus [116]. However, Prado and co-workers conducted their own study of perfusion of the PT tendon and concluded there was no zone of hypovascularity, and the vulnerability for rupture of this tendon in the retromalleolar region can be attributed to mechanical stresses which arise from a flatfoot deformity [119].

The retromalleolar region renders the posterior tibial tendon susceptible to degeneration and rupture for another reason. Apposition of a tendon against bone creates a "gliding zone" which has increased compression and friction forces [120, 121]. These forces develop from the gliding of the tendon against bone. Uchiyama and colleagues studied gliding resistance of the posterior tibial tendon in cadaver specimens [122]. Dorsiflexion of the ankle increased gliding resistance of the posterior tibial tendon, while plantarflexion decreased resistance. The investigators then created a flatfoot deformity by cutting key ligaments in the arch of the cadaver specimens and observed a 30% increase in friction or gliding resistance in the posterior tibial tendon. The authors concluded that a flatfoot deformity increased the risk of degenerative changes in the posterior tibial tendon due to creation of increased friction in the retromalleolar gliding zone.

A subsequent study by Arai and co-workers showed that motion of the foot in the frontal and transverse planes, rather than the sagittal plane, increased friction and gliding resistance of the posterior tibial tendon in the retromalleolar region [123]. This gliding resistance was increased with creation of a flatfoot deformity and increasing load in the cadaver models. This validates the role of obesity as a risk factor for AAF. Finally, a study by Fujii and co-workers concluded that the FDL and FHL tendons do not increase gliding resistance in a simulated flatfoot condition [124]. It appears that the boney apposition of the TP tendon in the retromalleolar groove provides the unique setting for increased friction of this tendon, particularly when the foot is moved in the direction of abduction and eversion.

Degeneration rather than inflammation appears to be the hallmark of the histopathology of posterior tibial tendon dysfunction. Mosier and colleagues studied surgical specimens of the posterior tibial tendon in 15 patients who had stage II AAF deformity [125]. Four types of histologic changes were found consistently: increased mucin content, hypercellularity, neovascularization, and chondroid metaplasia. No inflammatory infiltrates were seen. These findings led the authors to conclude that the pathology of the posterior tibial tendon represented a tendinosis, not a tendonitis, in stage II disease. Furthermore, the histologic findings showed marked disruption of orientation of collagen bundles, rendering the posterior tibial tendon vulnerable to rupture.

The authors could not determine if these findings were a cause or result of progressive flatfoot deformity.

Fowble and co-workers had similar findings when they studied the histopathology of patients with stage II and III PTTD [126]. The specimens taken from the damaged TP tendon showed mucoid change and tenocyte cellularity with chondrometaplasia. As with the Mosier study, Fowble et al. found no evidence of inflammation in the samples. Also, there was no significant difference in the histopathologic findings when comparing stage II and stage III patients.

Other studies have confirmed that as the posterior tibial tendon changes in PTTD, there is a loss of predominance of type I collagen normally seen in healthy tendons. Specimens taken from patients with PTTD show a reduction of type I collagen (40%) and increase in type V collagen (26%) and type III collagen (53%) [127, 128]. These findings are consistent with a progression toward degeneration and eventual rupture of a tendon.

Pathomechanics and Staging of Deformity

It was recognized in early reports that PTTD was a progressive pathology. In 1989 Johnson and Strom published a three-stage classification system of posterior tibial tendon dysfunction based upon condition of the posterior tibial tendon, position of the hindfoot, and flexibility of the deformity [36]. After the Johnson and Strom classification was published, considerable insight into the condition known as PTTD was gained from research over a 20-year period leading many to change the name of the condition to adult acquired flatfoot (AAF). This was primarily due to the recognition that AAF was not just the result of failure of the posterior tibial tendon. A shortcoming of the Johnson and Strom classification was attributing the condition solely to failure of the TP tendon. Furthermore this classification was based on subjective criteria and was never tested for validity and reliability. Inexplicably, the

Johnson and Strom classification became the gold standard for staging the AAF and has remained so over a 30-year period despite the publication of several more appropriate systems [39, 129–132].

Newer classification systems for the adult acquired flatfoot recognized that the deformity progresses to multiple joints, not just the hindfoot. Both the Bluman and Raikin classifications evaluate changes in the midfoot joints and the joints of the medial column which are the result of progressive ligament failure [131, 132]. These classification systems were designed to suggest treatment options for each level of deformity even though the reliability of the system to predict appropriate treatment has never been tested. Furthermore, all current systems for classification of the AAF still primarily rely on subjective criteria which lead to considerable variation of interpretation by the practitioner.

The classification system proposed by Bluman, Title, and Myerson will be used in this chapter to describe the pathomechanics of the adult acquired flatfoot [131] (Table 7.1). It is recognized that this system is based on the previous work of Myerson who based his classification upon the original work of Johnson and Strom [36, 39]. Thus, some of the shortcomings of that original system have been carried forward and are repeated in the newer system. Furthermore, the Bluman system is focused primarily on surgical treatment of the AAF from stage II onward, overlooking documented evidence of the efficacy of non-surgical interventions [133].

Notwithstanding, the Bluman system represents the best available system today to classify the stages of the AAF deformity.

Stage I Tenosynovitis of the TP Tendon Without Deformity

The description of stage I AAF by Bluman as well as Johnson and Strom clearly designates that there is no "deformity" in the affected foot. This contradicts the long-standing belief that AAF starts with a pre-existing flatfoot deformity [34, 35, 52]. Yet, Johnson and Strom, Bluman et al., and Raikin all propose a stage I level of AAF which has no evidence of flatfoot deformity. Instead, they propose that the AAF begins with tenosynovitis of the posterior tibial tendon in an otherwise structurally sound foot. Bluman et al. subdivide stage I, recognizing that some patients with tenosynovitis have underlying inflammatory joint disease such as rheumatoid arthritis or seronegative spondyloarthropathy and should be recognized as a separate group. Another subgroup of stage I has tenosynovitis of the posterior tibial tendon from unknown cause and have "normal hindfoot anatomy." Finally stage I-C has tenosynovitis with "mild hindfoot valgus" of 5 degrees or less. Bluman et al. recognize that this stage has a partial rupture of the TP tendon, yet the patient can still perform a single foot heel raise.

Much of the evidence for the cause-effect relationship of mechanical overload causing degeneration of the posterior tibial tendon is based upon a pre-existing pes planus foot deformity. Specifically, patients with pes planus have greater activation of the hindfoot inverters than patients with rectus foot alignment [106, 107]. Also, studies show that a flatfoot deformity experimentally produced in cadaver models

Table 7.1 Bluman classification of adult acquired flatfoot [131]

Classification of posterior tibial tendon rupture				
Stage	Substage	Most characteristic clinical findings	Most characteristic radiographic findings	Treatment
I	A	Normal anatomy	Normal	Immobilization, NSAIDs, cryotherapy, orthoses, tenosynovectomy
		Tenderness along PTT		
				± Systemic disease-specific pharmacotherapy
	B	Normal anatomy	Normal	Immobilization, NSAIDs, cryotherapy, orthoses Tenosynovectomy
		Tenderness along PTT		
	C	Slight HF valgus	Slight HF valgus	Immobilization, NSAIDs, cryotherapy, orthoses, tenosynovectomy
		Tenderness along PTT		
II	A1	Supple HF valgus	HF valgus	Orthoses
		Flexible forefoot varus	Meary's line disruption	Med. displ. calc. osteot.
		Possible pain along PTT	Loss of calcaneal pitch	TAL or Strayer and FDL transf. if deformity corrects only with ankle plantar flexion
	A2	Supple HF valgus	HF valgus	Orthoses
		Fixed forefoot varus	Meary's line disruption	Med. displ. calc. osteot. and FDL transf.
		Possible pain along PTT	Loss of calcaneal pitch	Cotton osteoctomy
	B	Supple HF valgus	HF valgus	Orthoses
		Forefoot abduction	Talonavicular uncovering	Med. displ. calc. osteot. and FDL transf.
			Forefoot abduction	Lateral column lengthening
	C	Supple HF valgus	HF valgus	Med. displ. calc. osteot. and FDL transf.
		Fixed forefoot varus	First TMT plantar gapping	Cotton osteot. or medial column fusion
		Medial column instability		
		First ray dorsiflexion with HF correction		
		Sinus tarsi pain		
III	A	Rigid HF valgus	Subtalar joint space loss	Custom bracing if not surgical candidate
		Pain in sinus tarsi	HF valgus	Triple arthrodesis

Table 7.1 (continued)

Classification of posterior tibial tendon rupture				
Stage	Substage	Most characteristic clinical findings	Most characteristic radiographic findings	Treatment
			Angle of Gissane sclerosis	
	B	Rigid HF valgus	Subtalar joint space loss	Custom bracing if not surgical candidate
		Forefoot abduction	HF valgus	Triple arthrodesis ± lateral column lengthening
		Pain in sinus tarsi	Angle of Gissane sclerosis	
			Forefoot abduction	
IV	A	Supple tibiotalar valgus	Tibiotalar valgus	Surgery for HF valgus and associated deformity
			HF valgus	Deltoid reconstruction
	B	Rigid tibiotalar valgus	Tibiotalar valgus	TTC fusion or pantalar fusion
			HF valgus	

Abbreviations: HF, hind foot; FDL transf., flexor digitorum longus transfer; med. displ. calc. osteot., medial displacement calcaneal osteotomy; NSAID, nonsteroidal anti-inflammatory drugs; PTT, posterior tibial tendon; TAL, tendo Achilles lengthening; TMT, tarsometatarsal joint; TTC, tibiotalocalcaneal.

significantly increases friction force on the TP tendon at the retromalleolar pulley [122, 123]. The classification systems proposed by Johnson and Strom, Bluman et al., and Raikin do not evaluate overall foot posture or arch profile in stage I, instead focusing only on rearfoot alignment in the frontal plane. Significant valgus deformity is not noted in the hindfoot until stage II, according to all classification systems, and this change will not occur without spring ligament disruption [134, 135].

Tenosynovitis of the TP tendon can occur from overuse, as seen in other tendons, without contributing mechanical factors of foot deformity [136, 137]. Tendinitis and tendinopathy can also occur from other risk factors including obesity, hyperlipidemia, and diabetes which are all related to circulating inflammatory markers seen in metabolic syndrome [30, 49, 50, 54–57]. Finally, genetic factors have been identified as a risk factor for tendinopathy [63, 64].

It is unclear which patients, untreated in stage I AAF with tenosynovitis, will progress to stage II which is characterized by rupture of the TP tendon and sequential rupture of key ligaments in the hindfoot. As discussed previously, a significant percentage of patients with flexible congenital flatfoot deformity remain asymptomatic all of their life and never undergo progression of deformity. Multiple factors must contribute to the etiology and progression of adult acquired flatfoot deformity besides the intrinsic risk factor of a pre-existing flatfoot.

Stage II: Rupture of Posterior Tibial Tendon with Flexible Flatfoot Deformity

The diagnosis of rupture of the TP tendon as originally proposed by Johnson and Strom and later supported by Bluman et al. is based upon physical examination. Specifically, the patient demonstrates "weakness with inversion" of the plantarflexed foot and is unable to perform a single foot heel rise. Also, a flatfoot deformity is now "clinically apparent."

What is missing from these classification systems is a recognition that AAF occurs unilateral over 80% of the time, making the unaffected foot an ideal comparison for weakness and deformity [138]. Therefore, the hallmark of stage II is asymmetry of flatfoot deformity, with the symptomatic side showing key clinical findings which differ from the asymptomatic side (Fig. 7.20).

An erroneous assumption made by Johnson and Strom and then carried forward by Bluman et al. is that the single foot heel rise test is diagnostic of rupture of the posterior tibial tendon. Studies have shown that patients who have lost function of their posterior tibial tendon due to surgical transfer can still perform a single foot heel rise [139–141]. The single foot heel rise requires stiff, stable midfoot joints with intact ligaments. With rupture of the spring ligament, excessive sagittal plane motion occurs at the talonavicular joint compromising the lever arm of the triceps surae to plantarflex the entire foot as a rigid body at the metatarsophalangeal joints. Thus, rupture of the spring ligament causes difficulty or inability to perform the single foot heel rise (Fig. 7.21).

The second error in these classification systems is attributing the structural changes seen in stage II to rupture of the posterior tibial tendon. Many studies published after the Johnson and Strom classification verified that isolated loss of the posterior tibial tendon will not cause a flatfoot deformity until multiple ligaments subsequently rupture [142–145]. It is recognized that spring ligament failure, rather than posterior tibial tendon failure, is the primary event leading to progressive adult acquired flatfoot deformity [83, 146–148].

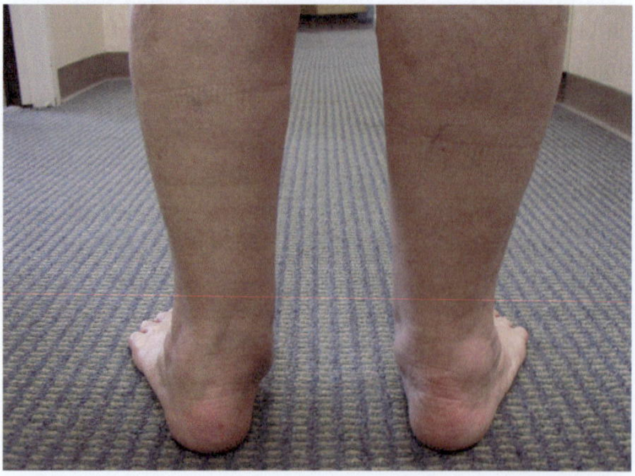

Fig. 7.20 Asymmetry with left foot demonstrating increased valgus alignment of the rearfoot and greater forefoot abduction is the hallmark of stage II adult acquired flatfoot

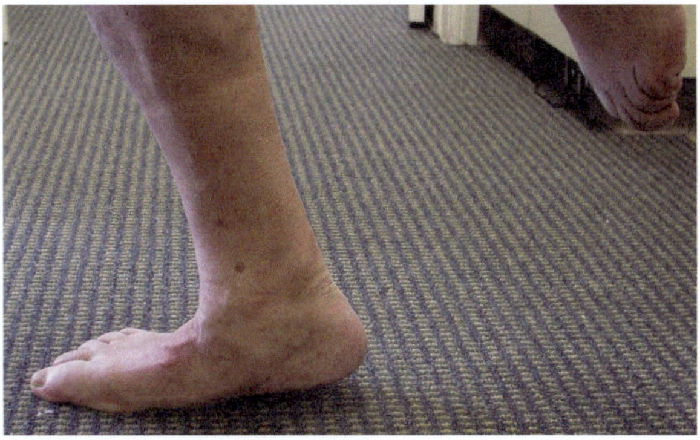

Fig. 7.21 Inability to complete the single foot heel rise is due to spring ligament deficiency causing plantarflexion moment of the triceps surae to be exerted across the unstable midfoot joints instead of at the metatarsophalangeal joints

In a landmark study, Jennings and Christensen identified the primary role of the spring ligament in AAF deformity. They sectioned the spring ligament in five cadaveric specimens and applied axial load to the tibia while also loading the posterior tibial tendon [148]. When the spring ligament was sectioned, a significant change of alignment associated with flatfoot deformity occurred which the posterior tibial tendon was unable to compensate.

These findings were further validated in a more recent study using finite analysis computer modeling to measure stress in key soft tissue structures in stage II AAF deformity [149]. The plantar aponeurosis and the spring ligament were identified to be the two primary structures which maintained integrity of the medial longitudinal arch, while the TP tendon played a secondary role. Furthermore, the plantar aponeurosis shows a primary role in preventing arch elongation. The spring ligament was found to have a primary function of controlling motion at the talonavicular joint and subtalar joint.

MacDonald and co-workers evaluated the effects of progressive sectioning of the spring ligament in nine cadaver specimens placed under physiologic loading. Motion capture demonstrated significant changes in alignment of the hindfoot and ankle with isolated spring ligament sectioning [150]. The talus was noted to undergo significant adduction and plantarflexion. The tibiotalar and subtalar joints underwent significant valgus rotation with spring ligament sectioning showing the global role of the spring ligament in providing stability to the ankle and hindfoot.

Rupture or deficiency of the spring ligament complex is the key event leading to stage II AAF deformity (Figs. 7.22 and 7.23). Deland studied 31 patients with documented tear or degeneration of the TP tendon [82]. The superomedial calcaneonavicular ligament was ruptured in 87% of the patients, while the interosseous talocalcaneal ligament was ruptured in 48% (Fig. 7.24). Of importance, the plantar fascia was intact in all these patients. This finding supported the previous work of Balen et al. who found spring ligament pathology in 92% of patients with TP tendon pathology but no evidence of plantar fascia pathology [151]. Williams found that spring ligament pathology was as frequent as posterior tibial tendon pathology in their MRI study of patients with AAF deformity [152].

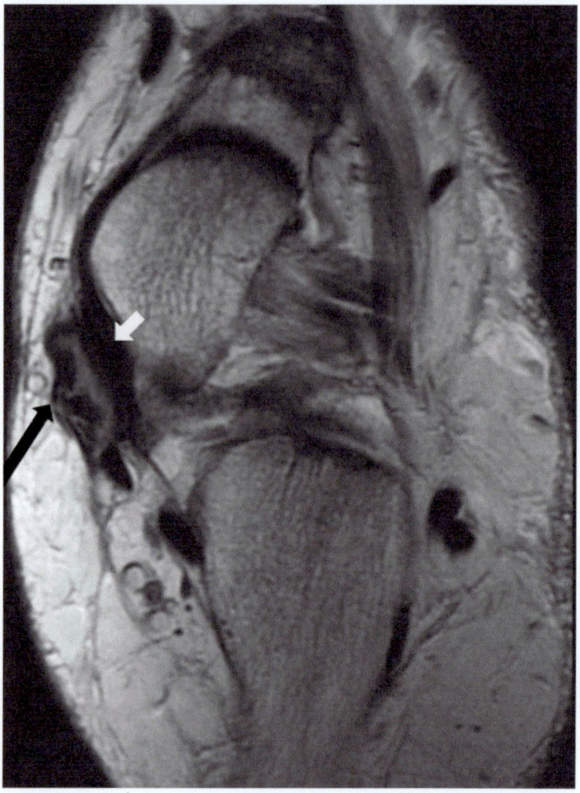

Fig. 7.22 PT tendon tear and spring ligament injury. (From: Omar et al. [83]). Axial PD-weighted image shows full-thickness longitudinal split tear of the PT tendon (long arrow) with tenosynovitis and thickening of the superomedial band of the spring ligament complex (small arrow)

The spring ligament and superficial deltoid ligament form a complex which contains multiple ligaments [91, 92]. There is evidence that rupture of one component of the complex may destabilize the remaining ligaments leading to progressive rupture [135, 153]. Ormsby and colleagues verified the sequential rupture of components of the spring-deltoid ligament complex in patients with stage II and III AAF deformity [154]. They found that the tibionavicular portion of the superficial deltoid ligament was attenuated or ruptured in more severe cases of AAF and this was almost always accompanied by attenuation of the spring ligament complex (Fig. 7.25). In less severe deformity, the tibionavicular ligament was intact, while the spring ligament complex was still attenuated or ruptured. The authors concluded that the superficial deltoid ligament will rupture after the spring ligament complex in the progression of the adult acquired flatfoot deformity.

The shortcoming of staging AAF according to the condition of the TP tendon was noted by Pasapula and colleagues who proposed a new classification system based upon spring ligament integrity [155]. They note the large body of evidence showing that spring ligament rupture is an essential part of the progressive AAF deformity and propose that attenuation of this ligament actually precedes the tenosynovitis seen in the TP tendon in stage I AAF deformity. Furthermore, Pasapula et al. propose that progressive collapse of the longitudinal arch of the foot will not

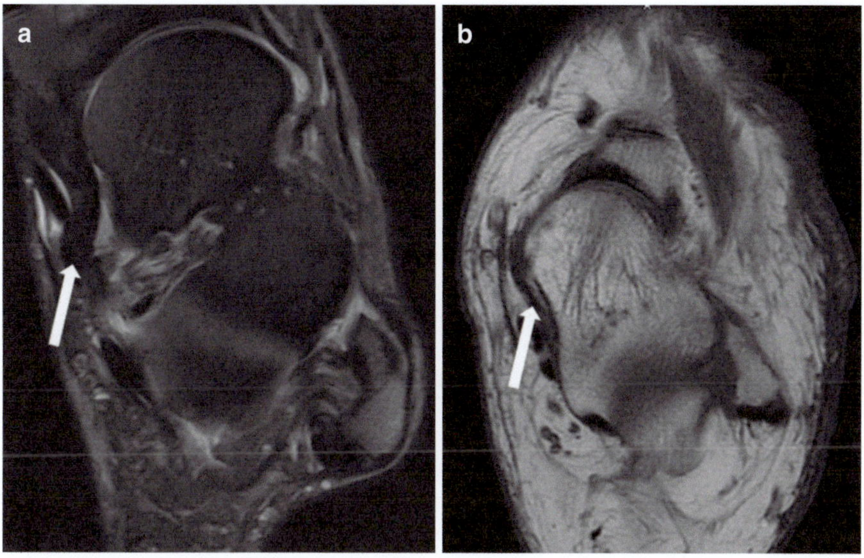

Fig. 7.23 Spring ligament degeneration. (From: Omar et al. [83]). (**a**) Axial 2D fat-suppressed PD-weighted image. Arrow shows thickening of the superomedial (SM) band of the spring ligament complex (SLC) consistent with degeneration. (**b**) Axial 2D PD-weighted image in a different patient shows diffuse thinning and irregularity of the SM band of SLC (arrow) consistent with chronic degenerative attenuation

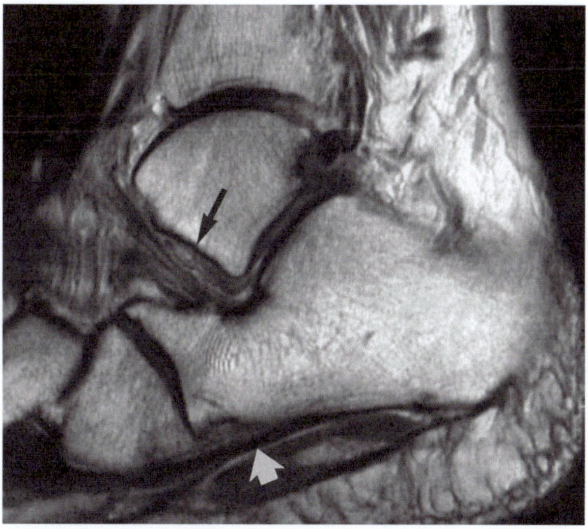

Fig. 7.24 Attenuation of the interosseous talocalcaneal ligament. (From: Deland et al. [82]). Figure 3 Sagittal MRI of a 69-year-old woman with stage II AAF demonstrates a thickened, hyperintense, torn (grade III) interosseous ligament (long arrow). Of note, the long plantar ligament is maintained (short arrow))

occur with isolated spring ligament failure but also requires acquired instability of the first TMT (metatarsocuneiform) joint. They observed that pronation of the hindfoot will be resisted by a stable first ray. However, with spring ligament failure, talonavicular joint instability will cause retrograde hindfoot valgus deformity which

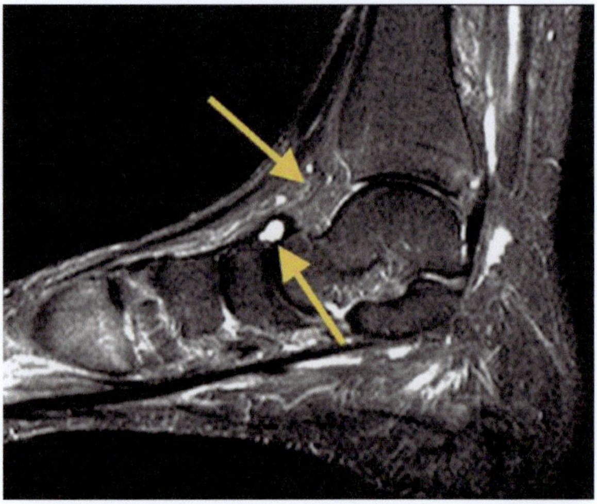

Fig. 7.25 Rupture of the tibionavicular ligament. (From: Ormsby et al. [154]. Sagittal T2-weighted image demonstrating attenuation and bulging of the TN ligament over the talonavicular joint. Arrows indicate dorsal and plantar margins of the TN ligament. (*PDFS* proton dense fat suppressed)

will progressively increase ground reaction forces on the first ray. The stabilizing structures of the first ray will eventually fail, causing dorsiflexion of the medial column upon weight bearing. Dorsiflexion of the medial column relative to the lateral column will result in a "supination deformity" of the forefoot.

Interestingly, despite attributing the deformity seen in stage II AAF to a rupture of the TP tendon rather than other ligaments, Bluman and colleagues acknowledge that supination deformity of the forefoot is an important part of the progression of flatfoot [131]. In fact, they break down stage II into four subdivisions depending upon the level and severity of forefoot-to-rearfoot deformity. In stage II-A, Bluman et al. attribute these changes of forefoot alignment to a valgus positioned hindfoot causing overload and adaptation in the medial column of the foot. Acquired supination deformity occurs as the forefoot inverts on the rearfoot to keep the metatarsals plantigrade. This is a normal compensatory motion of the forefoot on the rearfoot as originally described by Steindler and later by Saraffian (see Chap. 2). This supinated position of the forefoot is distinctly different from a congenital forefoot varus deformity which is not the result of, but can theoretically be the cause of, compensatory pronation of the rearfoot [138]. Forefoot supination or "supinatus" deformity is acquired from persistent valgus positioning of the rearfoot during weight bearing. Supination deformity is reducible in the early stage II-A-1 AAF and rigid in later stage II-A-2 AAF according to Bluman et al. [131] (Fig. 7.26). Another variation of deformity is abduction of the forefoot across the tarsometatarsal joints, stage II-B (Fig. 7.27). Finally, sagittal plane breakdown of the medial column across the talonavicular, naviculocuneiform, or first metatarsocuneiform joints will cause "medial column instability" viewed on a lateral weight-bearing radiograph and classified as stage II-C (Fig. 7.28).

What is missing from the Bluman classification is recognition of sequential ligament rupture causing the various changes in forefoot alignment seen in the four categories of stage II AAF deformity. Pasapula et al. attribute supination deformity

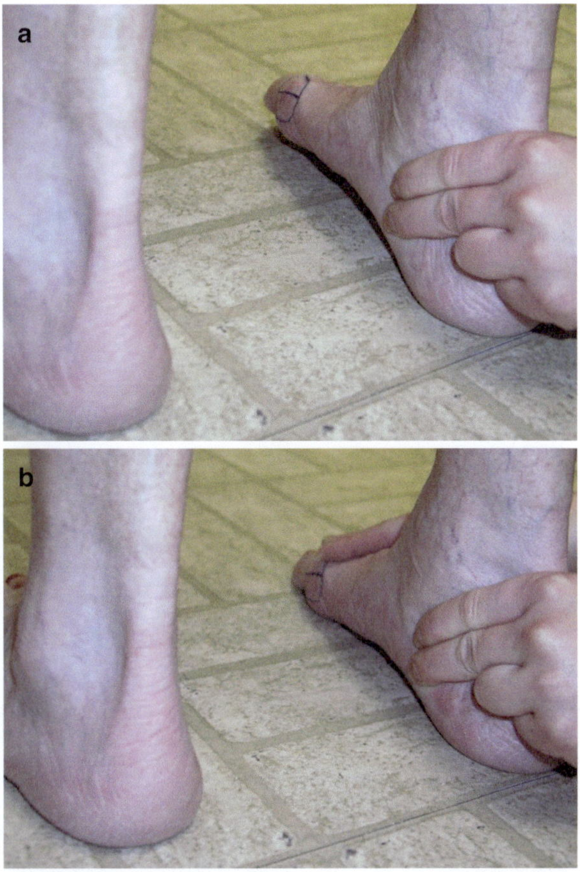

Fig. 7.26 (**a**) Acquired forefoot supination deformity. Supination is reducible in early stage II AAF deformity

and first ray instability to failure of the plantar ligaments of the first metatarsocuneiform joint, although many radiographic studies of AAF show that faulting or breakdown of the medial column can occur at multiple joints [72, 73, 156, 157]. Alsousou showed that when Meary's line is disrupted with breakdown of the medial column in flatfoot deformity, the naviculocuneiform joint is involved in 30% of the cases [156] (Fig. 7.29). Swanton and co-workers showed that the plantar naviculocuneiform ligament is actually an extension of the insertion of the posterior tibial tendon and is compromised in AAF and spring ligament failure [157].

Ligament Failure and Shift of Skeletal Segments

Rupture of the spring ligament and medial collateral ankle ligament complexes which occurs in stage II AAF deformity causes a global three-dimensional shift of alignment of skeletal segments in the ankle, hindfoot, and forefoot. A general observation in patients with stage II AAF is flattening of the medial longitudinal arch in

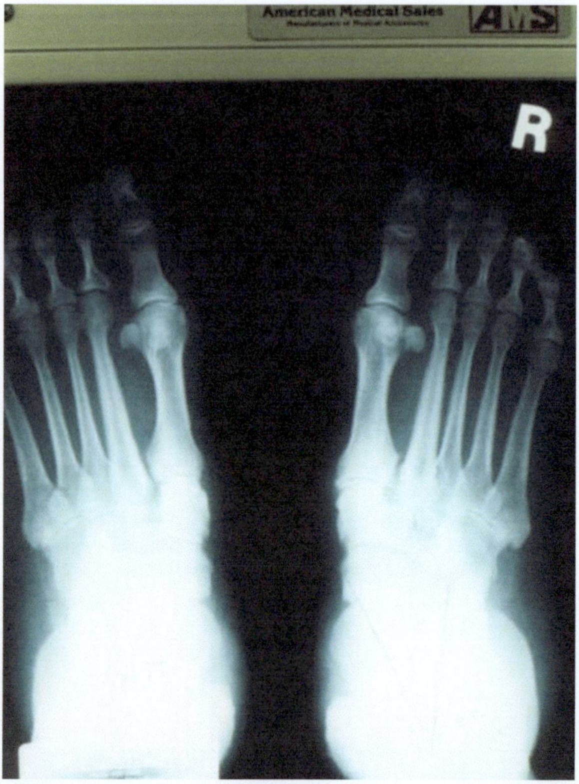

Fig. 7.27 Stage II-B AAF deformity-transverse plane subluxation across tarsometatarsal joints

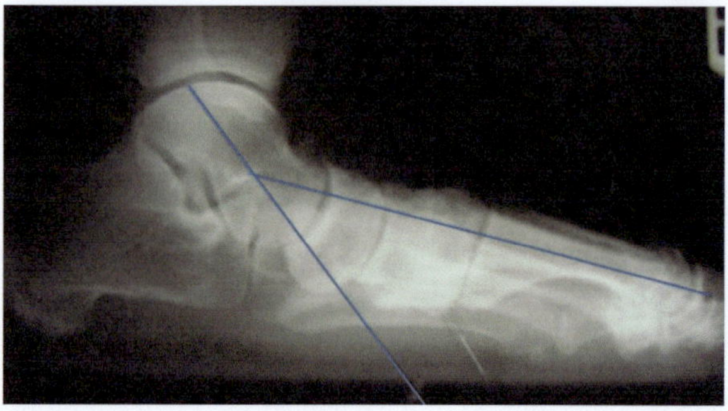

Fig. 7.28 Medial column subluxation at talonavicular joint in AAF contributing to forefoot supination deformity

the affected foot compared to the contralateral foot [39, 51]. Cadaver studies have shown that the plantar aponeurosis as well as the long and short plantar ligaments are the most important structures which maintain the integrity of the longitudinal arch of the foot [143, 158]. However, these ligament structures are unaffected in stage II AAF [82, 151]. Therefore, arch flattening is not the immediate change seen in stage II AAF. Instead, loss of the spring ligament allows abduction of the foot at the talo-calcaneonavicular joint, driven by the peroneal muscles. Abduction of the forefoot

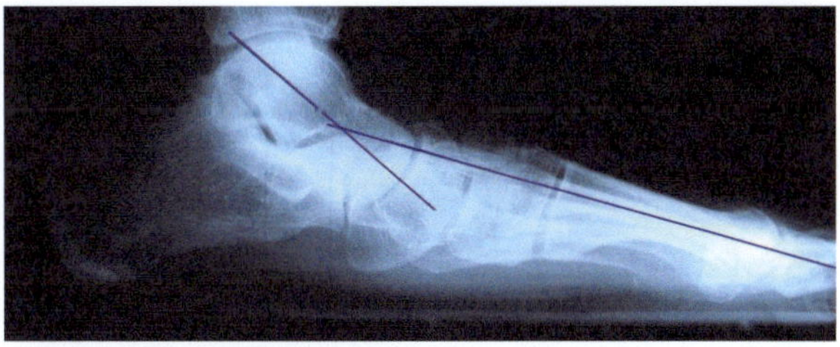

Fig. 7.29 Medial column subluxation at naviculocuneiform joint in AAF contributing to forefoot supination deformity

disrupts alignment of the medial truss structure of the longitudinal arch deviating and compromising the moment arm of the plantar aponeurosis. Spring ligament rupture causes significant plantarflexion of the talus and eventual attenuation of the remaining ligaments of the first ray leading to visible collapse of the longitudinal arch [52].

What is lacking in cadaver studies of arch stability is the reproduction of the effects of the peroneal musculature in AFF deformity. The peroneal muscles have been identified to be a significant deforming force in the propagation of adult acquired flatfoot [141]. This is validated by the fact that patients who undergo surgery for dropfoot, from common peroneal nerve palsy, do not develop a flatfoot deformity despite transfer of the attachment of the tibialis posterior tendon away from the navicular to the dorsum of the foot [139–141]. Clearly the lack of any deforming force from the peroneal musculature in these unique patients explains why they did not develop a flatfoot deformity. More importantly, these patients lost function of the TP tendon but did not develop ligamentous rupture.

The progression of the AAF deformity is due to isolated loss of the TP tendon. Even with the peroneal musculature acting un-opposed, flatfoot deformity will not progress unless key ligaments fail. The progressive AAF deformity is caused by rupture of the spring ligament causing sequential rupture of the medial collateral ankle ligaments and the interosseous talocalcaneal ligament [82]. Certainly the peroneal musculature contributes to the progression of deformity, but the sequence would not occur without ligamentous rupture.

The failure of the TCN ligament in the spring ligament and medial collateral ligament complexes results in translation of the head of the talus into a plantar-flexed and adducted position relative to the forefoot. This causes flattening of the medial longitudinal arch and also causes a retrograde motion of the calcaneus into the direction of eversion [91, 159]. While the plantar aponeurosis has a primary function of preventing arch elongation, sagittal plane collapse of the arch will occur once the spring ligament is lost and the talus plantarflexes relative to the forefoot [149].

The spring ligament provides stability to the calcaneus at the talocalcaneonavicular joint, while an important frontal plane restraint of the calcaneus is provided by the interosseous talocalcaneal ligament [160]. Belen found evidence of sinus tarsi abnormality suggesting injury to both the interosseous talocalcaneal ligaments and cervical ligaments in 72% patients with stage II PTTD [151]. Sinus tarsi syndrome and impingement of the calcaneus on the lateral malleolus are the result of extreme valgus rotation and lateral subluxation of the calcaneus which results from ligament rupture within the talocalcaneal joint [161–164] (Figs. 7.30 and 7.31).

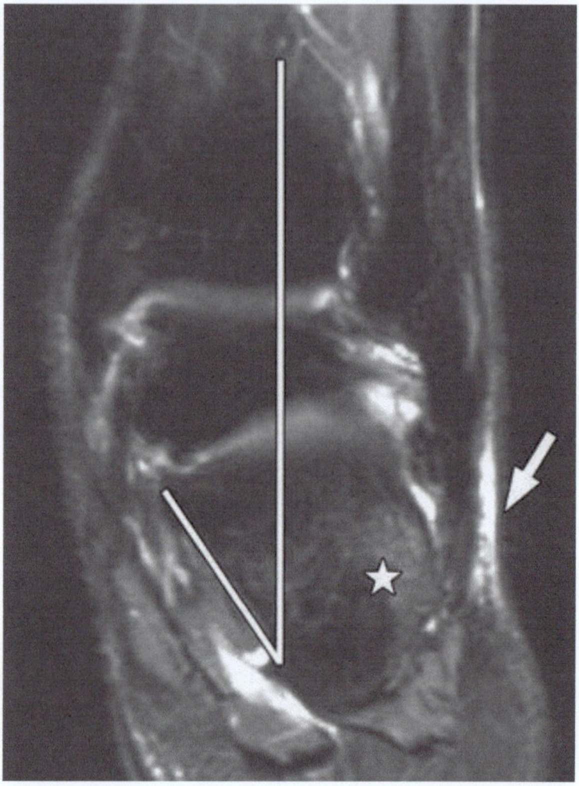

Fig. 7.30 Sinus tarsi and lateral impingement. (From: Donovan and Rosenberg [161]. Fig 2-c)

Standing radiographs will demonstrate all these alignment changes when the symptomatic foot is compared to the asymptomatic foot. On the anterior-posterior view, forefoot abduction will be evident with increased talonavicular angle and increased uncoverage of the head of the talus [165]. On the standing lateral radiograph, there will be a decrease in calcaneal pitch angle, increased Meary's angle (measuring alignment of the medial arch), and decreased height of the distance between the medial cuneiform and the fifth metatarsal [166, 167].

Three-dimensional computer modeling with CT imaging allows visualization of shifting of skeletal segments in AAF which are not seen on plain radiographic images. These studies show that the tibia and talus migrate medially while the navicular and calcaneus migrate laterally [168, 169] (Fig. 7.32). However, the cuboid does not shift relative to the calcaneus. This may be due to the anatomic locking of this joint, while no locking is available at the talonavicular joint.

The 3-D CT studies verify the notion of "peritalar subluxation" originally proposed by Hansen and colleagues, which describes the displacement of various skeletal segments in AAF (180,181). As seen in the CT imaging studies, the talus is stable in the ankle joint, while the bones of the foot which are distal to the talus undergo three-dimensional rotation due to ligament rupture (Fig. 7.33). This is a displacement or untwisting of the lamina pedis as originally described by

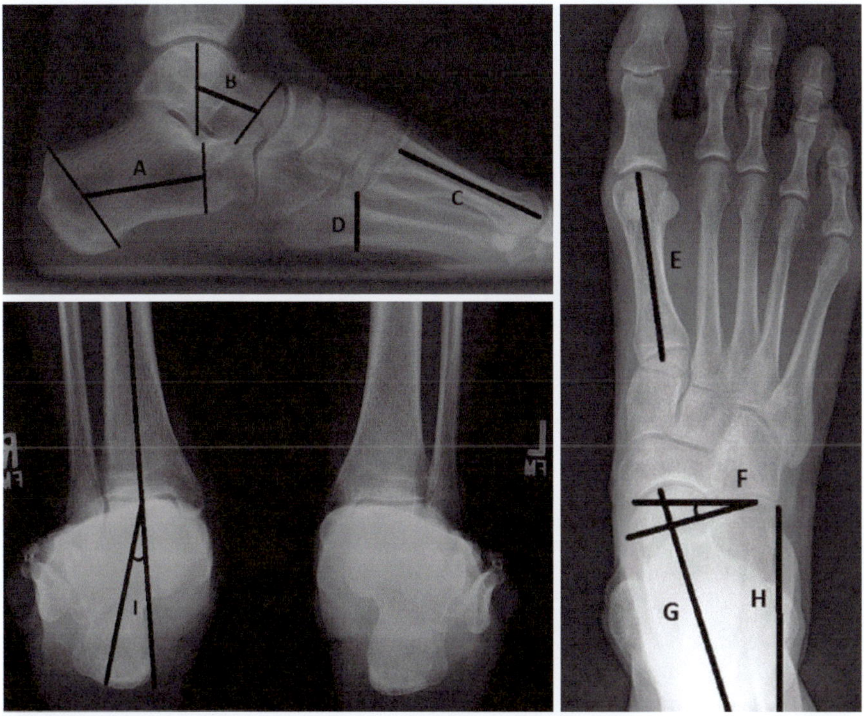

Fig. 7.31 Radiographic evaluation of AAF. (From: Cody et al. [53]). Standard weight-bearing lateral, AP, and hindfoot measurements demonstrate (**a**) calcaneal axis for calcaneal pitch and talocalcaneal angles; (**b**) talar axis for talocalcaneal angle and Meary's angle; (**c**) long axis of first metatarsal for Meary's angle; (**d**) medial column height; (**e**) AP first metatarsal axis for the talar-first metatarsal angle; (**f**) illustration of the AP talar coverage angle; (**g**) talar axis for the talocalcaneal angle and talar-first metatarsal angle; (**h**) calcaneal axis for the talocalcaneal angle; and (**i**) hindfoot alignment angle measured at the intersection of the long axes of the tibia and calcaneus

MacConnaill [76]. With untwisting of the lamina pedis which causes peritalar subluxation, the calcaneus moves proximal and plantarflexes, causing secondary Achilles and triceps contracture [171]. The calcaneus is already displaced laterally due to innate deformity of the talus with valgus angulation of the posterior facet [52, 53]. The navicular moves dorsal and lateral on the talus, and the medial column is now free to flatten and elongate due to loss of the truss mechanism. The lateral column has no ligament loss causing elongation, and the calcaneocuboid joint has osseous locking so it remains stable. The mobile, elongating medial column rotates into abduction, causing the entire forefoot to pivot laterally around the stable calcaneocuboid joint [170].

The rotation and displacement of skeletal segments within the foot affect the moment arm of key muscles as the AAF deformity progresses. As the foot moves into eversion and abduction, the moment arm of the invertors decreases (Fig. 7.18). Hinterman showed that inversion moment arm of the tibialis posterior is reduced

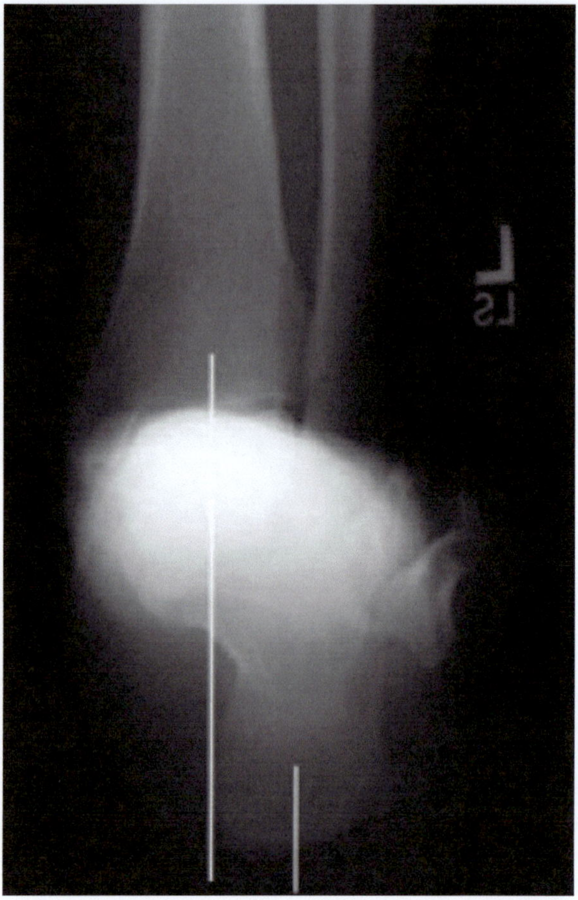

Fig. 7.32 Standing radiograph of stage III deformity shows significant medial shift of tibia and talus over the calcaneus

with eversion of the foot [95]. Lee and colleagues showed that the tibialis anterior has a significant inversion moment arm when the foot is in neutral or inverted, but this changes into an eversion moment arm when the foot is everted [172]. Therefore, as the foot moves into an everted position, the torque capacity of the two primary invertor muscles becomes compromised.

Lee et al. also showed that the gastrocnemius via the Achilles tendon has a small inversion moment arm on the rearfoot when the ankle is positioned at 90 degrees and the rearfoot is in a neutral position [172]. Eversion of the hindfoot creates a slight eversion moment arm of the Achilles. Displacement of the insertion of the Achilles with a medial displacement osteotomy increases the length of the inversion moment arm of the triceps surae acting on the subtalar joint [173].

As the calcaneus plantarflexes in stage II AAF deformity, the effective length of the Achilles tendon shortens which causes weakening due to compromise of the force/length relationship. Houck and colleagues showed that patients with stage II PTTD have delayed ankle joint plantarflexion and delayed heel off during walking

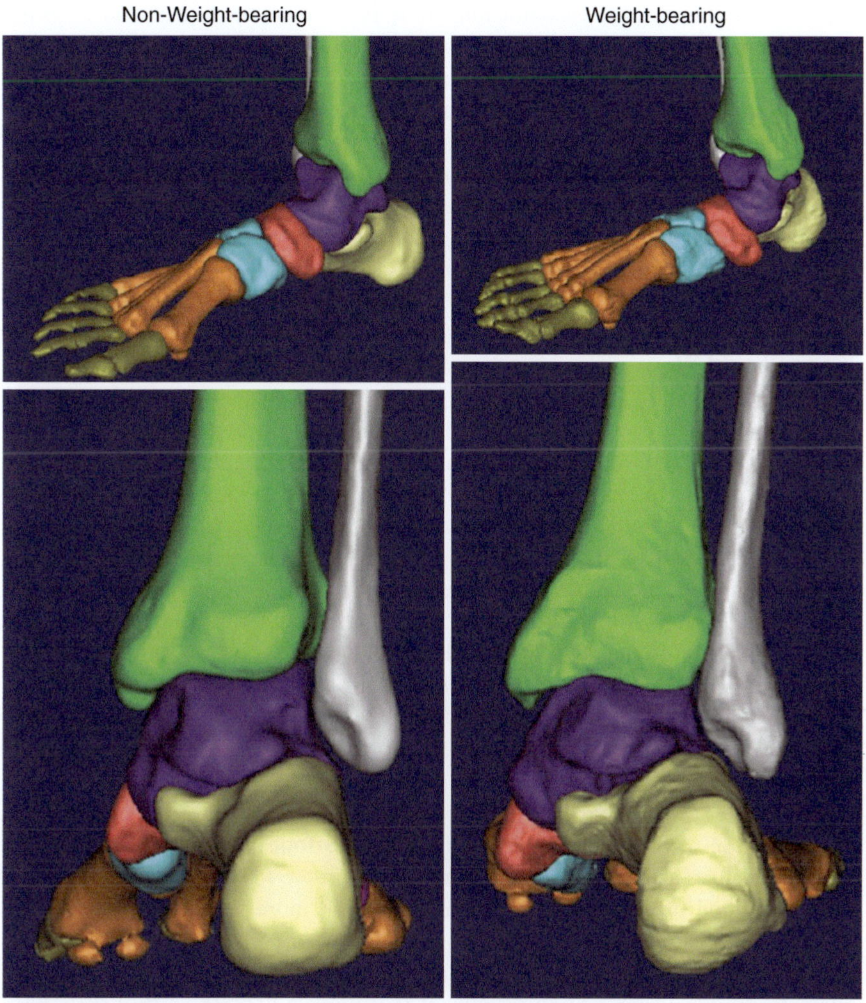

Fig. 7.33 Simulated weight-bearing CT imaging demonstrates medial shift of tibia and talus over calcaneus, eversion of calcaneus, supination, and abduction of forefoot. (From: Zhang et al. [169]. Figure 9 Lateral (top) and posterior (bottom) three-dimensional CT images of a stage II posterior tibial tendon dysfunction (PTTD) flatfoot subject's right foot under non-weight bearing (left) and full-body weight bearing (right))

gait [100]. The authors concluded that this finding was not only due to weakening of the triceps but also was a compensation which allowed early double limb support, thus decreasing the need for push off on the affected limb.

Plantarflexion of the calcaneus which causes shortening of the tendo Achilles will demonstrate a compensatory equinus contracture of the ankle joint upon examination of patients with adult acquired flatfoot deformity. Valgus moment arm increases in

AAF with increased Achilles contracture [174]. Equinus contracture of the triceps has been speculated as a contributing factor to the etiology of adult acquired flatfoot although no studies have confirmed this cause-effect relationship. While most patients with AAF deformity demonstrate equinus contracture on physical examination, there is a lack of research which can answer the question whether this contracture is a cause or effect of many structural changes seen in the foot and ankle in adult acquired flatfoot deformity. The triceps are recognized as a primary deforming force in the progression of the adult acquired flatfoot deformity, but the negative effects only occur when the stability of the midfoot is compromised by ligament failure.

Stage III and IV

In the final two stages of AAF, the foot is rigid due to degenerative arthritis and ankylosis of the hindfoot and ankle joints. As with stage I and II, there is opportunity for subjective interpretation when determining that a patient is in stage III or IV. Bluman et al. provide no objective criteria other than describing a "rigid hindfoot valgus" deformity. Stage IV differs from stage III with valgus deformity in the talocrural joint due to rupture of the deep tibiotalar section of the deltoid ligament. The clinical determination of a "flexible" vs "rigid" AAF deformity requires assessment of mobility of the hindfoot, while weight bearing, where the calcaneus is attempted to be reduced to neutral or aligned with the lower leg [138].

Gait Disturbance with AAF

The disruption of ligaments in the midfoot and rearfoot in adult acquired flatfoot deformity causes a shift of skeletal segments within the foot which are detected in static exam and in standing radiographs. However, more significant changes are seen during dynamic gait where excessive mobility of certain joints, loss of coupling between the foot and the leg, and loss of push off have been noted in several kinematic studies using 3-D multisegment foot models [97–100, 176–178]. These studies are remarkably similar in showing consistent changes.

Rattanaprasert studied a single patient who had suffered traumatic rupture of their posterior tibial tendon and then developed acquired flatfoot deformity [97]. Compared to a group of ten healthy subjects, the AAF patient showed significant dorsiflexion of the forefoot on the rearfoot during heel rise. Also there was excessive forefoot abduction in the AAF patient but not significant eversion of the rearfoot.

Tome and co-workers used a multisegment foot model to study gait in 12 patients in stage II PTTD and compared to 10 healthy control subjects [98]. Greater forefoot abduction was measured in the PTTD patients during the entire stance phase of walking gait. Also, the medial longitudinal arch was lower from loading response thru pre-swing in the PTTD patients. The authors concluded that after loss of the TP tendon, the other extrinsic arch supporting muscles are unable to compensate to provide stability. The excessive motion across the longitudinal arch suggested that an osseous locking mechanism does not exist in the foot as loss of soft tissue support leads to significant flexibility across the midfoot joints.

Ringleb and colleagues studied kinematics, plantar pressures, as well as EMG activity in five patients with PTTD deformity [99]. As with previous studies, the primary changes occurred in the sagittal and transverse plane across the midfoot joints, while hindfoot eversion motion did not differ between the two groups. Increased forefoot dorsiflexion across the midfoot joints was observed in the PTTD patients. The patient group demonstrated inversion of the forefoot during heel rise. The healthy controls demonstrated eversion of the forefoot during heel rise indicating healthy function of the first ray and engagement of the windlass mechanism. In this study, PTTD patients exhibited greater activity of the peroneus longus, perhaps a compensation attempting to evert the forefoot and plantarflex the first ray. Also, increased activity of the gastrocnemius was measured in the PTTD group suggesting excessive work to plantarflex the ankle with an unstable midfoot.

Also, functional shortening of the Achilles due to plantarflexion of the rearfoot weakens the tension producing capacity of the gastrocnemius which now must work harder to initiate heel rise [101].

Houck et al. [100] developed a multisegment foot model to evaluate rearfoot motion, midfoot motion, as well as first metatarsal motion in 30 patients with PTTD compared to 15 healthy controls. As with the study by Ringleb, Houck et al. found that the plantar-flexed alignment of the rearfoot caused shortening of the Achilles and weakness of the triceps which delayed heel rise. Also, greater flexibility across the midfoot joints compromised stiffening of the foot and delayed heel rise (Fig. 7.34). This flexibility was associated with lowering of the medial longitudinal

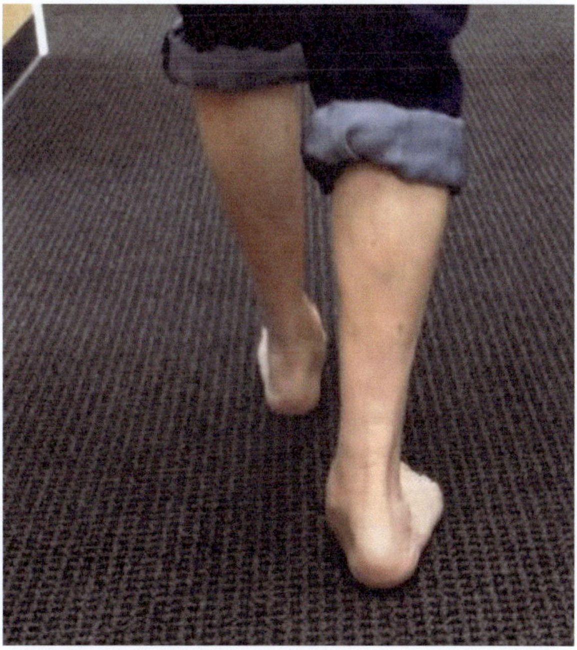

Fig. 7.34 Gait studies of patients in stage II AAF consistently show forefoot abduction on the affected side along with delayed heel rise due to midfoot instability. Right foot should have begun heel rise before heel contact by the left foot

arch and dorsiflexion of the first metatarsal. Unlike other studies, Houck et al. measured greater hindfoot eversion in the PTTD patient group compared to controls. Kinematic studies which do find greater hindfoot eversion with PTTD show a magnitude of increase in the realm of only 3 to 5 degrees [175, 176].

Neville et al. studied plantar loading patterns in subjects with PTTD using in shoe pressure measurement insoles [177]. A reduction in total loading of the forefoot during terminal stance in the patient group compared to healthy controls indicated loss of stability of the arch for push off. Furthermore, there was a loss of lateral forefoot loading in the patient group which would normally occur with hindfoot inversion. A shift of loading toward the midfoot and heel in the PTTD patient group was similar to what is seen in patients with rupture of the tendo Achilles. With loss of arch and midfoot stability, the Achilles appears ineffective in transferring load to the forefoot in patients with PTTD. This can be demonstrated when patients in stage II deformity are unable to perform a single foot heel rise (Fig. 7.22). In another study, this same research group determined that the loss of normal loading of the forefoot during terminal stance in PTTD is not due to muscle weakness but rather due to loss of ligament stability across the midfoot joints [176].

Van de Velde and co-workers were the first to measure coupling between key segments of the foot and leg [178]. Also, this was the first study to compare patients with stage II and stage III AAF deformity. As expected, the more rigid stage III deformity was associated with less motion across the medial arch and in the rearfoot compared to stage II. Both groups demonstrated decreased walking speed and delayed heel lift compared to healthy controls. Verifying previous studies, AAF is associated with reduced "twisting" of the forefoot on the rearfoot during terminal stance which is characterized by hindfoot inversion and forefoot eversion. Instead, patients with AAF demonstrated lack of eversion or pronation of the forefoot during terminal stance [178]. This study also showed decreased hallux dorsiflexion in the AAF patients during terminal stance, suggesting a lack of loading of the first ray and failure to engage the windlass mechanism during heel rise.

The most significant finding of the study by Van de Velde et al. was the disruption of coupling or synchronized motion between the rearfoot and the lower leg in PTTD patients, compared to healthy controls. Coupling is the mechanism where internal and external rotation of the leg is converted to inversion and eversion of the foot. Loss of coupling between skeletal segments has been theorized to be either the result or the cause of many musculoskeletal injuries [179]. A study of 11 injured runners showed out of phase coupling between the foot and leg which was improved with use of custom orthotic devices [180]. Hinterman, in a cadaver study, showed a significant loss of coupling between the foot and leg when the medial ankle ligaments and the interosseous talocalcaneal ligament are severed, identical to what is seen in stage III AAF deformity [181]. For the first time, an in vivo kinematic study from Van de Velde verified Hinterman's previous work, showing that the foot becomes mechanically disconnected from the leg when the medial ankle and arch ligaments are ruptured. The lack of connectivity or mechanical coupling between the foot and the leg poses one of several challenges in treating PTTD.

Kinematic studies verify a deterioration of specialization of the human foot as the adult flatfoot deformity takes on the features of the ape foot.

In the rearfoot, the calcaneus moves laterally and abuts on the distal fibula. The talus adducts and now lies adjacent to the calcaneus in the transverse plane.

The longitudinal arch is lost, and increased flexibility of the forefoot on the rearfoot in the sagittal and transverse planes occurs in the AAF deformity. As a result, heel rise is delayed, and a midfoot break is accentuated in AAF deformity. Eversion or pronation of the forefoot during heel rise is lost in AAF, and the first ray loses stability for loading. As a result, hallux dorsiflexion and engagement of the windlass are compromised. Due to all of these changes, the adult acquired flatfoot has now assumed the structure and function of the ape foot.

References

1. Tachdjian MO The child's foot. Philadelphia: WB Saunders Co.;1985
2. Evans AM, Rome KA. Cochrane review of the evidence for nonsurgical interventions for flexible pediatric flat feet. Eur J Phys Rehabil Med. 2011;47:69–89.
3. Staheli LT. Planovalgus foot deformity. Current status. J Am Podiatr Med Assoc. 1999;89(2):94–9.
4. El O, Akcali O, Kosay C, et al. Flexible flatfoot and related factors in primary school children: a report of a screening study. Rheumatol Int. 2006;26(11):1050–3.
5. Pfeiffer M, Kotz R, Ledl T, et al. Prevalence of flat foot in preschool-aged children. Pediatrics. 2006;118(2):634–9.
6. Rodriguez N, Volpe RG. Clinical diagnosis and assessment of the pediatric Pes Planovalgus deformity. Clin Podiatr Med Surg. 2010;27(1):43–58.
7. Cappello T, Song KM. Determining treatment of flatfeet in children. Curr Opin Pediatr. 1998;10(1):77.
8. Staheli L, Chew DE, Corbett M. The longitudinal arch. A survey of eight hundred and eighty-two feet in normal children and adults. J Bone Joint Surg Am. 1987;69(2):426–8.
9. Rome K, Ashford RL, Evans A. Non-surgical interventions for paediatric pes planus. Cochrane Database Syst Rev 2010;7.
10. Harris EJ, Vanore JV, Thomas JL, et al. Diagnosis and treatment of pediatric flatfoot. J Foot Ankle Surg. 2004;43(6):341–73.
11. Harris RI, Beath T. Hypermobile flatfoot with short tendo Achilles. J Bone Joint Surg Am. 1948;30:116–40.
12. Uden H, Scharfbillig R, Causby R. The typically developing paediatric foot:how flat should it be? A systematic review. J Foot Ankle Res. 2017;10:37.
13. Buldt AK, Murley GS, Butterworth P, Levinger P, Menz HB, Landorf KB. The relationship between foot posture and lower limb kinematics during walking: a systematic review. Gait Posture. 2014;38:363–72.
14. Kosashvili Y, Fridman T, Backstein D, Safir O, Ziv Y. The correlation between pes planus and anterior knee or intermittent low back pain. Foot Ankle Int. 2008;29(9):910–3.
15. Garrow A, Simlman A, Macfarlane G. The Cheshire foot pain and disability survey: a population survey assessing prevalence and associations. Pain. 2004;110:378–84.
16. Mølgaard C, Lundbye-Christensen S, Simonsen O. High prevalence of foot problems in the Danish population: a survey of causes and associations. Foot. 2010;20:7–11.
17. Nguyen US, Hillstrom HJ, Dufour AB, Kiel DP, Procter-Gray E, Gagnon MM, et al. Factors associated with hallux valgus in a population-based study of older women and men: the MOBILIZE Boston study. Osteoarthr Cartil. 2010;18:41–6.

18. Galica A, Hagedorn TJ, Dufour AB, Riskowski JL, Hillstrom HJ, Casey VA, Hannan MT. Hallux valgus and plantar pressure loading: the Framingham foot study. J Foot Ankle Res. 2013;6:1–18.
19. Shibuya N, Jupiter D, Ciliberti L, VanBuren V, La Fontaine J. Characteristics of adult flatfoot in the United States. J Foot Ankle Surg. 2010;49:236–368.
20. Shibuya N, Kitterman RT, LaFontaine J, Jupiter DC. Demographic, physical, and radiographic factors associated with functional flatfoot deformity. J Foot Ankle Surg. 2014;53:168–72.
21. Richie DH. Pathomechanics of the adult acquired flatfoot. In: Lepow GW, editor. Foot Ankle Q. Data Trace. 2005;17(4):109–24.
22. Kulowski J. Tendovaginitis (tenosynovitis): general discussion and report of one case involving posterior tibial tendon. J Missouri Med Assoc. 1936;33:135–7.
23. Fowler AW. Tibialis posterior syndrome. In: Proceedings of the Sought-West Orthopaedic Club. J Bone Joint Surg. 1955;37-B(3):520.
24. Williams R. Chronic nonspecific tendovaginitis of the tibialis posterior. J Bone Joint Surg. 1963;45-B(3):542–5.
25. Kettelkamp DB, Alexander HH. Spontaneous rupture of the posterior tibial tendon. J Bone Joint Surg. 1969;51-A:759–64.
26. Goldner JL, Keats PK, Bassett FH, et al. Progressive talipes equinovalgus due to trauma or degeneration of the posterior tibial tendon and medial plantar ligaments. Orthop Clin North Am. 1974;5(1):39–51.
27. Mann RA, Specht LH. Posterior tibial tendon ruptures and analysis of eight cases. Foot Ankle. 1982;2:350.
28. Jahss MH. Spontaneous rupture of the tibialis posterior tendon: clinical findings. Tenographic studies and a new technique of repair. Foot Ankle. 1982;3:158–66.
29. Jahss MH. Tendon disorders of the foot and ankle. In: Jahss M, editor. Disorders of the foot and ankle. 2nd ed. Philadelphia: W.B. Saunders; 1981. p. 1506.
30. Mann RA, Thompson FM. Rupture of the posterior tibial tendon causing flatfoot. J Bone Joint Surg. 1985;67A:556–61.
31. Fredenburg M, Tiley G, Yagoobian EM. Spontaneous rupture of the posterior tibial tendon secondary to chronic nonspecific tenosynovitis. J Foot Surg. 1983;22:198–202.
32. Mueller TJ. Ruptures and lacerations of the tibialis posterior tendon. J Am Podiatry Assoc. 1984:109–19.
33. Pomeroy GC, Manoli A. A new operative approach for flatfoot secondary to posterior tibial tendon insufficiency: a preliminary report. Foot Ankle Int. 1997;18(4):206–12.
34. Henceroth WD 2nd, Deyerle WM. The acquired unilateral flatfoot in the adult: some causative factors. Foot Ankle. 1982;2(5):304–8.
35. Funk DA, Cass JR, Johnson KA. Acquired adult flat foot secondary to posterior tibial-tendon pathology. J Bone Joint Surg Am. 1986;68(1):95–102.
36. Johnson KA, Strom DE. Tibialis posterior tendon dysfunction. Clin Orthop. 1989;239:196–206.
37. Gazdag AR, Cracchiolo A. Rupture of the posterior tibial tendon. Evaluation of injury of the spring ligament and clinical assessment of tendon transfer and ligament repair. J Bone Joint Surg Am. 1997;79A:675–81.
38. Deland JT, Arnoczky SP, Thompson FM. Adult acquired flatfoot deformity at the talonavicular joint: reconstruction of the spring ligament in an in vitro model. Foot Ankle Int. 1992;13:327–32.
39. Myerson MS. Adult-acquired flatfoot deformity: treatment of dysfunction of the posterior tibial tendon. Instr Course Lect. 1997;46:393–405.
40. Lee MS, Vanore JV, Thomas JL, Catanzariti AR, Kogler G, Kravitz SR, et al. Diagnosis and treatment of adult flatfoot. J Foot Ankle Surg. 2005;44(2):78–113.
41. Nair P, Deland J, Ellis SJ. Current concepts in adult acquired flatfoot deformity. Curr Orthop Pract. 2015;26(2):160–8.

42. Ross MH, Smith MD, Vicenzino B. Reported selection criteria for adult acquired flatfoot deformity and posterior tibial tendon dysfunction: are they one and the same? A systematic review. PLoS One. 2017;12(12):e0187201.
43. Downey DJ, Simpkin PA, Mack LA, Richardson ML, Kilcoyne RF, Hansen ST. Tibialis posterior tendon rupture: a cause of rheumatoid flat foot. Arthritis Rheumatol. 1988;31:441–4.
44. Michelson J, Easley M, Wigley FM, Hellmann D. Posterior tibial tendon dysfunction in rheumatoid arthritis. Foot Ankle Int. 1995;16:156–61.
45. Myerson M, Solomon G, Shereff M. Posterior tibial tendon dysfunction: its association with seronegative disease. Foot Ankle. 1989;9:219–25.
46. Kohls-Gatzoulis J, Haddad F, Livingston J, Berry G, Singh D, Angel JC. Posterior tibial tendon dysfunction: a frequent and treatable cause of adult acquired flatfoot. Br Med J. 2004;329:1328–33.
47. Kohls-Gatzoulis JA, Singh D. Tibialis posterior dysfunction as a cause of flatfeet in elderly patients. Foot. 2004;14:207–9.
48. Kohls-Gatzoulis J, Woods B, Angel JC, Singh D. The prevalence of symptomatic posterior tibialis tendon dysfunction in women over the age of 40 in England. Foot Ankle Surg. 2009; 15(2):75±81.
49. Holmes GB II, Mann RA. Possible epidemiological factors associated with rupture of the posterior tibial tendon. Foot Ankle. 1992;13:70–9.
50. Ross M, Smith M, Plinsinga ML, Vicenzino B. Self-reported social and activity restrictions accompany local impairments in posterior tibial tendon dysfunction: a systematic review. J Foot Ankle Res. 2018;11:49.
51. Dyal CM, Feder J, Deland JT, Thompson FM. Pes planus in patients with posterior tibial tendon insufficiency: asymptomatic versus symptomatic foot. Foot Ankle Int. 1997;18:85–8.
52. Probasco W, Haleem AM, Yu J, et al. Assessment of coronal plane subtalar joint alignment in peritalar subluxation via weight-bearing multiplanar imaging. Foot Ankle Int. 2015;36(3):302–9.
53. Cody EA, Williamson ER, Burket JC, Deland JT, Ellis SJ. Correlation of talar anatomy and subtalar joint alignment on weight bearing computed tomography with radiographic flatfoot parameters. Foot Ankle Int. 2016;37(8):874–81.
54. Ford ES, Giles WH, Dietz WH. Prevalence of the metabolic syndrome among US adults. Findings from the third national health and nutrition survey. JAMA. 2002;87(3):356–9.
55. Pidker PM. C-reactive protein: a simple test to help predict risk of heart attack and stroke. Circulation. 2003;108:e81–5.
56. Gaida JE, Alfredson L, Kiss ZS, Wilson AM, Alfredson H, Cook JL. Dyslipidemia in Achilles tendinopathy is characteristic of insulin resistance. Med Sci Sports Exerc. 2009;41(6):1194–7.
57. Shiri R, Viikari-Juntura E, Varonen H, Heliövaara M. Prevalence and determinants of lateral and medial epicondylitis: a population study. Am J Epidemiol. 2006;164(11):1065–74.
58. Mäntyselkä P, Miettola J, Niskanen L, Kumpusalo E. Persistent pain at multiple sites–connection to glucose derangement. Diabetes Res Clin Pract. 2009;84(2):e30–2.
59. Mäntyselkä J, Miettola L, Niskanen E. Kumpusalo. Glucose regulation and chronic pain at multiple sites. Rheumatology. 2008;47(8):1235–8.
60. Afifi AEA, Shaat RM, Gharbia OM, Boghdadi YE, Eshmawy MME, El-Emam OA. Osteoarthritis of knee joint in metabolic syndrome. Clin Rheumatol. 2018;37(10): 2855–61.
61. Scott A, Zwerver J, Grewal N, et al. Lipids, adiposity and tendinopathy: is there a mechanistic link? Critical review. Br J Sports Med. 2015;49(15):984–8.
62. Mokone GG, Schwellnus MP, Noakes TD, Collins M. The COL5A1 gene and Achilles tendon pathology. Scand J Med Sci Sports. 2006;16(1):19–26.
63. Posthumus M, September AV, O'Cuinneagain D, van der Merwe W, Schwellnus MP. The COL5A1 gene is associated with increased risk of anterior cruciate ligament ruptures in female participants. Am J Sports Med. 2009;37(11):2234–40.
64. September AWV, Schwellus MP, Collins M. Tendon and ligament injuries: the genetic component. Br J Sports Med. 2007;41:241–6.

65. Morimoto I. Notes on architecture of tibialis posterior muscle in man. Kaibogaku Zasshi. 1983;58:74–80.
66. Silver RL, de la Garza J, Rang M. The myth of muscle balance. J Bone Joint Surg. 1985;67-B:432–7.
67. Maharaj JN, Cresswell AG, Lichtwark GA. The mechanical function of the tibialis posterior muscle and its tendon during locomotion. J Biomech. 2016;49:3238–43.
68. Kelikian AS, Sarrafian SK. Sarrafian's anatomy of the foot and ankle: descriptive, topographic, functional. 3rd ed. Philadelphia: Wolters Kluwer Health/Lippincott Williams & Wilkins; 2011.
69. Petersen W, Hohmann G, Pufe T, et al. Structure of the human tibialis posterior tendon. Arch Orthop Trauma Surg. 2004;124(4):237–42.
70. Benjamin M, Ralphs JR. Fibrocartilage in tendons and ligaments and adaptation to compressive load. J Anat. 1998;193:481–94.
71. Olewnik L. A proposal for a new classification for the tendon of insertion of the tibialis posterior. Clin Anat. 2019;32:557–65.
72. Younger AS, Sawatzky B, Dryden P. Radiographic assessment of adult flatfoot. Foot Ankle Int. 2005;26(10):p820–5.
73. Sensiba PR, Coffey MJ, Williams NE, Mariscalco M, Laughlin RT. Inter and Intraobserver reliability in the radiographic evaluation of adult acquired flatfoot deformity. Foot Ankle Int. 2010;31(2):141–5.
74. O'Malley MJ, Deland JT, Lee KT. Selective hindfoot arthrodesis for the treatment of adult acquired flatfoot deformity: an in vitro study. Foot Ankle Int. 1995;16:412–7.
75. Nester CJ, Jarvis HL, Jones RK, Bowden PD, Anmin L. Movement of the human foot in 100 pain free individuals aged 18–45: implications for understanding normal foot function. J Foot Ankle Res. 2014;7:51.
76. MacConaill MA. The postural mechanism of the human foot. Proc R Ir Acad. 1944–1945;50B:265–78.
77. Jones FW. Structure and function as seen in the foot. London: Baillière, Tindall and Cox; 1944.
78. Pisani G. The coxa pedis. J Foot Ankle Surg. 1994;1:67–74.
79. Pisani G. Coxa pedis today. Foot Ankle Surg. 2016;22:78–84.
80. Pisani G. Peritalar destabilisation syndrome (adult flatfoot with degenerative glenopathy). Foot Ankle Surg. 2010;16:183–8.
81. Pisani G. About the pathogenesis of the so-called adult acquired pes planus. Foot Ankle Surg. 2010;16:1–2.
82. Deland JT, de Asla RJ, Sung IH, Ernberg LA, Potter HG. Posterior tibial tendon insufficiency: which ligaments are involved? Foot Ankle Int. 2005;26(6):427–35.
83. Omar H, Saini V, Wadhwa V, Liu G, Chhabra A. Spring ligament complex: illustrated normal anatomy and spectrum of pathologies on 3T MR imaging. Eur Jour Radiol. 2016;85:2133–43.
84. Davis WH, Sobel M, DiCarlo EF, et al. Gross, histological, and microvascular anatomy and biomechanical testing of the spring ligament complex. Foot Ankle Int. 1996;17(2):95–102.
85. Patil V, Ebraheim NA, Frogameni A, Liu J. Morphometric dimensions of the calcaneonavicular (spring) ligament. Foot Ankle Int. 2007;28(8):927–32.
86. Taniguchi A, Tanaka Y, Takakura Y, et al. Anatomy of the spring ligament. J Bone Joint Surg Am. 2003;85-A(11):2174–8.
87. Vadell AM, Peratta M. Calcaneonavicular ligament: anatomy, diagnosis, and treatment. Foot Ankle Clin. 2012;17:437–48.
88. Milner CE, Soames RW. The medial collateral ligaments of the human ankle joint: anatomical variations. Foot Ankle Int. 1998;19(5):289–92.
89. Savage-Elliott I, Murawski CD, Smyth NA, Golano P, Kennedy JG. The deltoid ligament: an in-depth review of anatomy, function, and treatment strategies. Knee Surg Sports Traumatol Arthrosc. 2013;21(6):1316–27.
90. Campbell KJ, Michalski MP, Wilson KJ, Goldsmith MT, Wijdicks CA, LaPrade RF, Clanton TO. The ligament anatomy of the deltoid complex of the ankle: a qualitative and quantitative anatomical study. J Bone Joint Surg. Am. 2014;A; 96:e62.

91. Cromeens BP, Kirchhoff CA, Patterson RM, Motley T, Stewart D, Fisher C, Reeves RE. An attachment-based description of the medial collateral and spring ligament complexes. Foot Ankle Int. 2015;36:710–21.
92. Amaha K, Nimura A, Yamaguchi R, Kampan N, et al. Anatomic study of the medial side of the ankle base on the joint capsule: an alternative description of the deltoid and spring ligament. J Exp Orthop. 2019;6(2):2–9.
93. Gray EG, Basmajian JV. Electromyography and cinematography of leg and foot ("normal" and flat) during walking. Anat Rec. 1968;161:1–15.
94. Mann, R. A.: Rupture of the tibialis posterior tendon. In Instructional Course Lectures, Am Acad Orthop Surg.1982; 33, pp. 302–309. St. Louis, C. V. Mosby, 1982.
95. Hinterman B, Nigg BM, Sommer C. Foot movement and tendon excursion: an in vitro study. Foot Ankle Int. 1994;15:386–95.
96. Klein P, Mattys S, Rooze M. Moment arm length variations of selected muscles acting on talocrural and subtalar joints during movement: an in vitro study. J Biomech. 1996;29:21–30.
97. Rattanaprasert U, Smith R, Sullivan M, Gilleard W. Three-dimensional kinematics of the forefoot, rearfoot, and leg without the function of tibialis posterior in comparison with normals during stance phase of walking. Clin Biomech. 1999;14:14–23.
98. Tome J, Nawoczenski DA, Flemister A, Houck J. Comparison of foot kinematics between subjects with posterior tibialis tendon dysfunction and healthy controls. J Orthop Sports Phys Ther. 2006;36:635–44.
99. Ringleb SI, Kavros SJ, Kotajarvi BR, Hansen DK, Kitaoka HB, Kaufman KR. Changes in gait associated with acute stage II posterior tibial tendon dysfunction. Gait Posture. 2007;25:555–64.
100. Houck JR, Neville CG, Tome J, Flemister AS. Ankle and foot kinematics associated with stage II PTTD during stance. Foot Ankle Int. 2009;30(6):530–9.
101. Ness ME, Long J, Marks R, Harris G. Foot and ankle kinematics in patients with posterior tibial tendon dysfunction. Gait Posture. 2008;27:331–9.
102. McCullough MB, Ringleb SI, Arai K, Kitaoka HB, Kaufman KR. Moment arms of the ankle throughout the range of motion in three planes. Foot Ankle Int. 2011;32(3):301–7.
103. Hui HJ, Beals TC, Brown NT. Influence of tendon transfer site on moment arms of the flexor digitorum longus muscle. Foot Ankle Int. 2007;28(4)
104. Wacker J, Calder JD, Engstrom CM, Saxby TS. MR morphometry of posterior tibialis muscle in adult acquired flat foot. Foot Ankle Int. 2003;24(4):354–7.
105. Angin S, Crofts G, Mickle KJ, Nester CJ. Ultrasound evaluation of foot muscles and plantar fascia in pes planus. Gait Posture. 2014;40:48–52.
106. Keenan MA, Peabody TD, Gronley JK, Perry J. Valgus deformities of the feet and characteristics of gait in patients who have rheumatoid arthritis. J Bone Joint Surg Am. 1991;73:237–47.
107. Hunt AE, Smith RM. Mechanics and control of the flat versus normal foot during the stance phase of walking. Clin Biomech. 2004;19(4):391–7.
108. Murley GS, Menz HB, Landorf KB. Foot posture influences the electromyographic activity of selected lower limb muscles during gait. J Foot Ankle Res. 2009;2(1):35.
109. Kehl AS, Corr M, Weisman MH. Review: enthesitis: new insights into pathogenesis, diagnostic modalities, and treatment. Arthritis Rheumatol. 2016;68(2):312–22.
110. Yang J, Zhang L, Yu C, Yang X-F, Wang H. Monocyte and macrophage differentiation: circulation inflammatory monocyte as biomarker for inflammatory diseases. Biomark Res. 2014;2(1):1–9.
111. Benjamin M, Ralphs JR. The cell and developmental biology of tendons and ligaments. Int Rev Cytol. 2000;196:85–130.
112. Benjamin M, McGonagle D. The anatomical basis for disease localisation in seronegative spondyloarthropathy at entheses and related sites. J Anat. 2001;199(Pt 5):503–26.
113. Moriggl B, Kumai T, Milz S, et al. The structure and histopathology of the enthesis organ at the navicular insertion of the tendon of the tibialis posterior. J Rheumatol. 2003;3:508–17.
114. Schweitzer ME, Caccese R, Karsick D, et al. Posterior tibial tendon tears: utility secondary signs for MR imaging diagnosis. Radiology. 1993;188:655–9.

115. Kiter E, Erdaq N, Karatosun V, et al. Tibialis posterior tendon abnormalities in feet with accessory bone and flatfoot. Acta Orthop Scand. 1999;70(6):618–21.
116. Petersen W, Hohmann G, Stein V, et al. The blood supply of the posterior tibial tendon. J Bone Joint Surg Br. 2002;84:141–4.
117. Stepien M. The sheath and arterial supply of the tendon of the posterior tibial muscle in man. Folia Morphol (Warsz). 1973;32:51–62.
118. Frey C, Shereff M, Greenidge N. Vascularity of the posterior tibial tendon. J Bone Joint Surg. 1990;72-A:884–8.
119. Prado MP, Carvalho AE, Rodrigues CJ, et al. Vascular density of the posterior tibial tendon: a cadaver study. Foot Ankle Int. 2006;27(8):628–31.
120. An KN, Berglund L, Uchiyama S, Coert JH. Measurement of friction between pulley and flexor tendon. Biomed Sci Instrum. 1993;29:1–7.
121. Zhao C, Amadio PC, Zobitz ME, Momose T, Couvreur P, An KN. Gliding resistance after repair of partially lacerated human flexor digitorum profundus tendon in vitro.2001. Clin Biomech. 16:696–701.
122. Uchimaya E, Kitaoka HB, Fujii T, et al. Gliding resistance of the posterior tibial tendon. Foot Ankle Int. 2006;27(9):723–6.
123. Arai K, Ringleb SI, Zhao KD, Gerglund LJ, Kitaoka HB, Kaufman KR. The effect of flatfoot deformity and tendon loading on the work of friction measured in the posterior tibial tendon. Clin Biomech. 2007;22:592–8.
124. Fujii T, Uchiyama E, Kitaoka HB, Luo ZP, Zhao K, et al. The influence of flatfoot deformity on the gliding resistance of tendons about the ankle. Foot Ankle Int. 2009;30:11.
125. Mosier SM, Lucas DR, Pomeroy G, et al. Pathology of the posterior tibial tendon in posterior tibial tendon insufficiency. Foot Ankle Int. 1998;19:520–4.
126. Fowble A, Vigorita V, Bryk E, et al. Neovascularity in chronic posterior tibial tendon insufficiency. Clin Orthop Relat Res. 2006;450:225.
127. Satomi E, Teodoro WR, Parra ER, et al. Changes in histoanatomical distribution of types I, III and V collagen promote adaptative remodeling in posterior tibial tendon rupture. Clinics (Sao Paulo). 2008;63:9.
128. Goncalves-Nieto J, Witzel S, Teodoro WR, et al. Changes in collagen matrix composition in human posterior tibial tendon dysfunction. Joint Bone Spine. 2002;69:189.
129. Deland JT, Arnoczky SP, Thompson FM. Adult acquired flatfoot deformity at the talonavicular joint: reconstruction of the spring ligament in an in vitro model. Foot Ankle. 1992;13:327–32.
130. Mankey MG. A classification of severity with an analysis of causative problems related to the type of treatment. Foot Ankle Clin. 2003;8:461–71.
131. Bluman EM, Title CI, Myerson MS. Posterior tibial tendon rupture: a refined classification system. Foot Ankle Clin. 2007;12:233–49.
132. Raikin SM, Winters BS, Daniel JN. The RAM classification: a novel, systematic approach to the adult-acquired flatfoot. Foot Ankle Clin. 2012;17:169–81.
133. Richie D. Biomechanics and orthotic treatment of the adult acquired flatfoot. Clin Podiatr Med Surg. 2020;37(1):71–89.
134. McCormack AP, Ching RP, Sangeorzan BJ. Biomechanics of procedures used in adult flatfoot deformity. Foot Ankle Clin. 2001;6:15–23.
135. Reeck J, Felten N, McCormack AP, et al. Support of the talus: a biomechanical investigation of the contributions of the talonavicular and talocalcaneal joints, and the superomedial calcaneonavicular ligament. Foot Ankle Int. 1998;19:674–82.
136. Backman C, Boquist L, Frid'en J, et al. Chronic achilles paratendonitis with tendinosis: an experimental model in the rabbit. J Orthop Res. 1990;8:541–7.
137. Enwemeka CS. Inflammation, cellularity, and fibrillogenesis in regenerating tendon: implications for tendon rehabilitation. Phys Ther. 1989;69:816–25.
138. Richie DH. Biomechanics and clinical analysis of the adult acquired flatfoot. Clin Podiatr Med Surg. 2007;24:617–44.
139. Yeap JS, Sing D, Birch R. Tibialis posterior tendon dysfunction: a primary or secondary problem? Foot Ankle Int. 2001;22:51–5.

140. Yeap JS, Birch R, Singh D. Long-term results of tibialis posterior tendon transfer for drop-foot. Int Orthop. 2001;25:114–8.
141. Mizel M, Temple H, Scranton PJ, Gellman R, Hecht PJ, et al. Role of the peroneal tendons in the production of the deformed foot with posterior tibial tendon deficiency. Foot Ankle Int. 1999;20:285–9.
142. Tao K, Ji W-T, Wang D-M, Wang C-T, Wang X. Relative contributions of plantar fascia and ligaments on the arch static stability: a finite element study. Biomed Tech Biomed Eng. 2010;55:265–71.
143. Huang CK, Kitaoka HB, An KN, Chao EYS. Biomechanical evaluation of longitudinal arch stability. Foot Ankle. 1993;14(6):353–7.
144. Kitaoka HB, Luo ZP, An KN. Three-dimensional analysis of flatfoot deformity: cadaver study. Foot Ankle Int. 1998;19(7):447–51.
145. Chu IT, Myerson MS, Nyska M, Parks BG. Experimental flatfoot model: the contribution of dynamic loading. Foot Ankle Int. 2001;22(3):220–5.
146. Pasapula C, Devany A, Fischer N, Wijdicks C, Hübner T, et al. The resistance to failure of spring ligament reconstruction. Foot. 2017;33:29–34.
147. Borton DC, Saxby T. Tear of the plantar calcaneonavicular (spring) ligament causing flatfoot. A case report. J Bone Joint Surg Br. 1997;79:641–3.
148. Jennings MM, Christensen JC. The effects of sectioning the spring ligament on rearfoot stability and posterior tibial tendon efficiency. J Foot Ankle Surg. 2008;47:219–24.
149. De la Portilla CC, Larrainzar-Garijo R, Bayod J. Biomechanical stress analysis of the main soft tissues associated with the development of adult acquired flatfoot deformity. Clin Biomech. 2019;61(1):163–71.
150. MacDonald A, Cifo D, Knapp E, Awad H, Ketz J, Flemister A. Peritalar kinematic changes associated with increased spring ligament tear in cadaveric flatfoot model. Foot Ankle Orthop. 2018:3(3).
151. Balen PF, Helms CA. Association of posterior tibial tendon injury with spring ligament injury, sinus tarsi abnormality, and plantar fasciitis on MR imaging. AJR Am J Roentgenol. 2001;176(5):1137–43.
152. Williams G, Widnall J, Evans P, Platt S. Could failure of the spring ligament complex be the driving force behind the development of the adultflatfoot deformity? J Foot Ankle Surg. 2014;53(2):152–5.
153. McCormack AP, Niki H, Kiser P, Tencer AF, Sangeorzan BJ. Two reconstructive techniques for flatfoot deformity comparing contact characteristics of the hindfoot joints. Foot Ankle Int. 1998;19(7):452–62.
154. Ormsby N, Jackson G, Evans P, Platt S. Imaging of the tibionavicular ligament and its potential role in adult acquired flatfoot deformity. Foot Ankle Int. 2018;39(5):629–35.
155. Pasapula C, Cutts S. Modern theory of the development of adult acquired flat foot and an updated spring ligament classification system. Clin Res Foot Ankle. 2017;5(3):247.
156. Alsousou J. Identification of the medial column line collapse variation is crucial in flatfoot management. Bristol: BOFAS; 2016.
157. Swanton E, Fisher L, Fisher A, Molloy A, Mason L. An anatomic study of the naviculocuneiform ligament and its possible role maintaining the medial longitudinal arch. Foot Ankle Int. 2019;40(3):352–5.
158. Sharkey NA, Linda F, Donahue SW. Biomechanical consequences of plantar fascial release or rupture during gait. Part I. disruption in longitudinal arch conformation. Foot Ankle. 1998;18:812–20.
159. Shibuya N, Ramanujam CL, Garcia GM. Association of tibialis posterior tendon pathology with other radiographic findings in the foot: a case-control study. J Foot Ankle Surg. 2008;47:546–53.
160. Kitaoka HB, Ahn T, Luo ZP, An K. Stability of the arch of the foot. Foot Ankle Int. 1997;18:644–8.
161. Donovan A, Rosenberg ZS. Extraarticular lateral hindfoot impingement with posterior tibial tendon tear: MRI correlation. AJR. 2009;193:672–8.

162. Kjaersgaard-Andersen P, Wethelund J, Helmig P, Soballe K. The stabilizing effect of the ligamentous structures in the sinus and canalis tarsi on the movements in the hindfoot. Am J Sports Med. 1988;16:512–6.
163. Beltran J. Sinus tarsi syndrome. Magn Reson Imaging Clin N Am. 1994;2:59–65.
164. Klein MA, Spreitzer AM. MR imaging of the tarsal sinus and canal: normal anatomy, pathologic findings, and features of the sinus tarsi syndrome. Radiology. 1993;186:233–40.
165. Ellis SJ, Yu JC, Williams BR, Lee C, Chiu YL, Deland JT. New radiographic parameters assessing forefoot abduction in the adult acquired flatfoot deformity. Foot Ankle Int. 2009;30:1168–76.
166. Lin YC, Mhuircheartaigh JN, Lamb J, Kung JW, Yablon CM, Wu JS. Imaging of adult flatfoot: correlation of radiographic measurements with MRI. AJR. 2015;204:354–9.
167. Younger AS, Sawatzky B, Dryden P. Radiographic assessment of adult flatfoot. Foot Ankle Int. 2005;26(10):820–p825.
168. Xu J, Zhang Y, Muhammad H, Wang X, Huang J, Chao Z, et al. In vivo three-dimensional analysis of hindfoot kinematics in stage II PTTD flatfoot. J Orthop Sci. 2015;20:488–97.
169. Zhang Y, Xu J, Wang J, Huang J, Zhang C, et al. An in vivo study of hindfoot 3D kinetics in stage II posterior tibial tendon dysfunction (PTTD) flatfoot based on weight-bearing CT scan. Bone Joint Res. 2013;2:255–63.
170. Hansen ST. Progressive symptomatic flat foot (lateral peritalar subluxation). In: Hansen ST, editor. Functional reconstruction of the foot and ankle. Philadelphia: Lippincott Williams & Wilkins; 2000. p. 195–207.
171. Van Boerum DH, Sangeorzan BJ. Biomechanics and pathophysiology of flat foot. Foot Ankle Clin. 2003;8:419–30.
172. Lee SM, Piazza SJ. Inversion-eversion moment arms of gastrocnemius and tibialis anterior measured in vivo. J Biomech. 2008;41:3366–70.
173. Sung IH, Lee S, Otis J, Deland J. Posterior Tibial Tendon Force Requirement in Early Heel Rise After Calcaneal Osteotomies. Foot Ankle Int. 2002;23:842–9.
174. Arangio G, Rogman A, Reed J. Hindfoot alignment valgus moment arm increases in adult flatfoot with Achilles tendon contracture. Foot Ankle Int. 2009;30(11):1078–82.
175. Niki H, Ching RP, Kiser P, Sangeorzan BJ. The effect of posterior tibial tendon dysfunction on hindfoot kinematics. Foot Ankle Int. 2001;22:292–300.
176. Neville C, Flemister AS, Houck JR. Deep posterior compartment strength and foot kinematics with stage II posterior tibial tendon dysfunction. Foot Ankle Int. 2010;31(4):320–8.
177. Neville C, Flemister AS, Houck J. Total and distributed plantar loading in subjects with stage II tibialis posterior tendon dysfunction during terminal stance. Foot Ankle Int. 2013;34:131–9.
178. Van de Velde M, Matricali GA, Wuite S, Roels C, Staes F, Deschamps K. Foot segmental motion and coupling in stage II and III tibialis posterior tendon dysfunction. Clin Biomech. 2017;45:38–42.
179. Chuter VH, Janse de Jonge XAK. Proximal and distal contributions to lower extremity injury: a review of the literature. Gait Posture. 2012;36:7–15.
180. Ferber R, McClay Davis I, Williams DS III. Effect of foot orthotics on rearfoot and tibia joint coupling patterns and variability. J Biomech. 2005;38:477–83.
181. Hinterman B, Sommer C, Nigg B. Influence of ligament transection on tibial and calcaneal rotation with loading and dorsiflexion – plantarflexion. Foot Ankle. 1995;16:567–71.
182. Semple R, Murley G, Woodburn J, Turner D. Tibialis posterior in health and disease: a review of structure and function with specific reference to electromyographic studies. J Foot Ankle Res. 2009;2,2(1).

Plantar Heel Pain

8

It is reasonably certain that a condition which has so many different theories of etiology and treatment does not have valid proof of any one cause. – Snook and Chrisman: A study of plantar heel pain in 1972. [1]

Key Points

- Plantar heel pain is a common disorder of the foot which involves multiple anatomic structures.
- It is widely accepted that inflammation followed by degeneration of the plantar aponeurosis at its attachment to the calcaneus is the most common pathology associated with plantar heel pain.
- During walking and running, the plantar aponeurosis is subjected to loads which approach the maximal limit of load to failure.
- Entrapment of the inferior calcaneal nerve can be a primary or secondary cause of chronic plantar heel pain.
- Increased body mass index and reduced ankle joint dorsiflexion are the two most common risk factors for plantar heel pain.
- Deficits in toe flexor strength have been identified in patients with plantar heel pain.
- There is conflicting evidence verifying the role of hindfoot pronation and plantar heel pain.
- Dorsiflexion of the digits increases strain in the plantar aponeurosis by 70%, while Achilles tendon tension increases strain by 30%.
- Kinematic studies of patients fail to demonstrate a consistent pattern of foot mechanics which predispose to the development of plantar heel pain.
- Plantar fascia thickening rather than calcaneal spurs is more commonly associated with plantar heel pain.

D. H. Richie Jr, *Pathomechanics of Common Foot Disorders*,
https://doi.org/10.1007/978-3-030-54201-6_8

Introduction

Heel pain is probably the most common condition affecting the human foot. However, it is under reported. This may be due to the simple fact that incidence rates of foot conditions are largely based upon a number of visits to doctors, yet with plantar heel pain, a larger portion of patients never seek treatment. It is reported that over 2 million people per year in the United States seek treatment for plantar heel pain and that the condition will affect 10% of the population in their lifetime [2]. However, recent surveys of incidence of foot pain suggest that plantar heel pain may be more prevalent than these estimates. A population-based survey of 3206 people age 20 and older in Australia revealed that 4% of the subjects had plantar heel pain at the time of taking the survey. Using this percent incidence in the United States, which has a population of 253,768,092 adults age 18 years old or older, 10 million of them could be expected to have plantar heel pain at the same time [3].

The term "plantar heel pain" is descriptive and encompasses a range of diagnoses [4]. Conditions which can cause pain on the plantar aspect of the calcaneus include plantar fasciitis, calcaneal spur, subcalcaneal bursitis, calcaneal periostitis, calcaneal stress fracture, fat pad syndrome, entrapment of Baxter's nerve, and radiculopathy [5]. There is general consensus that plantar fasciitis is the most common cause of plantar heel pain [5–10].

The term plantar fasciitis has fallen under some scrutiny as the name implies an inflammatory condition. Histopathologic studies of patients with chronic plantar heel pain demonstrate degenerative changes in the plantar fascia rather than inflammatory infiltrates, which has prompted some authors to use the term "plantar fasciopathy" [11]. However, the term "plantar fasciitis" continues to be preferred by most clinicians when diagnosing the common presentation of plantar heel pain. The pathomechanics of plantar heel pain is a continuum from inflammation to degeneration of the plantar fascia in most patients. This process will be discussed in depth in this chapter justifying the use of the term plantar fasciitis to describe the common cause of plantar heel pain. We will also discuss the relationship between plantar fasciitis and entrapment of the inferior calcaneal nerve which is another cause of plantar heel pain.

Anatomy

The anatomy of the plantar aponeurosis is summarized according to the detailed descriptions which are provided by Sarrafian [12]. The plantar aponeurosis has three components: central, lateral, and medial (Fig. 8.1). The medial and lateral components are thin fascial coverings for the medial and lateral intrinsic musculature and do not play a role in plantar heel pain. The central component of the plantar aponeurosis is a strong fibrous sheet of tissue which extends from the heel to the ball of the foot. Its origin is primarily from the posteromedial calcaneal tuberosity, approximately 1.5–2 cm in width. In this chapter, the term plantar fascia is synonymous with the term plantar aponeurosis, and both terms refer to the central band of this structure rather than the superficial medial and lateral bands.

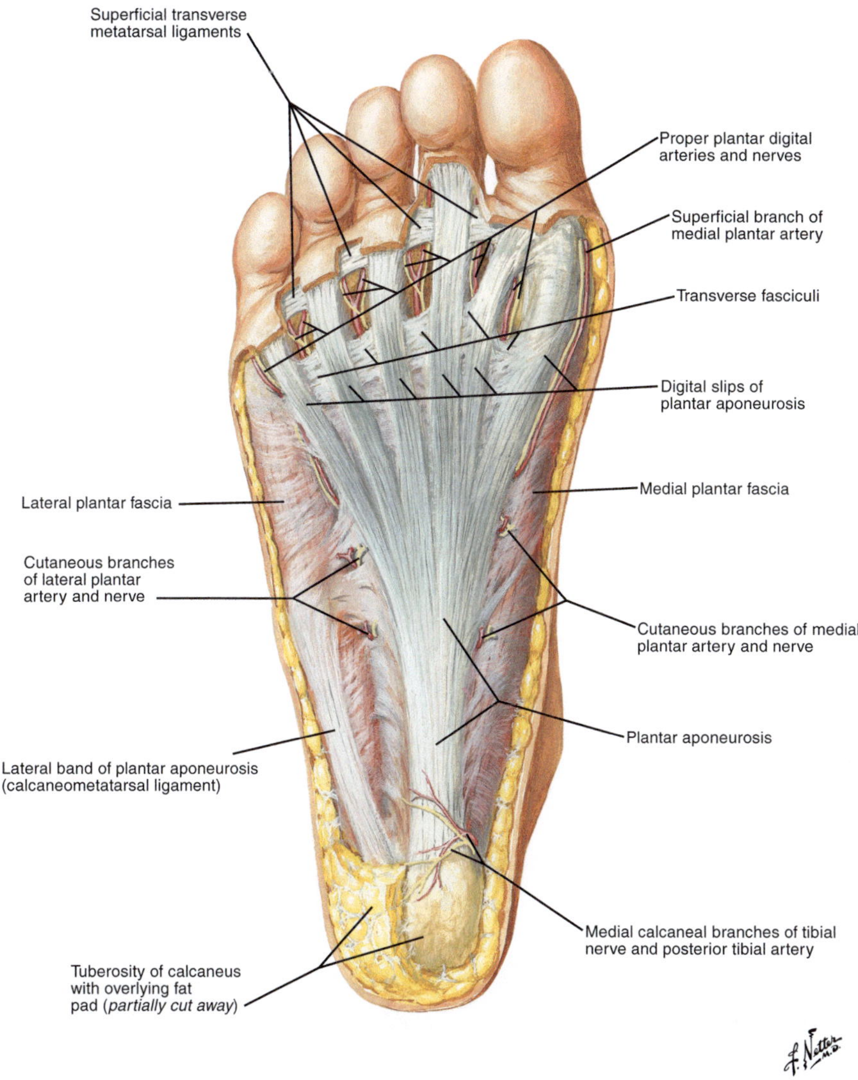

Fig. 8.1 The plantar fascia: medial, central, and lateral bands. (Re-printed with permission from Netterimages)

Anatomic and clinical articles often describe a connection between the insertion of the Achilles tendon and the origin of the plantar aponeurosis. In the Italian literature, there is a description of the Achilles-calcaneal-plantar arch system [13]. In the English literature, there is a reference to the "foot and ankle linkage system" describing an anatomic connection between the Achilles and the plantar aponeurosis [14]. However, in a cadaver study of 10 adults, three neonates, and two fetal feet, Snow and co-workers showed that, while a connection between the Achilles and the plantar aponeurosis was clearly evident in the neonate, it becomes separated by periosteum in the adult foot [15]. The authors concluded that while the linkage between the Achilles and the plantar aponeurosis is strongly mechanical, it is not anatomic.

As the plantar aponeurosis is followed from its origin from the calcaneus into the foot, it is observed to divide at the mid-metatarsal level. The division provides five longitudinally oriented segments which then divide into three superficial and five deep tracts. The three central superficial tracts insert into the skin and into the superficial transverse metatarsal ligament, also known as the plantar interdigital ligament (Fig. 8.2). The five deep tracts form ten sagittal septa which pass along each side of the long flexor tendons. Each sagittal septa inserts in the interosseous fascia, the fascia of the transverse head of the adductor hallucis, the deep transverse metatarsal ligament, and the plantar plate of the metatarsophalangeal joint.

In the great toe joint, the medial sagittal septa inserts on the plantar plate of the 1st MTPJ and the medial sesamoid as well as the medial head of the flexor hallucis brevis. The lateral septum of the aponeurotic band of the great toe inserts on the transverse metatarsal ligament, the plantar plate, and the lateral sesamoid and connects with the fascia of the lateral head of the flexor hallucis brevis muscle. The distal insertions of the plantar aponeurosis provide essential stabilization of the digits. A lesser recognized function of the plantar aponeurosis is stabilization of the

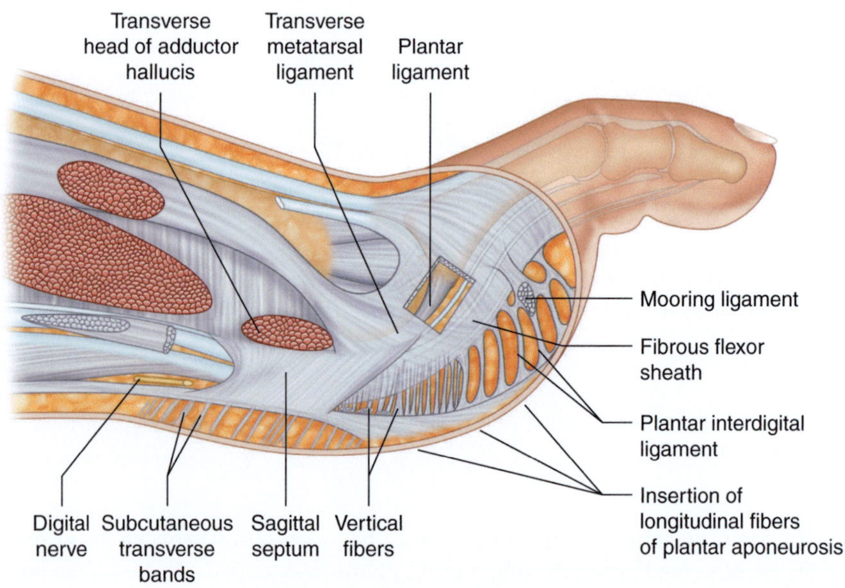

Fig. 8.2 Distal insertions of the plantar aponeurosis: each sagittal septa inserts in the interosseous fascia, the fascia of the transverse head of the adductor hallucis, the deep transverse metatarsal ligament, and the plantar plate of the metatarsophalangeal joint. The three central superficial components insert into the skin and into the superficial transverse metatarsal ligament, also known as the plantar interdigital ligament. (From: Bojsen-Moller and Flagstad [145]. Figure 6)

plantar fat pad and plantar integument distal to the MTPJs particularly during terminal stance and pre-swing.

The Nerves and Plantar Heel Pain

There is much confusion and misuse of terminology regarding the nerves passing along the medial and inferior aspect of the calcaneus. The first branch of the lateral plantar nerve has been implicated by many authors to be a potential cause of plantar heel pain [16, 18–22]. This nerve, also known as the nerve to the abductor digiti quinti, originates from the lateral plantar nerve at the level of the bifurcation of the tibial nerve into the medial and lateral plantar nerves [17]. The first branch of the lateral plantar nerve passes thru a discreet corridor bordered medially by the abductor hallucis muscle and laterally by the medial head of the quadratus plantae muscle (Fig. 8.3). At the lower border of the abductor hallucis, the nerve makes an abrupt lateral turn passing above the thick fascia of the medial intermuscular septum. At this lateral turn, located across the bony inferior medial ridge of the calcaneus, the first branch of the lateral plantar nerve is in a vulnerable position for entrapment. This nerve crosses beneath the quadratus plantae muscle obliquely giving sensory branches to the periosteum of the medial tuberosity of the calcaneus, the long plantar ligament, and motor branches to the lateral head of the quadratus plantae, the flexor digitorum brevis, and finally divides into two branches which insert into the abductor digiti quinti [17]. The first branch of the lateral plantar nerve is thus encased in a tight tunnel as it passes from medial to lateral across the plantar surface of the calcaneus (Fig. 8.4). The plantar fascia and flexor digitorum muscle form the floor of the tunnel, while the calcaneus, the quadratus plantae, and the long plantar ligament form the roof. The roof and floor of this tunnel converge proximally to meet at the anterior margin of the calcaneus which forms a proximal wall. The first branch of the lateral plantar nerve has potential for entrapment from multiple sources: the deep fascia or medial intermuscular septum separating the abductor hallucis from the quadratus plantae, the bony inferior medial ridge of the calcaneus,

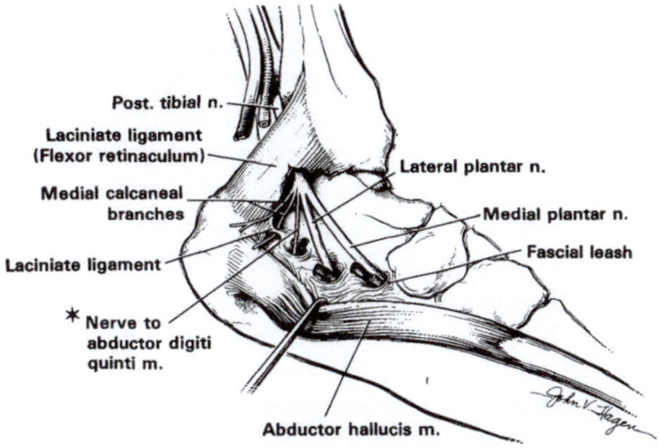

Fig. 8.3 First branch of the lateral plantar nerve (nerve to the abductor digiti minimi). (From: Baxter and Thigpen [146]. Figure 2)

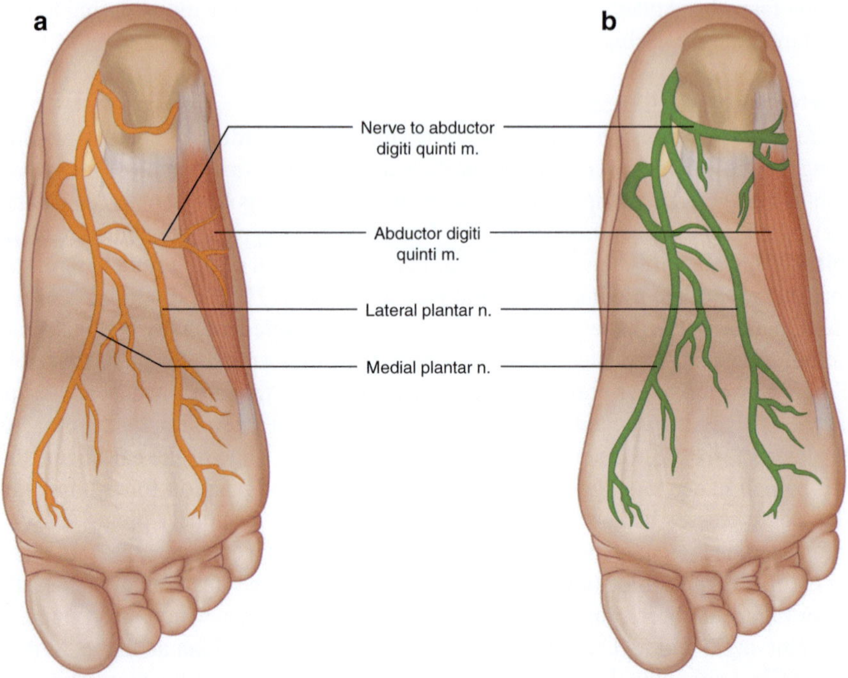

Fig. 8.4 Baxter and Thigpen showed that, contrary to most anatomy texts depicting a distal location of the nerve to the abductor digiti quinti muscle (**a**), the nerve actually passes along the distal plantar margin of the medial calcaneal tubercle (**b**) where entrapment may occur along the corridor for passage. (Baxter and Thigpen [146]. Fig. 1)

and the direct pressure from either the plantar aponeurosis or the flexor digitorum brevis, or both. With degeneration and thickening of the proximal plantar fascia, the inferior calcaneal nerve, lying directly above, can become entrapped against a "wall" which is the inferior surface of the body of the calcaneus.

Roegholt was the first to propose a mechanism whereby the first branch of the lateral plantar nerve could become entrapped, causing plantar heel pain [18]. He renamed the first branch of the lateral plantar nerve the *inferior calcaneal nerve* and speculated that its position could be entrapped by a calcaneal spur. Tanz evaluated the first branch of the lateral plantar nerve and also observed that there was potential for entrapment by a calcaneal spur, inflammation, or venous engorgement [19].

Aronson et al. studied 30 cadaver feet and confirmed previous reports describing the course of the inferior calcaneal nerve with sensory branches to the periosteum of the calcaneus [20]. Przylucki and Jones dissected the inferior calcaneal nerve in four patients with heel pain and performed histologic examination of the specimens, documenting changes consistent with entrapment neuropathy [21]. They proposed a mechanism for entrapment at the medial edge of the calcaneus.

In their seminal article, Baxter and Thigpen described neurolysis of the "mixed nerve supplying the abductor digiti quinti muscle" in 26 patients [22]. The authors proposed two sites of potential entrapment: the fascia of the abductor hallucis muscle and the medial edge of the calcaneus (Figs. 8.3 and 8.4). After this article was published, entrapment of the first branch of the lateral plantar nerve became known as "Baxter's neuropathy" despite the fact that the condition had been described by several authors in previous publications [23, 24].

It is appropriate that Saraffian adopted the name *inferior calcaneal nerve* to describe the first branch of the lateral plantar nerve [17]. The term "first branch of the lateral plantar nerve" is a misnomer because the nerve is not always the first branch off of the lateral plantar nerve. The term "nerve to abductor digiti quinti" is misleading because it neglects the important motor input that the nerve provides to the flexor digitorum brevis and quadratus plantae muscles. The term inferior calcaneal nerve is appropriate in describing its unique location and its potential role in plantar heel pain.

In comparison, the medial calcaneal nerve is a pure sensory nerve which has no innervation to the plantar fascia or periosteum of the calcaneus [25]. This nerve originates from the posterior tibial nerve usually posterior and proximal to the first branch of the lateral plantar nerve [17] (Fig. 8.5). The medial calcaneal nerve divides into three branches at the level of the tarsal tunnel. The posterior branch innervates the skin covering the medial aspect of the Achilles tendon as well as the integument of the medial and posterior portion of the heel. The middle and anterior branches course thru the fat pad of the heel innervating the skin on the plantar heel and proximal arch area. There is no connection of this nerve to the deep fascia, the periosteum of the calcaneus, or the plantar aponeurosis. The medial calcaneal nerve is a sensory nerve which innervates the fat pad and bursa structures which are superficial to the calcaneus. Entrapment of the medial calcaneal nerve can be involved with a variation of plantar heel pain not caused by impingement of the structures surrounding the plantar fascia [17, 25].

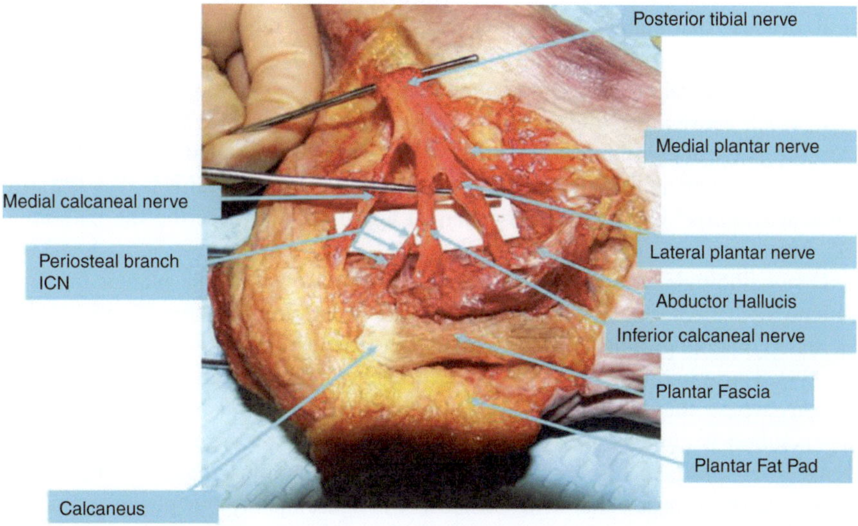

Fig. 8.5 Medial plantar nerve innervates fat pad and integument under calcaneus. Inferior calcaneal nerve passes above plantar aponeurosis and abductor hallucis

Histology

The enthesis or proximal attachment of the plantar fascia to the calcaneus is a direct bridge of fibrocartilage rather than an indirect periosteal attachment [26–28]. Fibrocartilage appears to be located predominantly at sites within tendons and ligaments that are subjected to bending, shear or compressive forces, or a combination of all three forces [29]. It is noted that tensile forces are not associated with formation of fibrocartilage at tendon insertions.

Wearing and co-workers make an argument that the fibrocartilage enthesis of the plantar aponeurosis, void of periosteal attachment to the calcaneus, is not susceptible to direct traction and detachment [30]. Traction in the plantar fascia is commonly blamed as the cause of an inferior calcaneal spur, yet there is very little evidence to support this theory. Tensile strain may be a mechanism for plantar aponeurosis injury located more distal to the enthesis. Indeed most authorities attribute degeneration of the plantar fascia to abnormal tensile loads. Hence, many studies have been conducted to determine what causes these abnormal loads. This will be discussed further in the section covering function of the plantar aponeurosis.

Wearing et al. describe four zones of the fibrocartilaginous enthesis which allow a gradual transition from uncalcified fibrocartilage to partially calcified cartilage which transitions to the bone. According to Wearing et al., by incorporating calcified and uncalcified fibrocartilaginous zones, direct attachments provide a gradual transition from hard to soft tissue, which helps to dissipate stress evenly [30].

The fibrocartilaginous insertion of the plantar aponeurosis into the calcaneus is prone to degenerative change [31]. This leads to ossification within the fibrocartilage and the formation of a spur which is typically found deep to the plantar fascia at the dorsal surface of the enthesis [31, 32] (Fig. 8.6). The changes seen in the fibrocartilage of the plantar aponeurosis may also be a response to compression, bending, and shear forces for which this structure is designed to dissipate. Wearing et al. propose that plantar heel spurs may provide a buttressing effect to the fascia by effectively minimizing bending at the insertion [30].

There is general consensus that plantar fasciitis is the result of mechanical overload which causes micro-tears in the fascia leading to an inflammatory response [33, 34]. A chronic inflammatory process results from continued repetitive trauma causing breakdown of tissue which exceeds tissue remodeling. This is why the term "plantar fasciitis" has traditionally seemed appropriate to describe the pathology [35, 36].

Histopathologic studies of surgical specimens taken from the plantar fascia in patients with plantar heel pain show collagen degeneration with fiber disorientation, increased mucoid ground substance, angiofibroblastic hyperplasia, and calcification [35–37]. These findings are consistent with the same degenerative changes seen in tendons where the condition is commonly termed "tendinosis" [38, 39]. While degeneration is evident, inflammatory infiltrates were not commonly seen in the tissue specimens taken during heel spur surgery or plantar fasciotomy. Hence, some authors have proposed the term "fasciosis" or "fasciopathy" in place of fasciitis when describing a common cause of plantar heel pain [11, 40].

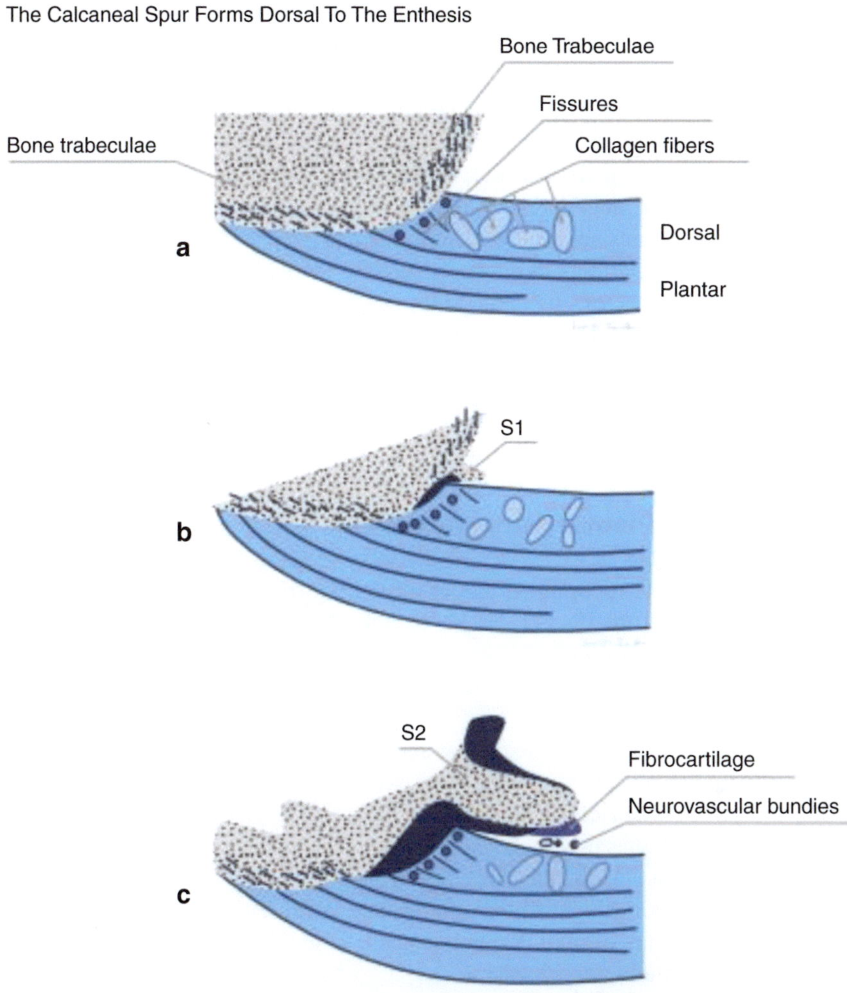

Fig. 8.6 Stages of formation of the calcaneal spur, dorsal to the plantar aponeurosis. Three possible stages in the formation of a heel spur. Each drawing shows two groups of trabeculae at the plantar fascia enthesis. The distal trabeculae are oriented along the direction of pull of the fascia and the proximal trabeculae along the line of force transmission that is associated with weight bearing. (**a**) Prior to spur formation, degenerative changes occur in the dorsal enthesis fibrocartilage, near the bony interface. (**b**) The development of a small spur (S1) is associated with subchondral bone sclerosis (shaded area), on the plantar side of the spur (arrow). (**c**) In larger spurs (S2), the sclerosis (shaded area) is more obvious and is present on both its dorsal and plantar sides (arrows). The tip of the spur (particularly its plantar side) is covered with a pad of fibrocartilage and between the pad and the plantar fascia itself are neurovascular bundles. (Adapted from: Kumai and Benjamin [147])

It is important to note that these histopathologic studies evaluated specimens taken from patients undergoing surgery for chronic heel pain [35–37, 40]. Perhaps these specimens demonstrate the final end stage of a continuum of tissue response to repetitive microtrauma. There is experimental evidence that tendons undergo a period of inflammatory change before developing degenerative changes seen in tendinosis [41, 42]. On the other hand, there are theories that tendons can undergo degenerative change without any preceding inflammation and that the two entities are independent processes [43]. What is needed are more histology studies of patients in the early stage of plantar heel pain to compare to the existing studies of patients with long-standing chronic heel pain. Notwithstanding, the term "plantar fasciitis" continues to be the preferred term by most health professionals as well as the lay public to describe the condition of chronic heel pain and therefore will be used in preference to "plantar fasciopathy" in this chapter.

Function of the Plantar Aponeurosis

The plantar fascia is recognized as the most important arch supporting structure of the human foot [8–11]. However, the mechanism explaining how the plantar aponeurosis supports the arch of the foot, and when this arch supporting mechanism is most critical during the gait cycle has been open to debate and conjecture.

Insight into the function of the plantar aponeurosis can be gained with studies of effects of surgical plantar fasciotomy. Several important studies were conducted during the 1990s when endoscopic plantar fasciotomy procedures were gaining popularity. These studies used static cadaver models which were positioned to simulate loads in the foot during midstance. Later, cadaver studies positioned the foot to mimic terminal stance which is more appropriate, given that maximal loads on the plantar aponeurosis are exerted during this phase of gait when heel rise occurs with activation of the windlass mechanism.

Huang et al. evaluated cadaver specimens under vertical loading and sequentially sectioned the plantar fascia, the spring ligament, and the long and short plantar ligaments [44]. The plantar fascia provided the most significant contribution to arch stability, and when sectioned, a decrease in arch stiffness by 25% was observed. An important finding from this study was that after complete release of all three structures, i.e., the plantar fascia, the spring ligament, and the long and short plantar ligaments, the longitudinal arch still retained 63% of its stiffness. Therefore, other static support mechanisms also contribute to arch stability. Later in this chapter, the transverse longitudinal arch contribution to foot stiffness will be discussed.

In a study of six cadaveric specimens, Murphy et al. showed that a partial, medial transection of the plantar fascia caused an 18% drop in the medial longitudinal arch and an 8% drop in the lateral longitudinal arch [45]. When a complete release of the plantar fascia was performed, the medial longitudinal arch dropped 29%, and the lateral longitudinal arch dropped 18%. This static cadaver model also positioned the foot flat on the supportive surface, duplicating midstance.

Thordarson studied the effects of sequential release of the plantar aponeurosis in eight cadaver specimens mounted on a loading frame [46]. The medial arch of the foot was observed to lower and elongate with progressive sectioning of the plantar fascia. In addition, the forefoot was noted to abduct, while the navicular rotated into inversion indicating triplane instability throughout the foot. Furthermore, the windlass mechanism was significantly compromised with complete sectioning of the plantar fascia.

In another cadaver study, Kitaoka et al. measured three-dimensional rotation of multiple joints of the midfoot and hindfoot when a complete plantar fasciotomy was performed [47]. Along the medial column, there was motion of the joints in the direction of dorsiflexion, abduction, and eversion suggesting an overall destabilization of the arch in three planes. Kitaoka and Daly reported the results of a mean 8-year follow-up of 12 patients who had undergone plantar fasciotomy for plantar heel pain [48]. There was evidence of flattening of the longitudinal arch and a loss of push off power indicating less efficient propulsion during walking.

The effect of plantar fascia release on stability of the digits was demonstrated in a study by Sharkey and co-workers [49]. Using cadaver specimens positioned in terminal stance, complete plantar fasciotomy shifted plantar pressure away from the toes to the metatarsal heads. This resulted in significant increase in strain and bending of the second metatarsal.

The previous described studies utilized static or quasi-dynamic cadaver models. These models do not replicate all the forces and movements occurring in the human foot during walking or running. Erin Ward and colleagues were among the first researchers to develop and study a dynamic cadaver model which was able to simulate human walking. Ward and co-workers studied the effects of sequential release of the plantar fascia on the kinematics of the foot [50]. They were also able to measure tensile strain or loads in the plantar aponeuronsis during the walking gait cycle. The force in the plantar fascia begins just after heel strike and peaks during midstance with a second peak during terminal stance. Force was greater in the medial band of the plantar aponeurosis compared to the lateral band.

Ward and co-workers also discovered a previous unreported function of the plantar fascia to facilitate supination of the foot during late midstance [50]. As the plantar fascia was sectioned, the rearfoot was unable to supinate from a pronated position during late midstance. This suggests a role of the plantar aponeurosis to facilitate the dynamic mechanisms which stiffen the foot for push off. Thus, the plantar fascia is not just a simple arch supporting structure. For example, a computerized tomography study verified that after plantar fasciotomy, cadaver specimens demonstrated increased strain in the subtalar joint ligaments [51]. Therefore, the plantar aponeurosis appears to play a role in stability of the subtalar joint in addition to supporting the longitudinal arch.

Behavior of the Plantar Fascia Under Loading

Studies of the effects of plantar fasciotomy provide a partial insight into the function of the plantar aponeurosis. By combining the data from these studies of cadaver models, we can see that complete plantar fasciotomy will result in an average decrease in arch height of 7.4 ± 4.1 mm and in an average 15% elongation of the arch of the foot [45–47]. Consistently, these studies also show that complete plantar fasciotomy does not cause complete collapse of the longitudinal arch of the foot. Other passive mechanisms such as the bone configuration, the articular surfaces of the joints spanning the arch, as well as other ligamentous structures can partially compensate and maintain arch stability in spite of loss of the plantar aponeurosis.

Since tensile strain has been implicated as the primary deforming force causing plantar fasciitis, numerous studies have been undertaken to learn how and when strain is developed in the plantar aponeurosis during standing and walking. The challenge when conducting these studies is performing direct measure of strain in the plantar aponeurosis, as this would require invasive measures to implant strain gauges; an intervention which cannot be carried out on human subjects. When studying cadaveric specimens, strain gauges can be inserted directly into the plantar aponeurosis, but simulating the loads imposed upon the foot during dynamic gait is challenging. Notwithstanding, valuable insight has been gained by several studies which are presented here to better understand the mechanics of plantar fascia overload.

Carlson studied the effects of Achilles tendon and windlass engagement on strain in the plantar fascia [52]. Eight cadaver specimens were mounted on a frame which positioned the foot to simulate terminal stance or heel rise. Strain in the plantar fascia was measured with an extensometer. At a low magnitude of force application of the Achilles tendon, there was almost a 1:1 transfer of force to the plantar fascia. As higher force approaching body weight was applied to the Achilles, the load transfer to the plantar fascia diminished. A tensile force in the Achilles of 500 Newtons produced 314 N of tensile force in the plantar fascia at zero degrees dorsiflexion of the toes. With 45 degrees dorsiflexion of the toes, the plantar fascia tension increased to 511 Newtons, a 1.6 times increase due to the windlass mechanism. A regression analysis of the results revealed that the toe angles were more important in transferring force to the plantar fascia than Achilles tendon force. *Overall force in the plantar fascia was influenced 84% by dorsiflexion of the digits and only 15% by Achilles tendon force.*

Further insight into the role of the Achilles and windlass on strain of the plantar fascia provided by a study by Cheng and co-workers [53]. Using a three- dimensional finite element model of the foot, they studied the behavior of the plantar fascia under stretch induced by dorsiflexion of the toes or tensioning of the Achilles tendon. Finite element method (FEM) is a computational tool which uses computer simulations to quantify mechanical responses of a model. With FEM, the researchers can manipulate certain parameters while holding others constant to evaluate the influence of the particular parameter. This model allowed computation of strain in all five slips of the plantar fascia. The results showed a transfer of strain from toe dorsiflexion angle and the Achilles in similar magnitude as reported by Carlson et al. The strain was most concentrated near the attachment of the plantar fascia at the medial calcaneal tubercle. Strain was higher beneath the first ray and decreased moving lateral toward the fifth ray. The predominant load bearing of the plantar fascia under the first ray increased as Achilles tendon and toe dorsiflexion movement was applied. The authors attributed the location of strain medially in the plantar aponeurosis to the enhanced efficiency of the windlass mechanism at the 1st MTPJ made possible by the enlarged first metatarsal head and sesamoid apparatus. *Dorsiflexion of the digits contributed 66% of the strain to the plantar fascia, while Achilles tendon force accounted for 33% of the strain in the plantar fascia.*

Geffen used fluoroscopy and plantar pressure mapping to determine arch configuration and plantar fascia length in two human subjects while walking [54]. The plantar fascia undergoes rapid lengthening during weight acceptance thru midstance while undergoing slower elongation during push off. Geffen attributes the lengthening of the plantar fascia to the histology of the structure which demonstrates both elastic and collagen fibers. The elastic fibers appeared to provide structural support during initial loading of the foot thru midstance allowing the plantar fascia to elongate. The collagen fibers were placed under load during heel rise and push off, providing stiffness to the plantar fascia. This study revealed the important finding that the plantar aponeurosis is certainly not a rigid, non-yielding structure. Its elastic fibers allow stretching, reaching a maximum of 9–12% elongation between midstance and pre-swing.

Erdemir and co-workers developed a dynamic cadaver simulator to study the effects of Achilles tendon load transfer to the forefoot during walking gait [55]. A fiber optic transducer was used to measure plantar aponeurosis force which was shown to peak in late stance. The force approached 96% of body weight. There was good correlation between Achilles tension/force and plantar fascia tension/force. The authors concluded that the plantar aponeurosis plays a critical role in transmitting Achilles tendon force to the forefoot during terminal stance.

Another finite element model was constructed by Cheung and co-workers to study the effect of Achilles tendon loads on plantar fascia strain [56]. Measured strain increased linearly with increasing vertical load placed upon the foot. By comparison, Achilles tendon load produced two times the strain in the plantar fascia than vertical loading. During bilateral standing, the plantar fascia bears 44% of the total weight applied to the foot. During terminal stance, the plantar fascia bears 77% of body weight. The large load borne by the plantar fascia confirms its role as a major arch supporting mechanism of the human foot. The finite element model predicted that during bilateral quiet standing, the Achilles tendon generates a force approximating 75% body weight or 350 Newtons on the standing foot.

The Kogler Studies

Kogler and co-workers conducted a series of studies which provided significant insight into the role of arch mechanics as well as interventions to reduce plantar fascia strain.

[57–59] Cadaveric lower limb specimens were positioned on a supportive surface simulating midstance, and a loading frame was applied to the tibia which then provided cyclical loading of the foot starting a 0 Newtons and progressing to 900 N. Strain in the plantar aponeurosis was measured with a differential variable reluctance transducer imbedded in the central band of the plantar aponeurosis, 2.5 cm distal to the medial calcaneal tubercle. The model did not engage the windlass mechanism and did not tension the Achilles beyond its resting length.

In the first study, Kogler et al. compared five different designs of foot orthoses in their ability to reduce strain in the plantar aponeurosis with loading of the foot [57]. Three of the devices significantly reduced strain in the plantar aponeurosis while two devices did not. Kogler et al. determined that the three devices which effectively reduced strain in the plantar aponeurosis did so by elevating the apex of the truss configuration of the longitudinal arch of the foot, thus relieving strain in the tie rod or the plantar aponeurosis. The authors speculated that the shape of the three devices effectively enhanced the "twisted plate" mechanism of the human foot as proposed by MacConnail and Saraffian to decrease strain in the plantar aponeurosis [60, 61]. However the authors made this correlation based upon the assumption that a foot orthosis design which most accurately contoured to the "apical" region of the longitudinal arch of the foot would reduce strain in the plantar fascia. This apical region was described as the area spanning the base of the first metatarsal, the medial cuneiform, the navicular, the sustentaculum, and the talus. By elevating the apical region, the foot moves into a position similar to what happens when it is "twisted" according to the original description provided by MacConnail [60]. The twisted plate mechanism elevates the apex of the longitudinal arch, but not via a direct support to the bony architecture as theorized by Kogler et al. Instead, the plate is twisted by everting the forefoot and/or inverting the rearfoot which then secondarily raises the longitudinal arch (Fig. 8.7). Kogler et al. acknowledge this principle in concluding

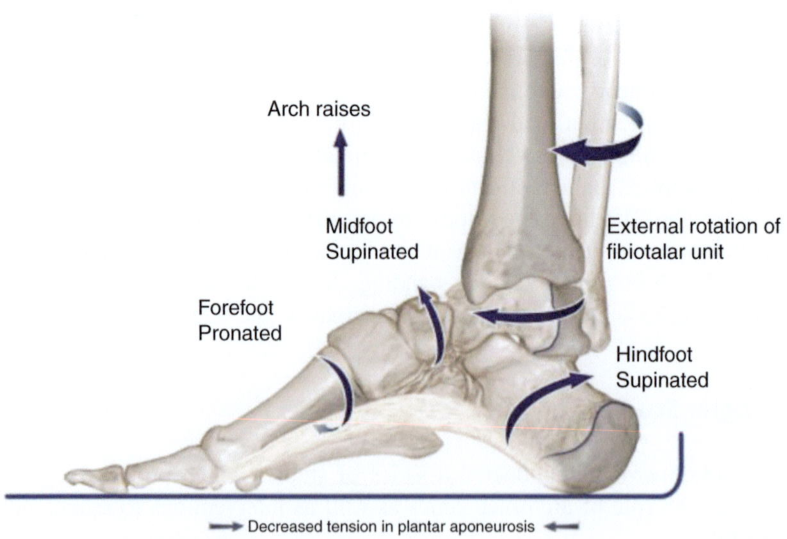

Fig. 8.7 Twisting the footplate to offload the plantar fascia

that the ideal movement of the foot to decrease strain in the plantar aponeurosis is a combination of hindfoot supination, midfoot supination, and forefoot pronation. When the rearfoot and midfoot are supinated, the medial aspect of the foot remains in contact with the ground due to pronation of the forefoot [12].

It was the second study by Kogler et al. which has provided a new and unexpected insight into the role of foot position and loading of the plantar fascia. They used the same experimental set up to measure strain in the plantar aponeurosis under axial load when 6 degree medial or lateral wedges were placed under the foot [58]. A *medial* wedge placed under the forefoot significantly *increased* strain in the plantar aponeurosis, while a *lateral* wedge placed under the forefoot significantly *decreased* strain in the plantar aponeurosis. Wedges placed under the calcaneus, whether medial or lateral had no significant influence on strain in the plantar aponeurosis.

Kogler et al. concluded that a lateral forefoot wedge would "stress shield" the plantar aponeurosis by two mechanisms. First, a lateral wedge would pronate the forefoot on the rearfoot thus facilitating the locking mechanism of the calcaneocuboid joint described by Bojsen-Moller [62]. Conversely, a medial forefoot wedge would supinate the forefoot and unlock the calcaneocuboid joint. Locking of the calcaneocuboid joint will stiffen the lateral column to accept load while reducing load at the medial longitudinal arch. A second mechanism of the lateral forefoot wedge is its ability to unload the medial truss of the longitudinal arch (Fig. 8.8). A lateral wedge placed under the forefoot will increase ground reaction force under the lateral forefoot and decrease force under the medial forefoot (Fig. 8.9). A medial forefoot wedge would increase loading of the medial forefoot (Fig. 8.10). This basically increases ground reaction forces under the distal strut of the truss of the arch of the foot: the first, second, and third metatarsals. This dorsiflexion moment applied

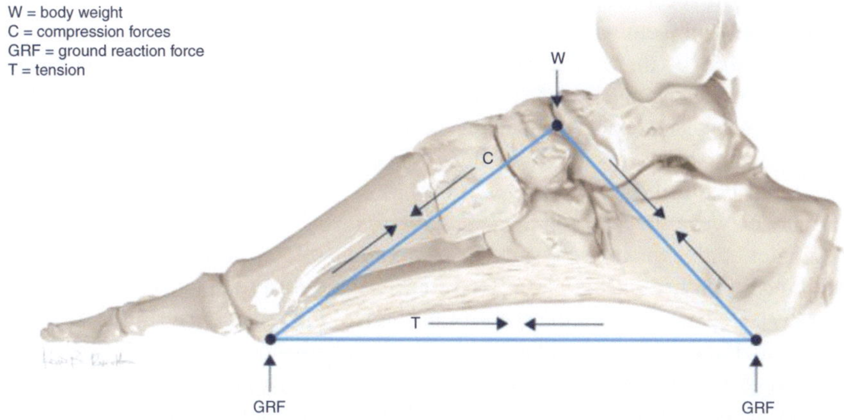

Fig. 8.8 The truss mechanism of the medial longitudinal arch. Struts under compression: Proximal strut = talus and calcaneus. Distal strut = first ray. Tie Rod under tension: Plantar aponeurosis

Fig. 8.9 Lateral forefoot wedge increases ground reaction force under lateral strut; decreases load under medial tie rod (plantar fascia)

Fig. 8.10 Medial forefoot wedge increases ground reaction force under medial strut; increases strain in the medial tie rod (plantar fascia)

to the distal strut would increase strain in the tie rod, i.e., the plantar aponeurosis. A lateral forefoot wedge would increase loading of the lateral column and off-load the medial column or medial truss mechanism. Interestingly, Kogler et al. do not discuss the twisted plate theory in their second paper although the effects of forefoot wedging to pronate the forefoot would account for the decreased strain in the plantar aponeurosis which occurred. Pronation of the forefoot will twist the plate of bones (lamina pedis) favorably to decrease strain in the plantar aponeurosis (see Chap. 1). Finally, the failure of medial wedging either in the forefoot or rearfoot to influence strain in the plantar aponeurosis defies a long held theory that pronation of the subtalar joint is a primary factor in plantar fascia overload. The subtalar joint by itself appears to have very little influence directly on plantar fascia strain as illustrated in this study by Kogler et al. as well as subsequent investigations which will be presented further in this chapter.

In their final investigation, Kogler et al. studied the effects of various heel elevations on strain in the plantar aponeurosis [59]. Two methods of elevating the heel of the cadaveric specimens were implemented. When simple rectangular blocks of 2 cm, 4 cm, and 6 cm were inserted under the heel of the specimens, no significant change of strain in the plantar fascia was measured with application cyclical loading from 0 to 450 Newtons. When a contoured platform, mimicking the shank profile of footwear was placed under the heel of the specimens, significant reduction of strain in the plantar fascia occurred, in similar magnitude for all three elevations. The contoured platform had a curve or arch profile with apex of the arch positioned halfway between the heel and the metatarsal heads.

Kogler et al. explain the results based upon the effects of the contoured support combined with heel elevation on load transmission thru the plantar surface of the foot [59]. Depending upon the shape of the arch of the subject, the sloped surface of the shank of the shoe may contact the lateral arch of the foot but not the medial arch. The shank of the shoe creates a curvature which may contact the lateral column of the foot accurately, while the higher medial arch of the subject is not in contact with the shank of the shoe. Thus, the area of the calcaneocuboid joint would have increased load bearing, while the medial truss of the foot would have decreased load. In patients with extreme cavus deformity, there may be minimal contact of any part of the arch of the foot with the slope or shank curvature of the shoe so no change in load transmission would be anticipated. Similarly, in pes planus, both the medial and lateral arches would contact the shank curvature so there may not be predominant load shift to the lateral column as seen in more normal arch shapes. Indeed, Kogler et al. noted that only certain cadaver specimens responded to the elevated shank profile platforms with reduced strain in the plantar aponeurosis, while other specimens showed minimal change [59].

It is interesting that the application of rectangular blocks of various elevation had no influence on strain in the plantar aponeurosis with loading of the cadaveric feet. This is likely due to the lack of loading of the Achilles tendon in the experimental model created by Kogler et al. Had the Achilles been loaded with tension normally seen in quiet stance (350N), elevation of the heel would be expected to offload both the Achilles and the plantar fascia. A block under the calcaneus will increase plantar pressure under the posterior strut of the truss of the arch of the foot. Loading of the strut will increase tensile strain in the tie rod or the plantar aponeurosis which spans the medial truss structure of the foot (Fig. 8.8). Kogler et al. point out that a simple truss model of the foot is a triangular structure in equilibrium. Application of increased force under the calcaneus would apply dorsiflexion moment to the posterior strut which would increase tension in the tie rod, i.e., the plantar aponeurosis. Yet in their study, Kogler et al. recognize that rectangular heel blocks did not increase strain in the plantar aponeurosis. The authors explain that the truss mechanism is a simple two-dimensional model which fails to account for the complex contribution of other ligaments and the interaction of multiple joint surfaces which are involved in the truss itself. The truss mechanism is isolated to the medial column of the foot itself and does not take into account the contribution of the lateral column to modify overall load transmission thru the foot. Kogler et al. propose that a complex three-dimensional interaction occurs in the human foot which directs force transmission thru the foot [59]. They cite their research on application of medial and lateral wedges under the forefoot where several mechanisms with both medial and lateral support were discovered. The research from Kogler et al. adds validity to the twisted plate mechanism of the foot whereby raising and lowering of the longitudinal arch is not achieved by a simple elevation of the medial truss, but rather by a complex three-dimensional rotation of the forefoot upon the rearfoot.

Based on the research from Kogler and co-workers, there appears to be a load-sharing mechanism operating between the medial column of the foot and the lateral column. The central band of the plantar aponeurosis originates from the medial calcaneal tubercle and has a direct course to the 1st and 2nd MTPJs. The long plantar ligament is situated to provide similar longitudinal support to the lateral column, just as as the central band of the plantar aponeurosis provides support to the medial column (Fig. 8.11). However, the long plantar ligament does not extend the full length of the lateral column as it inserts on the base of the lateral metatarsals. Kogler's work shows that when the lateral column is loaded, strain on the central band of the plantar aponeurosis is reduced. Loading the lateral column theoretically increases strain in the long plantar ligament while decreasing strain in the central band of the plantar aponeurosis. A well-known complication of plantar fasciotomy is the development of lateral column pain, centered under the cuboid. This pain likely eminates from the

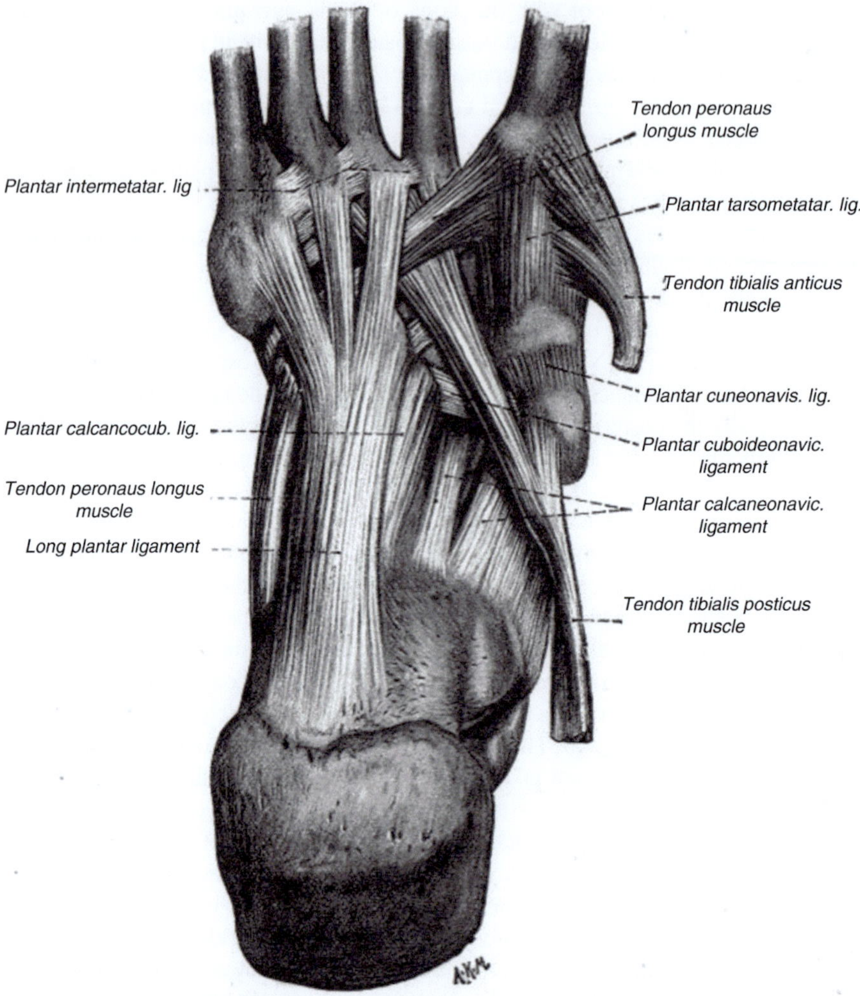

Fig. 8.11 The long plantar ligament provides a truss support to the lateral column. (From: https://wikimili.com/en/Long_plantar_ligament)

long plantar ligament which is now placed under increased strain due to loss of function of the central band of the plantar aponeurosis. Surgical procedures which lengthen the lateral column of the foot have a known risk of lateral column pain postoperative which may be due to overload of the long plantar ligament.

The Plantar Fascia and the Arch Spring Mechanism

Caravaggi and co-workers used a multisegment foot model to measure arch movement during walking in 10 human subjects and then calculated strain in the plantar fascia [63]. Deformation of the medial longitudinal arch began at heel strike and continued to terminal stance. Faster walking caused larger arch deformation. Tension in the plantar aponeurosis increased with arch deformation with maximal tension occurring at terminal stance. Tension in the plantar aponeurosis was greatest medially and diminished in the lateral slips. Increased walking speed caused increased strain in the medial section of the plantar aponeurosis, and increased walking speed increased overall dorsiflexion of the digits at the MTPJs verifying the role of the windlass to increase loads in the plantar fascia.

Most important, Caravaggi and co-workers describe a preloading phenomenon of the plantar aponeurosis which occurs during swing phase and may be due to certain muscle activity. In a previous study, Caravaggi et al. proposed that a preloading mechanism of the plantar aponeurosis prior to heel strike would take up slack and stiffen the foot for load acceptance at touchdown [64]. This mechanism is the result of activation of the toe and ankle joint dorsiflexors (tibialis anterior, extensor digitorum longus) providing clearance of the foot during swing phase and creating antagonistic passive tension in the Achilles tendon. Dorsiflexion of the forefoot against the rearfoot, resisted by the Achilles tendon in the sagittal plane, will elongate the arch and pretension the plantar aponeurosis. This muscle activity will increase with faster walking speed according to Caravaggi et al.

In their second study, Caravaggi et al. observed that there was larger arch deformation and larger strain medially in the plantar aponeurosis with faster walking speed [63]. They speculated that some type of intrinsic mechanism operates within the foot which regulates muscle activity to modulate arch stiffness during walking, particularly in response to faster speed and increased ground reaction force. Regulation of activity of the ankle and toe dorsiflexors as well as activity of the triceps surae can affect pre-activation and stiffness of the plantar aponeurosis to affect compliance of the longitudinal arch in response to ground reaction force. This allows the plantar aponeurosis to provide early resistance to arch deformation which will increase with greater ground reaction force. Increased strain in the plantar aponeurosis was noted by Caravaggi and co-workers to correlate with increased walking speed and was noted to occur predominantly along the medial section of the plantar aponeurosis. They also noted that the longitudinal arch height actually increased in terminal stance thru pre-swing, and the magnitude of arch raising was dependent upon walking speed. Faster walking increased arch height during push off. Since the plantar fascia tension remained constant during this phase of the gait cycle, Caravaggi and co-workers theorized that the plantar intrinsic muscles must play a role in raising the arch during push off. In fact, plantar fascia tension was noted to drop off to almost zero at toe off. This indicated that the plantar fascia would be ineffective as an arch supporting mechanism at the pre-swing phase of the gait cycle. They concluded that the most likely contribution to arch raising during terminal stance and pre-swing was the plantar intrinsic muscles as well as the extrinsic arch supporting muscles (tibialis posterior and flexor digitorum longus) which begin contracting at midstance.

Kelly and co-workers conducted a series of studies which verified some of Caravaggi's proposals about the contribution of the plantar intrinsic muscles to stability of the arch of the foot. Two studies showed that the abductor hallucis, the flexor digitorum brevis, and the quadratus plantae muscles are all active during the stance phase of gait and increase activity as increase in speed of walking and running occurs [65, 66]. These plantar intrinsic muscles appear to regulate stiffness of the longitudinal arch in response to mechanical load and reduce strain in the plantar aponeurosis. However, two subsequent studies by Kelly's research team contradicted their earlier conclusions and added further insights into the role of passive and dynamic support mechanisms operating within the human foot during ambulation.

Welte et al. studied nine human subjects using a custom apparatus which simulated midstance and applied loads similar to what is seen during walking [67]. The researchers measured longitudinal arch compression and elongation when the 1st MTPJ was moved from 30 degrees plantarflexion, to neutral and then to 30 degrees dorsiflexion. Contrary to previous beliefs, the study showed that when the windlass mechanism was engaged with dorsiflexion of the hallux, the longitudinal arch elongated more and dissipated more energy under loading. In the unloaded condition, dorsiflexion of the hallux caused an elevation of longitudinal arch height and decreased length of the arch, as expected. However, with loading and engagement of the windlass, the arch elongated, but the height of the arch was unaffected. The marker for arch height was the navicular, so clearly other bones may have dropped closer to the ground as the arch elongated. The authors speculated that shortening the arch length probably placed other ligaments (spring, long and short plantar ligaments) closer to their resting lengths. The arch could then undergo greater excursion with loading. In other words, engagement of the windlass raised the arch which then allowed more available range for flattening under loading. This is similar to coiling a spring (See Chap. 3 Fig. 3.23). This study did not evaluate arch stiffness during dynamic gait when other factors such as extrinsic muscle activity would potentially stiffen the arch. The results also suggest that the plantar aponeurosis undergoes lengthening during loading and is not completely rigid. This has been confirmed by other studies [52, 53, 63].

Welte et al. also observed that when the hallux was dorsiflexed, the forefoot moved relative to the rearfoot in a different direction than seen when the hallux was plantarflexed. Dorsiflexion of the hallux pushes the first metatarsal plantar, but this is resisted by the supportive surface. Therefore, in order for the hallux to dorsiflex, the first metatarsal must move proximal which explains shortening of the longitudinal arch. Welte et al. also observed that the entire forefoot moved on the rearfoot about an inclined axis which allowed transverse plane rotation of the forefoot in the direction of external rotation (abduction). This verifies that the windlass engagement followed the twisted plate theory by pronating the forefoot.

Farris and co-workers studied 12 human subjects before and after performing a tibial nerve block to eliminate function of the plantar intrinsic muscles [68]. Contrary to expectation, the plantar intrinsic muscles did not appear to contribute to stability of the longitudinal arch during dynamic loading from initial contact thru midstance while walking and running. Although previous studies had shown that the plantar intrinsic muscles are active during the stance phase of gait, the current study showed that the overall contribution of these muscles to arch stiffness was negligible. The passive structures such as the plantar aponeurosis and the plantar ligaments appear to play the critical role in arch stiffening which is part of an energy storage mechanism. However, at terminal stance thru pre-swing, the plantar intrinsic muscles appear to play an important role in stiffening the MTPJs. This mechanism is critical to the positive work created by the ankle during push off [69, 70]. In their study, Farris et al. noted that subjects without function of the intrinsic muscles of the foot showed less mechanical work produced at the ankle and also showed reduced propulsive impulse during terminal stance and pre-swing. They attributed this finding to the important role of the plantar intrinsic muscles to generate plantar flexion moments at the MTP joints during push off. Furthermore, based upon previous research, Farris et al. note that the windlass mechanism is incapable by itself to stiffen the forefoot during push off. Finally, Farris et al. allude to Bojsen-Moller and propose that the abductor hallucis muscle acting on an adducted hallux promotes a high gear push off, a feature unique to human gait.

These studies underscore the importance of the plantar aponeurosis as not only an arch supporting mechanism but also as an *elastic* energy storage- and- release mechanism which ultimately provides *metabolic* energy savings.

During walking gait, the plantar aponeurosis elongates up to 12% of its resting length [54]. Stearne and co-workers showed that during walking, the average arch compression is 7 mm, while during running, this arch compression increases to 11 mm [71]. The elastic energy return of all of the passive arch structures during running was estimated by Stearne et al. to provide approximately 8% of the mechanical energy required to run at 2.7 meters per second [71]. In another study, Ker and Bennett estimated that the combined elastic energy of the plantar aponeurosis and plantar arch ligaments provides approximately 17% of the total mechanical energy expended when running at 4.5 meters per second [72]. Clearly the "arch spring" of the human foot is the most important passive energy storage release mechanism operating in the lower extremity during running.

However, as seen in the studies from Welte and Farris, extension of the digits which engage the windlass mechanism can actually result in decreased strain in the plantar aponeurosis [67, 68]. This may represent an energy absorption mechanism, rather than release of energy for positive work.

McDonald and co-workers studied 18 subjects running on a force plate-instrumented treadmill and used three-dimensional motion capture to measure the effects of extension of the MTPJ as well as arch depression on loading of the plantar aponeurosis [73]. While running at a speed of 2.7 meters per second, strain in the plantar aponeurosis reached a peak at 60% of stance. This is the point of maximal energy storage by the plantar aponeurosis which is then returned during the final 40% of stance, when the longitudinal arch undergoes recoil from previous compression. In this study, arch compression had the primary influence on strain in the plantar aponeurosis, while motion of the MTPs had minimal effect. With extension of the MTPJs in terminal stance and pre-swing, this study showed that 11.7% of total lower limb joint energy absorption occurred at the 1st MTPJ. McDonald et al. proposed that some of this absorbed energy was transferred by the plantar aponeurosis between the MTPJ and the recoiling longitudinal arch to assist with power generation.

Pathomechanics of Plantar Heel Pain

Mechanical overload of the plantar fascia has been the most common accepted mechanism causing plantar heel pain [6–9, 74]. Kitaoka and co-workers evaluated the material properties of the plantar aponeurosis in 12 cadaver specimens [75]. The average load to failure was 1189N which is 1.8 times body weight. Studies presented previously in this chapter reveal that the average load transmitted to the plantar aponeurosis during walking and running can approximate 1.5 times body weight which is close to the limit of load failure [54, 55]. Therefore, repetitive loading at a magnitude near load failure is speculated to be the primary cause of degeneration of the plantar fascia in plantar heel pain pathology [74, 76]. Failure of the tissue to repair under continued mechanical load leads to chronic inflammation and eventual degeneration [77, 78].

Many extrinsic and intrinsic factors have been identified which may contribute to overload of the plantar fascia and the development of plantar heel pain. Intrinsic risk factors relate to anatomic or functional abnormalities such as foot and leg alignment, range of motion, muscle strength, and body mass. Extrinsic factors relate to lifestyle and environment such as time spent standing, exercise, footwear, and floor surfaces.

Over the past 40 years, many studies have been undertaken to understand the cause of plantar heel pain. There are two excellent reviews of published research on this subject from Irving et al. [79] and Sullivan et al. [80]. Also, a study of a large group of patients with plantar heel pain was conducted by Sullivan et al. which focused on the contribution of both mechanical factors and activity factors [81]. These three publications provided the bulk of evidence which will be presented here regarding the intrinsic and extrinsic causes of plantar heel pain.

Excessive Loading

Excessive loading of the plantar surface of the heel during walking and running has been suggested as a cause of plantar heel pain [82, 83]. It is intuitive that impact shock or excessive pressure on the enthesis of the plantar fascia attachment on the calcaneus would contribute to degenerative changes.

However, studies of patients with plantar heel pain fail to validate this cause-effect relationship.

Liddle and co-workers found no difference between symptomatic and asymptomatic limbs in terms of magnitude and timing of the vertical heel strike transient or first vertical force maximum in patients with heel pain while walking with preferred speed [84]. It is reasonable to expect compensation in the painful foot to offload the calcaneus in these studies, yet most studies of patients with symptomatic plantar fasciitis show that regional loading of the heel is unaffected by the condition [85–87]. However, more recent studies have shown that patients with plantar heel pain engage a compensatory loading mechanism to off-load the plantar fascia.

Wearing et al. studied 16 patients with plantar heel pain and compared to 16 healthy controls while walking over a pressure measurement platform [88]. In comparison with the control subjects, patients with plantar fasciitis had reduced forces beneath the heel and forefoot. The patients spent less time on the rearfoot and developed peak pressure in the forefoot later, indicating less activation of the triceps which will diminish load on the plantar aponeurosis. Conversely, this study showed greater pressure under the toes of the symptomatic patients, which may indicate a splinting mechanism to again off load the plantar fascia. Greater toe flexion will reduce dorsiflexion motion which increases strain in the plantar aponeurosis.

Sullivan and co-workers collected plantar pressure data from 198 patients with plantar heel pain and compared to 70 asymptomatic controls [89].

The results showed a strategy to reduce loading beneath the painful heel which was characterized by lower peak pressure under the posterolateral heel and lower maximum force throughout the entire heel during walking. There was also a decreased maximum force in the medial forefoot combined with a higher force time integral in the lateral forefoot of the plantar heel pain patients. This suggested to the authors that there was a strategy to reduce plantar fascia loading at the 1st MTPJ which would decrease engagement of the windlass operating during late stance. By reducing maximum force across the 1st MTPJ, reduced tension and pain in the plantar fascia could be the result of reduced windlass function.

1st MTPJ Range of Motion

The work from Sullivan et al. as well as cadaveric studies from Carlson and Cheng discussed earlier suggest that the most significant contribution to plantar fascia strain is from the windlass mechanism initiated by dorsiflexion of the toes across the MTPJs [52, 53, 89]. Therefore, it is reasonable to consider a link between range of dorsiflexion range of motion of the 1st MTP and risk of developing plantar heel pain. Two possibilities exist regarding the effects of range of motion of the 1st MTPJ and tension in the plantar aponeurosis. One assumption would be that increased dorsiflexion range of motion of the 1st MTPJ would put greater tensile strain on the plantar aponeurosis via the windlass mechanism. The opposite finding of reduced range of motion of the 1st MTPJ could be from a tight, inflexible plantar fascia which would then be vulnerable to increased tensile strain during gait [90].

Two studies have shown no difference in 1st MTPJ dorsiflexion range of motion comparing groups of subjects with and without plantar heel pain [81, 90]. In contrast, a study showed reduced range of extension of the 1st MTPJ in 50 patients with plantar heel pain compared to 50 matched controls [91]. These studies utilized static measures of passive range of motion of the 1st MTPJ, rather than dynamic measures during gait.

A study by Creighton and Olson measured active and passive range of motion of the 1st MTPJ of six patients with plantar heel pain compared to healthy controls [92]. A significant reduced range of both active and passive extension of the 1st MTPJ was found in the patient group with plantar heel pain. Wearing et al.

examined dynamic range of 1st MTPJ motion in ten subjects with plantar heel pain compared to a control group [93]. The patient group with heel pain had a mean dynamic extension of the 1st MTPJ which was significantly (4.6 degrees) less than the control group.

These studies dispel the notion that a full functioning windlass mechanism, placing maximal tensile strain on the plantar fascia, is a risk factor for plantar heel pain. Rather, it appears that loss of extension of the 1st MTPJ, whether being the cause or result of reduced flexibility of the plantar fascia is more closely associated with plantar heel pain.

Strength Deficits

Toe flexor strength deficits have been identified in three different studies of patients with plantar heel pain [81, 90, 94]. Although these studies used different methods to measure toe flexor force, the findings suggest that toe flexor weakness may be either the cause or the result of plantar heel pain. From a mechanical standpoint, the intrinsic muscles and the extrinsic flexor digitorum longus muscle are most active during late midstance and terminal stance which coincides with maximal strain developed in the plantar aponeurosis [55, 65, 66]. There is a growing body of evidence that the toe flexors modulate stiffness across the MTPJs depending upon tension produced by the plantar aponeurosis [63, 67, 68]. Failure of timing and power of toe flexion strength could logically place increased tensile load on the plantar aponeurosis during a critical time during the stance phase of gait.

Sullivan and co-workers carried out a study of a large group of patients which plantar heel pain to determine if there was commonality in musculoskeletal risk factors [83]. Compared to healthy controls, toe flexor and ankle eversion strength were reduced in patients with plantar heel pain. The peroneus longus provides the most powerful eversion moment to the foot during terminal stance [95, 96]. Loss of timing or strength of the peroneus longus could result in failure to shift load medially to the forefoot and engage the windlass mechanism. Furthermore, the peroneus longus acts to stabilize the first ray against ground reaction forces [97, 98]. Loss of stability of the first ray will also compromise the windlass mechanism and normal extension of the MTPJ during push off [99].

Loss of calf muscle strength has been demonstrated in patients with plantar heel pain. Runners with plantar heel pain showed deficits in plantar flexion strength on the involved limb compared to the asymptomatic contralateral limb [83]. Another study which isolated motion at the ankle joint showed reduced endurance of the calf musculature in 27 patients with plantar heel pain compared to matched controls [100]. In contrast, two studies showed no significant difference in plantar flexion endurance comparing patients with plantar heel pain and healthy controls [81, 101]. Since numerous studies have shown that the calf muscles, via the Achilles tendon exert significant tensile load on the plantar aponeurosis, evidence of weakness in these muscles in patients with plantar heel pain may suggest a compensatory rather than a causative mechanism.

Pronation and Arch Mechanics

The single mechanism most commonly proposed causing plantar heel pain is instability of the longitudinal arch of the foot [102–104].

Closely linked to arch instability and also commonly assumed to be a contributing factor for plantar heel pain is excessive pronation of the foot, presumably at the subtalar joint [105–107]. If the longitudinal arch has a truss mechanism, lowering the arch will move the struts into a less perpendicular position and form a more acute angle to the supportive surface [61]. Simple geometry will dictate that tension will increase in the tie rod, i.e., the plantar aponeurosis as the height of the arch decreases (Figs. 8.12 and 8.13). The tie rod of the lateral longitudinal arch, i.e., the long plantar ligament, can be expected to develop increased tensile strain as the height of the lateral arch lowers.

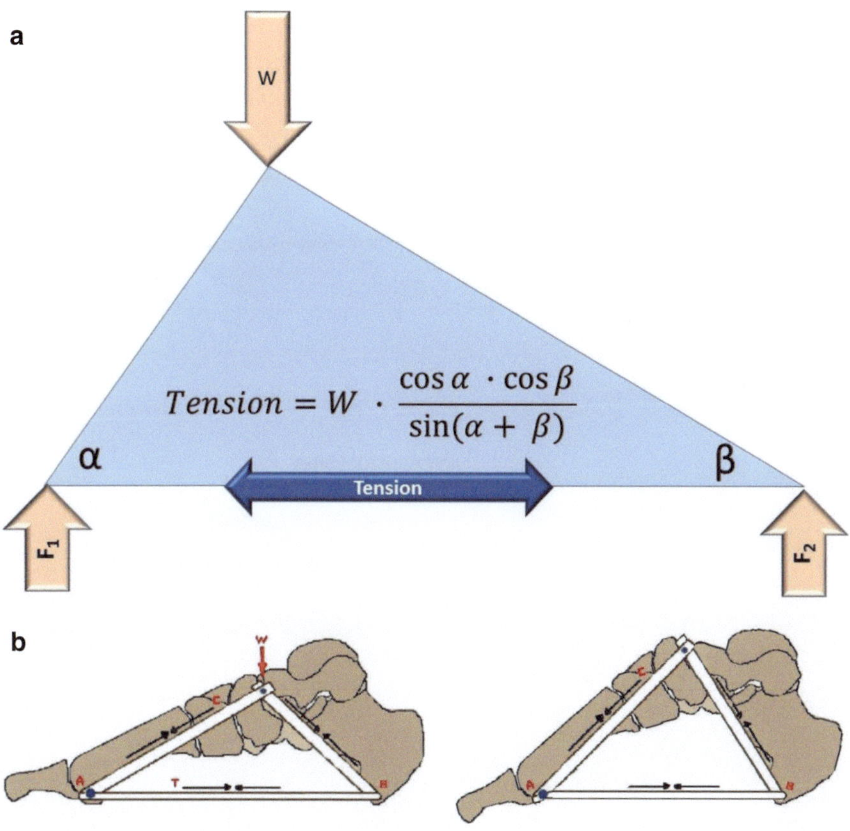

Fig. 8.12 Arch height will affect tension in the plantar aponeurosis. (**a**) Geometry of the truss mechanism. (**b**) Arch height and calcaneal pitch directly influences tension in the plantar aponeurosis. (**c**) Increased calcaneal pitch will decrease strain in plantar aponeurosis (courtesy of Robert D. Phillips)

c

Calcaneal Inclination	Plantar Fascia Strain (lb)
10	682.9
20	334.4
30	214.8

Fig. 8.12 (continued)

Fig. 8.13 Twisting the calcaneopedal unit (CPU) influences the calcaneal inclination angle to the ground

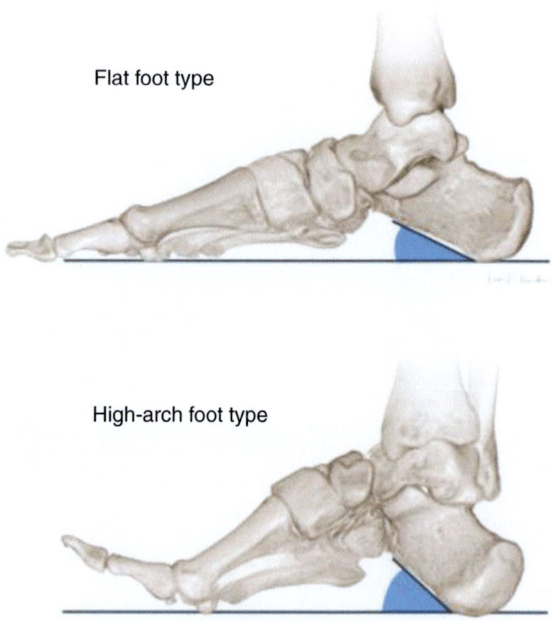

While cadaveric models have shown a consistent and significant relationship between medial longitudinal arch mobility and plantar fascia strain, clinical studies fail to verify a cause/effect relationship when evaluating patients with plantar heel pain. A challenge when interpreting these studies is the wide range of methods to measure foot posture, arch profile, and magnitude of pronation.

Published reports studying the relationship between foot alignmnet and arch height while standing have used various measures such as rearfoot angle, navicular height, as well as global foot posture. Radiographic studies have been published evaluating patients with plantar heel pain which utilize various measures of arch height and truncated foot length. Another popular method is the use of footprint measures, most commonly the arch index. However, this method has been questioned for accuracy in predicting true arch height and alignment [108, 109].

Prichasuk and Subhadrabandhu measured calcaneal pitch on radiographs of 82 patients with plantar heel pain and concluded that pes planus was a risk factor for the condition [110]. Similarly, Shama et al. evaluated static radiographs of 52 subjects with plantar heel pain and concluded that pronation was a contributing factor to the condition in 82% of the symptomatic patients [111]. Conversely, Rano et al. found no difference in radiographic parameters of arch height when comparing 49 patients with heel pain to a control group [112].

Surface markers have been used to measure arch height and rearfoot position during quiet standing in patients with heel pain. Rome et al. found no difference in navicular height when comparing 36 patients with plantar heel pain to a control group [113]. Studies measuring the rearfoot angle in static stance show no differences comparing patients with heel pain to a control group [90, 114, 115]. A study of 20 patients with plantar heel pain compared to 20 matched controls used the longitudinal arch angle which are lines drawn from the medial malleolus to the navicular tuberosity and to the first metatarsal head and found no difference in the angle comparing the two groups [90].

The Foot Posture Index (FPI) is a well accepted and reliable method of assessing global foot alignment [116]. Using the FPI, two studies showed that patients with plantar heel pain had a more pronated foot posture than healthy controls, although the entire subject pool did not have significantly pronated feet when compared to normative data [100, 117]. In a larger study of 202 patients with plantar heel pain, the FPI did not show any evidence of pronation compared to the control group, and both groups had scores falling within the normal range of foot posture [81].

Measures of static, standing arch height may not accurately depict dynamic arch height during gait and resultant strain on the plantar aponeurosis. Saraffian proposes that pronation of the rearfoot indirectly increases strain on the plantar fascia by increasing ground reaction forces medially on the forefoot resulting in supination of the forefoot on the rearfoot [61]. Nakamura and Kakurai showed that maximal rearfoot eversion was associated with reduction of dynamic arch height during gait [117]. Lee and co-workers measured static arch height, dynamic arch height, rearfoot pronation, and plantar fascia tension in 28 healthy subjects running on a treadmill [118]. There was no relationship between static arch height and dynamic arch height during running. Only dynamic arch height predicted strain in the plantar aponeurosis. By itself, the magnitude of rearfoot pronation during running was not a predictor of plantar fascia tension. However, an indirect effect of rearfoot pronation on plantar fascia strain was detected when magnitude of pronation was combined with dynamic arch height. This verifies the fact that rearfoot pronation by itself cannot influence strain on the plantar aponeurosis. Other movements in the midfoot must take place in response to rearfoot pronation before strain is increased in the plantar aponeurosis. This is in accordance with the twisted plate mechanism of plantar fasica strain proposed by Saraffian [61].

Wearing et al. used digital fluoroscopy to evaluate dynamic movement of the arch during walking gait studying 10 patients with plantar heel pain compared to 10 healthy controls [93]. Dynamically lower arch shapes were correlated with thicker plantar fascia in the patient groups. However, there was no correlation with arch height or dynamic arch movement when comparing patients with heel pain to the control group. The authors concluded that plantar fasciitis is not associated with a lower arch or excessive arch movement during walking. Later, Wearing et al. speculated that other forces on the plantar fascia besides tensile strain might be responsible for the degenerative changes seen in chronic plantar heel pain patients [119]. These would include bending, shear, and compressive forces which would cause "stress shielding" as seen in insertional tendinopathy [120–122].

Adding to the debate about the role of foot pronation and arch movement in causing plantar fasciitis, Chang and co-workers published results of a high- quality study of 22 patients with chronic plantar fasciitis [123]. This study used a multi-segment foot model to measure three-dimensional motion of the shank, rearfoot, medial forefoot, and hallux while walking. Patients with plantar fasciitis demonstrated greater peak dorsiflexion of the 1st MTPJ (53.3 degrees vs 49 degrees) compared to healthy controls. Overall, rearfoot range of motion was greater in the plantar fasciitis patients, but there was no difference in magnitude of peak rearfoot eversion when compared to controls. There was a small but statistically significant increase of sagittal plane motion in the medial arch of one degree in the plantar fasciitis patients compared to the control group. When comparing overall rearfoot motion and medial arch motion, the difference between patients with plantar fasciitis and healthy controls is apparent when evaluating the alignment of the foot at touchdown. Patients with plantar fasciitis land with their feet slightly more inverted in the rearfoot and slightly more plantarflexed in the forefoot which then allows greater range of motion during midstance. The authors noted that a greater range of pronation, from a more inverted rearfoot position, and greater dorsiflexion of the forefoot after foot strike may cause greater loading of the plantar fascia. They also acknowledge that these findings may simply represent compensation for pain in the patients with plantar fasciitis.

Ankle Joint Range of Motion

Multiple mechanisms have been proposed describing how restricted ankle joint dorsiflexion could increase tensile strain in the plantar aponeurosis during walking and running. During midstance, restriction of ankle joint dorsiflexion prevents tibial progression which is compensated for by excessive sagittal plane dorsiflexion motion across the midfoot joints [124]. This in turn would increase tensile strain in the plantar aponeurosis via the medial truss mechanism of the longitudinal arch [61]. The structure most likely to restrict motion during this critical phase of the gait cycle is the gastrocnemius which is placed under maximal tension with knee extension at midstance. The Achilles tendon has a direct connection to the origin of the plantar aponeurosis in infants, but this connection is lost by adulthood [15]. Notwithstanding, the Achilles tendon has a significant indirect effect on loading the plantar aponeurosis during stance [56].

Gastrocnemius contracture, measured with a goniometer with the knee extended, has been determined to be a contributing factor for plantar heel pain in a number of studies on athletic individuals [83] as well as the general population [2, 125]. Sullivan et al. found that patients with plantar heel pain demonstrate restricted ankle joint dorsiflexion with both a bent knee lunge test and a straight knee lunge test [81]. The authors concluded that structures other than the gastrocnemius could have caused restricted ankle joint dorsiflexion in the bent knee lunge test. In their recent review of current literature studying mechanical factors of heel pain, Sullivan et al. concluded that restriction of ankle joint dorsiflexion is a likely intrinsic risk factor for developing plantar heel pain [80]. This conclusion was also reached by Irving et al. in their systematic review of factors related to heel pain [79].

Body Mass Index

Increase body mass directly translates to increased load borne by the foot during weight bearing [126, 127]. Riddle found that a body mass index (BMI) of greater than 30 kg/m^2 was associated with greater risk of plantar heel pain. Similar findings were reported by Sullivan et al. where patients with plantar heel pain had a BMI of 28 kg/m^2, and the control group had a BMI of 25 kg/m^2, providing the strongest association of any musculoskeletal risk factor in their study for developing heel pain [81]. Four other studies have shown a correlation with BMI scores and plantar heel pain [100, 110, 112, 128]. The only study which contradicts this association was performed by Rome et al. who found no difference in BMI in a group of 36 runners with heel pain compared to asymptomatic controls [113]. Clearly, among running athletes, other risk factors for plantar heel pain come into play besides BMI which is still the largest risk factor in the general population. A systematic review verified that BMI is the most consistent and valid reported intrinsic risk factor for plantar heel pain [129].

Calcaneal Spur

Prichasuk and Subhadrabandhu found that a spur, projecting horizontally from the plantar calcaneal tubercle on plain radiographs, was present in 66% of patients with plantar heel pain and was found in 16% of the control subjects. Calcaneal spurs were found in 28% of the feet of runners with plantar heel pain [83] and 44% of patients from the general population with heel pain [112]. Ozdemir et al. found that radiographic evidence of calcaneal spurs are found in 45% of patients with symptoms of plantar fasciitis and are also found in 20% of patients without symptoms of heel pain [128].

Other imaging studies using MRI and ultrasound find associated findings with calcaneal spur including thickening and hypoechogenicity of the plantar fascia, peri-fascial edema, and calcaneal bone marrow edema [130, 131]. Plantar fascia thickening as well as calcaneal spurs have showed a strong association with plantar heel pain [127, 131]. This raises a fundamental questions regarding the role of calcaneal spurs as a risk factor for plantar heel pain: Is the spur a cause of or a result of plantar fasciopathy? Or does the spur cause a variation of heel pain without plantar fasciopathy?

Menz et al. studied a large number of patients with plantar heel pain (n = 530) and used plain radiography and ultrasound to determine the association between calcaneal spurs, plantar fascia thickening, and plantar heel pain [132].

Plantar fascia thickening was more closely associated with plantar heel pain (47% of patients) than calcaneal spurs (26%). Plantar calcaneal spurs and plantar fascial thickening were commonly found together (30%), while isolated calcaneal spurs without fascial thickening was rare (4%). Recent onset of plantar heel pain less than 1 month duration was commonly associated with fascial thickening but not with calcaneal spurs. The authors concluded that fascial thickening, greater than 4 mm, is more closely associated with plantar heel pain than are calcaneal spurs. This same conclusion was found in a previous study [130].

The conclusion which could be made is that fascial thickening precedes the formation of calcaneal spurs and not vice versa. Calcaneal spurs appear later in the pathlology of plantar heel pain possibly acting as a mechanism to offload the enthesis of the plantar aponeurosis from compressive loads [31]. Finally, plantar calcaneal spurs are reported to occur in 10–60% of asymptomatic individuals making this finding an unlikely contributor to the cause of plantar heel pain [133].

Other Factors

Prolonged standing has been implicated as an extrinsic risk factor for plantar heel pain, although published studies show conflicting results. Two studies of factory workers have shown increased incidence of heel pain in subjects who stand for prolonged periods compared to those who do not [134, 135].

Riddle et al. showed that subjects who stand all day have a three times greater risk for developing plantar heel pain than those who do not [2]. In contrast, two other studies found no correlation between incidence of plantar heel pain with standing during the day [81, 100].

Both Beeson [11] and Wearing [136] suggest that genetics, which have already proven to be associated with tendinopathy and ligament injuries, might also be involved in the pathological changes seen in plantar fasciopathy. Certain genes have been identified as important regulators of fibrillogenesis and collagen cross bridging which are critical to establishing integrity to tendons and ligaments [137–139]. These genetic factors may interact with other intrinsic and extrinsic risk factors to contribute to degeneration of the plantar aponeurosis.

Sequence of Events

Scott Wearing, who has published multiple studies of plantar heel pain, suggests a newer algorithm for the pathomechanics of plantar fascia degeneration [136, 140]. He refers to studies of tendinopathy which show that repeated mechanical loading cannot explain the degenerative changes seen histologically in injured tendons [141–143]. In tendons, other intrinsic or extrinsic factors must come into play before degeneration occurs [140]. To add further confusion, over half of patients with plantar heel pain experience the condition in only 1 foot, yet both feet are exposed to basically the same internal and external risk factors [132]. Wearing argues that there is very little evidence that any of the previously discussed risk factors are actually the *cause* of plantar heel pain, but rather, these factors act to modulate the levels of pain once it is present [88, 119, 144].

Wearing proposes that plantar fasciopathy begins with an underlying defect in the plantar aponeurosis which may be genetic and which leads to asymptomatic thickening with repetitive loads. Alternatively, an underlying neuromuscular deficit causing strength deficits in the intrinsic or extrinsic muscles of the foot can cause overload and asymptomatic thickening of an otherwise healthy plantar aponeurosis. Subsequently, when an extrinsic factor such as unaccustomed prolonged activity or faulty footwear comes into play, repair of the thickened plantar fascia begins to lag behind, and a degenerative process ensues. Wearing points out that these degenerative changes, in the absence of inflammation in the plantar fascia cannot explain the pain experienced by the patient. His review of all possibilities shows no logical or consistent explanation why plantar fasciopathy is painful [136].

One cannot ignore the contribution of the inferior calcaneal nerve in generating and transmitting the pain of plantar fasciopathy. As noted in the beginning of this chapter, the inferior calcaneal nerve is located in a precarious position sandwiched between the calcaneus and the enthesis of the plantar aponeurosis. There is very little room in this corridor to accommodate the thickening of the plantar aponeurosis which is seen in over 50% of patients with plantar heel pain. The inferior calcaneal nerve may be an innocent bystander in the pathomechanics of degeneration of the plantar aponeurosis, but it then becomes the primary generator of pain and disability as the condition progresses.

Ultimately, plantar heel pain may be the result of a nerve entrapment which is secondary to degeneration and thickening of the planar aponeurosis. What triggers this degenerative process is probably multi-factoral as no single dominant causative factor has yet to be identified.

References

1. Snook GA, Chrisman OD. The management of subcalcaneal pain. Clin Orthop Relat Res. 1972;82:163–8.
2. Riddle DL, Pulisic M, Pidcoe P, et al. Risk factors for plantar fasciitis: a matched case-control study. J Bone Joint Surg. 2003;85(5):872–7.
3. https://www.census.gov/programs-surveys/acs/news/data-releases/2018/release.html.
4. Riel H, Cotchett M, Delahunt E, Skovdal R, Vicenzino B, Weir A, Landorf KB. Is 'plantar heel pain' a more appropriate term than 'plantar fasciitis'? Time to move on. Br J Sports Med. 2017;51(22):1576–7.
5. Michelson JD. Heel pain: when is it plantar fasciitis? J Musculoskelet Med. 1995;12:22–6.
6. Rome K. Anthropometric and biomechanical risk factors in the development of plantar heel pain—a review of the literature. Phys Ther Rev. 1997;2(3):123–34.
7. Barrett P, Black M. In response to: anthropometric and biomechanical risk factors in the development of plantar heel pain—a review of the literature [Rome, Phys Ther Rev 1997;2:123–34]. Phys Ther Rev. 1998;3(1):55–6.
8. Singh D, Angel J, Bentley G, Trevino SG. Plantar fasciitis-a clinical review. Br Med J. 1997;315:172–5.
9. Buckbinder R. Plantar fasciitis. N Engl J Med. 2004;350:2159–66.
10. David JA, Chatterjee A, Macaden AS, Sankerapandian V, Christopher PRH. Injected corticosteroid for treating plantar heel pain in adults. Cochrane Database Syst Rev. 2011;10:1–10.
11. Beeson P. Plantar fasciopathy: revisiting the risk factors. Foot Ankle Surg. 2014;20:160–5.
12. Sarrafian SK, Kelikian AS. "Retaining systems and compartments" in Sarrafian's anatomy of the foot and ankle. 3rd ed. Philadelphia: Wolters Kluwer; 2011. p. 120–63.
13. De Palma L, Coletti V, Santuccl A, Tulli A. L'Embriogenesi del sistema Achilleo- calcaneo-plantare. Archivio "Putti" di chirurgia degli organi di movimento. 1985;35:135–41.
14. Waller JF Jr, Maddalo A. The foot and ankle linkage system. In: Nicholas JA, Hershman EB, editors. The lower extremity and spine in sports medicine. St. Louis: Mosby; 1986.
15. Snow SW, Bohne WH, DiCarlo E, et al. Anatomy of the Achilles tendon and plantar fascia in relation to the calcaneus in various age groups. Foot Ankle Int. 1995;16:418–21.
16. Rondhuis JJ, Huson A. The first branch of the lateral plantar nerve and heel pain. Acta Morphol Neerl Scand. 1986;24:269.
17. Sarrafian SK, Kelikian AS. Nerves in Sarrafian's anatomy of the foot and ankle. 3rd ed. Philadelphia: Wolters Kluwer; 2011. p. 381–428.
18. Roegholt MN. Een nervus calcaneus inferior als overbrenger van de pijn bij calcaneodynie of calcanensspoor en de daaruit volgende therapie. Ned Tijdschr v Geneeskd. 1940;84:1898.
19. Tanz SS. Heel pain. Clin Orthop. 1963;28:169.
20. Arenson DJ, Cosentina GL, Suran SM. The inferior calcaneal nerve, an anatomical study. J Am Podiatr Assoc. 1980;70:552.
21. Przylucki H, Jones CL. Entrapment neuropathy of muscle branch of lateral plantar nerve. J Am Podiatr Assoc. 1981;71:119.
22. Baxter DE, Thigpen CM. Heel pain-operative results. Foot Ankle. 1984;5(1):16.
23. Moroni S, Zwierzina M, Starke V, Moriggl B, et al. Clinical-anatomic mapping of the tarsal tunnel with regard to Baxter's neuropathy in recalcitrant heel pain syndrome: part 1. Surg Radiol Anat. 2018;41(1):29–41.
24. Hornick B, Amabile AH. Baxter's nerve impingement and other neuropathies of the foot: implications for physical therapy. FASEB J. 2017;4:31.
25. Louisia S, Masquelet AC. The medial and inferior calcaneal nerves: an anatomic study. Surg Radiol Anat. 1999;21:169–73.
26. Benjamin M, McGonagle D. The anatomical basis for disease localisation in seronegative spondyloarthropathy at entheses and related sites. J Anat. 2001;199:503–26.
27. McGonagle D, Marzo-Ortega H, O'Connor P, et al. Histological assessment of the early enthesitis lesion in spondyloarthropathy. Ann Rheum Dis. 2002;61:534–7.

28. Merkel KH, Hess H, Kunz M. Insertion tendopathy in athletes: a light microscopic, histochemical and electron microscopic examination. Pathol Res Pract. 1982;173:303–9.
29. Benjamin M, Ralphs JR. Fibrocartilage in tendons and ligaments: an adaptation to compressive load. J Anat. 1998;193:481–94.
30. Wearing SC, Smeathers JE, Urry SR, Henning EM, Hills AP. The pathomechanics of plantar fasciitis. Sports Med. 2006;36(7):585–611.
31. Kumai T, Benjamin M. Heel spur formation and the subcalcaneal enthesis of the plantar fascia. J Rheumatol. 2002;29:1957–64.
32. Abreu MR, Chung CB, Mendes L, et al. Plantar calcaneal digital enthesophytes: new observations regarding sites of origin based on radiographic, MR imaging, anatomic, and palepossible mechanism for tendon degeneration. Skelet Radiol. 2003;32:13–21.
33. Kaya BK. Plantar fasciitis in athletes. J Sport Rehabil. 1996;5:305–20.
34. Warren BL. Plantar fasciitis in runners: treatment and prevention. Sports Med. 1990;10:338–45.
35. Schepsis AA, Leach RE, Gorzyca J. Plantar fasciitis: etiology, treatment, surgical results, and review of the literature. Clin Orthop. 1991;266:185–96.
36. LeMelle DP, Kisilewicz P, Janis LR. Chronic plantar fascial inflammation and fibrosis. Clin Podiatr Med Surg. 1990;7:385–9.
37. Snider MP, Clancy WG, McBeath AA. Plantar fascia release for plantar fasciitis in runners. Am J Sports Med. 1983;11:215–9.
38. Astrom M, Rausing A. Chronic Achilles tendinopathy: a survey of surgical and histopathologic findings. Clin Orthop. 1995;316:151–64.
39. Kannus P. Etiology and pathophysiology of chronic tendon disorders in sports. Scand J Med Sci Sports. 1997;7:78–85.
40. Lemont H, Ammirati KM, Usen N. Plantar fasciitis: a degenerative process (fasciosis) without inflammation. J Am Podiatr Med Assoc. 2003;93:234–7.
41. Backman C, Boquist L, Frid'en J, et al. Chronic achilles paratendonitis with tendinosis: an experimental model in the rabbit. J Orthop Res. 1990;8:541–7.
42. Enwemeka CS. Inflammation, cellularity, and fibrillogenesis in regenerating tendon: implications for tendon rehabilitation. Phys Ther. 1989;69:816–25.
43. Kannus P, Jo'zsa L. Histopathological changes preceding spontaneous rupture of a tendon: a controlled study of 891 patients. J Bone Joint Surg Am. 1991;73:1507–25.
44. Huang CK, Kitaoka HB, An KN, Chao YS. Biomechanical evaluation of longitudinal arch stability. Foot Ankle Int. 1993;14(6):353–7.
45. Murphy GA, Pneumaticos SG, Kamaric E, et al. Biomechanical consequences of sequential plantar fascia release. Foot Ankle Int. 1998;19:149–52.
46. Thordarson DB, Kumar PJ, Hedman TP, Ebramzadeh E. Effect of partial versus complete plantar fasciotomy on the windlass mechanism. Foot Ankle Int. 1997;18(1):16–20.
47. Kitaoka HB, Luo ZP, An KN. Mechanical behavior of the foot and ankle after plantar fascia release in the unstable foot. Foot Ankle Int. 1997;18:8–15.
48. Daly PJ, Kitaoka HB, Chao EY. Plantar fasciotomy for intractable plantar fasciitis: clinical results and biomechanical evaluation. Foot Ankle. 1992;13:188–95.
49. Sharkey NA, Donahue SW, Ferris L. Biomechanical consequences of plantar fascial release or rupture during gait, part II: alterations in forefoot loading. Foot Ankle Int. 1999;20:86–96.
50. Ward ED, Smith KM, Cocheba JR, Patterson PE, Phillips RD. In vivo forces in the plantar fascia during the stance phase of gait: sequential release of the plantar fascia. J Am Podiatr Med Assoc. 2003;93(6):429–42.
51. Anderson DJ, Fallat LM, Savoy-Moore RT. Computer-assisted assessment of lateral column movement following plantar fascial release: a cadaveric study. J Foot Ankle Surg. 2001;40:62–70.
52. Carlson RE, Fleming LL, Hutton WC. The biomechanical relationship between the tendon achilles, plantar fascia and metatarsophalangeal joint dorsiflexion angle. Foot Ankle Int. 2000;21:18–25.

53. H-YK C, Lin C-L, Wang H-W, Chou S-W. Finite element analysis of plantar fascia under stretch—the relative contribution of windlass mechanism and Achilles tendon force. J Biomech. 2008;41:1937–44.
54. Gefen A. The in vivo elastic properties of the plantar fascia during the contact phase of walking. Foot Ankle Int. 2003;24:238–44.
55. Erdemir A, Hamel AJ, Fauth AR, Piazza SJ, Sharkey NA. Dynamic loading of the plantar aponeurosis in walking. J Bone Joint Surg Am. 2004;86-A:546–52.
56. Cheung J, Zhang M, An K-N. Effect of achilles tendon loading on plantar fascia tension in the standing foot. Clin Biomech. 2006;21:194–203.
57. Kogler GF, Solomonidis SE, Paul JP. Biomechanics of longitudinal arch support mechanisms in foot orthoses and their effect on plantar aponeurosis strain. Clin Biomech. 1996;11(5):243–52.
58. Kogler GF, Veer FB, Solomonidis SE, Paul JP. The influence of medial and lateral orthotic wedges on loading of the plantar aponeurosis. Book of Abstracts from the 9th World Congress of the International Society for Prosthetics and Orthotics 1998; 579–581.
59. Kogler GF, Veer FB, Verhulst SJ, Solomonidis SE, Paul JP. The effect of heel elevation on strain within the plantar aponeurosis: in vitro study. Foot Ankle Int. 2001;22:433. https://doi.org/10.1177/107110070102200513.
60. MacConaill MA. The postural mechanism of the human foot. Proc Roy Irish Acad. 1944-1945;50B:265–78.
61. Sarrafian SK. Functional characteristics of the foot and plantar aponeurosis under tibiotalar loading. Foot Ankle. 1987;8:4–18.
62. Bojsen-Møller F. Calcaneocuboid joint and stability of the longitudinal arch of the foot at high and low gear push off. J Anat. 1979;129:165.
63. Caravaggi P, Pataky T, Gu¨nther M, Savage R, Crompton R. Dynamics of longitudinal arch support in relation to walking speed: contribution of the plantar aponeurosis. J Anat. 2010;217:254–61.
64. Caravaggi P, Pataky T, Goulermas JY, Savage R, Crompton R. A dynamic model of the windlass mechanism of the foot: evidence for early stance phase preloading of the plantar aponeurosis. J Exp Biol. 2009;212:2491–9.
65. Kelly LA, Cresswell AG, Racinais S, Whiteley R, Lichtwark G. Intrinsic foot muscles have the capacity to control deformation of the longitudinal arch. J R Soc Interface. 2014;11:20131188.
66. Kelly LA, Lichtwark G, Cresswell AG. Active regulation of longitudinal arch compression and recoil during walking and running. J R Soc Interface. 2014;12:1076.
67. Welte L, Kelly LA, Lichtwark GA, Rainbow MJ. Influence of the windlass mechanism on arch-spring mechanics during dynamic foot arch deformation. J R Soc Interface. 2018;15:20180270.
68. Farris DJ, Kelly LA, Cresswell AG, Lichtwark GA. The functional importance of human foot muscles for bipedal locomotion. Proc Natl Acad Sci. 2019;116(5):1645–50.
69. Takahashi KZ, Gross MT, van Werkhoven H, Piazza SJ, Sawicki GS. Adding stiffness to the foot modulates soleus force-velocity behaviour during human walking. Sci Rep. 2016;6:29870.
70. Zelik KE, Honert EC. Ankle and foot power in gait analysis: implications for science, technology and clinical assessment. J Biomech. 2018;75:1–12.
71. Stearne SM, McDonald KA, Alderson JA, North I, Oxnard CE, Rubenson J. The foot's arch and the energetics of human locomotion. Sci Rep. 2016;6:19403.
72. Ker RF, Bennett MB, Bibby SR, Kester RC, Alexander RM. The spring in the arch of the human foot. Nature. 1987;325(6100):147–9.
73. McDonald KA, et al. The role of arch compression and metatarsophalangeal joint dynamics in modulating plantar fascia strain in running. PLoS One. 2016;11:e0152602.
74. Aquino A, Payne C. Function of the plantar fascia. Foot. 1999;9:73–8.
75. Kitaoka HB, Luo ZP, Growney ES, Berglund BS, An KN. Material properties of the plantar aponeurosis. Foot Ankle Int. 1994;15(10):557–60.

76. Carlstsdt CA, Nordin M. Biomechanics of tendons and ligaments. In: Nordin M, Frankel VH, editors. Biomechanics of tissues and structures of musculoskeletal system. London: Lea and Feibiger; 1980. p. 59–74.
77. Laurent GJ, McAnulty RJ, Gibson J. Changes in collagen synthesis and degradation during skeletal muscle growth. Am J Physiol. 1985;249:C352–5.
78. McAnulty RJ, Laurent GJ. Collagen synthesis and degradation in vivo. Evidence for rapid rates of collagen turnover with extensive degradation of newly synthesized collagen in tissues of the adult rat. Coll Relat Res. 1987;7:93–104.
79. Irving DB, Cook JL, Menz HB. Factors associated with chronic plantar heel pain: a systematic review. J Sci Med Sport. 2006;9(1–2):11–22.
80. Sullivan J, Pappas E, Burns J. Role of mechanical factors in the clinical presentation of plantar heel pain: implications for management. Foot. 2020;42:101636.
81. Sullivan J, Burns J, Adams R, Pappas E, Crosbie J. Musculoskeletal and activity-related factors associated with plantar heel pain. Foot Ankle Int. 2015;36(1):37–45.
82. League AC. Current concepts review: plantar fasciitis. Foot Ankle Int. 2008;29:358–6.
83. Kibler WB, Goldberg C, Chandler J. Functional biomechanical deficits in running athletes with plantar fasciitis. Am J Sports Med. 1991;19:66–71.
84. Liddle D, Rome K, Howe T. Vertical ground reaction forces in patients with unilateral plantar heel pain: a pilot study. Gait Posture. 2000;11:62–6.
85. Kelly AJ, Wainwright AM, Winson IG. Plantar pressures are normal in plantar fasciitis. Foot Dis. 1995;11:129–32.
86. Katoh Y, Chao EYS, Laughman RK, Schneider E, Morrey BF. Biomechanical analysis of foot function during gait and clinical applications. Clin Orthop. 1983;177:23–33.
87. Bedi HS, Love BRT. Differences in impulse distribution patterns in patients with plantar fasciitis. Foot Ankle Int. 1998;19:153–6.
88. Wearing SC, Smeathers JE, Urry SR. The effect of plantar fasciitis on vertical foot-ground reaction forces. Clin Orthop Relat Res. 2003;17:175–85.
89. Sullivan J, Burns J, Adams R, Pappas E, Crosbie J. Plantar heel pain and foot loading during normal walking. Gait Posture. 2015;41:688–93.
90. Allen RH, Gross MT. Toe flexors strength and passive extension range of motion of the first metatarsophalangeal joint in individuals with plantar fasciitis. J Orthop Sports Phys Ther. 2003;33:468–78.
91. Aranda Y, Munuera PV. Plantar fasciitis and its relationship with hallux limitus. J Am Podiatr Med Assoc. 2014;104:263–8.
92. Creighton DS, Olson VL. Evaluation of range of motion of the first metatarsophalangeal joint in runners with plantar fasciitis. J Orthop Sports Phys Ther. 1987;8(7):357–61.
93. Wearing SC, Smeathers JE, Yates B, Sullivan PM, Urry SR, Dubois P. Sagittal movement of the medial longitudinal arch is unchanged in plantar fasciitis. Med Sci Sports Exerc. 2004;36:1761–7.
94. McClinton S, Collazo C, Vincent E, Vardaxis V. Impaired foot plantar flexor muscle performance in individuals with plantar heel pain and association with foot orthosis use. J Orthop Sports Phys Ther. 2016;46:681–8.
95. Perry J. Gait analysis: normal and pathological function. 2nd ed. Thorofare: SLACK Inc.; 1995. p. 2010.
96. McCullough MB, Ringleb SI, Arai K, Kitaoka HB, Kaufman KR. Moment arms of the ankle throughout the range of motion in three planes. Foot Ankle Int. 2011;32:3.
97. Johnson C, Christensen JC. Biomechanics of the first ray part 1. The effects of the peroneus longus function. A three dimensional kinematic study on a cadaver model. J Foot Ankle Surg. 1999;38(5):313–21.
98. Perez HR, Reber LK, Jeffrey CC. The effect of frontal plane position on first ray motion: forefoot locking mechanism. Foot Ankle Int. 2008;29(1):72–6.
99. Roukis TS, Scherer PR, Anderson CF. Position of the first ray and motion of the first metatarsophalangeal joint. J Am Pod Med Assoc. 1996;86:538–46.

100. McPoil TG, Martin RL, Cornwall MW, et al. Heel pain–plantar fasciitis: clinical practice guidelines linked to the international classification of function, disability, and health from the orthopaedic section of the American Physical Therapy Association. J Orthop Sports Phys Ther. 2008;38(4):A1–A18.
101. Irving DB, Cook JL, Young MA, Menz HB. Obesity and pronated foot type may increase the risk of chronic plantar heel pain: a matched case-control study. BMC Musculoskelet Disord. 2007;8:41.
102. Gill LH. Plantar fasciitis: diagnosis and conservative management. J Am Acad Orthop Surg. 1997;5:109–17.
103. Kwong PK, Kay D, Voner RT, et al. Plantar fasciitis: mechanics and pathomechanics of treatment. Clin Sports Med. 1988;7:119–26.
104. Huang YC, Wang LY, Wang HC, et al. The relationship between the flexible flatfoot and plantar fasciitis: ultrasonographic evaluation. Chang Gung Med J. 2004;27:443–8.
105. Ross M. Use of the tissue stress model as a paradigm for developing an examination and management plan for a patient with plantar fasciitis. J Am Podiatr Med Assoc. 2002;92:499–506.
106. Taunton JE, Clement DB, McNicol K. Plantar fasciitis in runners. Can J Appl Sport Sci. 1982;7:41–4.
107. Cornwall MW, McPoil TG. Plantar fasciitis: etiology and treatment. J Orthop Sports Phys Ther. 1999;29:756–60.
108. Wearing SC, Hills AP, Byrne NM, Hennig EM, McDonald M. The arch index: a measure of flat or fat feet? Foot Ankle Int. 2004;25:575–81.
109. McCrory JL, Young MJ, Boulton AJM, Cavanagh PR. Arch index as a predictor of arch height. Foot. 1997;7:79–81.
110. Prichasuk S, Subhadrabandhu T. The relationship of pes planus and calcaneal spur to plantar heel pain. Clin Orthop. 1994;306:92–6.
111. Shama SS, Kominsky SJ, Lemont H. Prevalence of non painful heel spur and its relation to postural foot position. J Am Podiatry Assoc. 1983;73:122–3.
112. Rano JA, Fallet LM, Savoy-Moore RT. Correlation of heel pain heel spur and its relation to postural foot position. J Am with body mass index and other characteristics of heel pain. J Foot Ankle Surg. 2001;40:351–6.
113. Rome K, Howe T, Haslock I. Risk factors associated with the development of plantar heel pain in athletes. Foot. 2001;11(3):119–25.
114. Pohl MB, Hamill J, Davis IS. Biomechanical and anatomic factors associated with a history of plantar fasciitis in female runners. Clin J Sport Med. 2009;19:372–6.
115. Ribeiro AP, Trombini-Souza F, Tessutti V, Rodrigues Lima F, Sacco Ide C, Joao SM. Rearfoot alignment and medial longitudinal arch configurations of runners with symptoms and histories of plantar fasciitis. Clinics. 2011;66:1027–33.
116. Redmond AC, Crosbie J, Ouvrier RA. Development and validation of a novel rating system for scoring standing foot posture: the foot posture index. Clin Biomech (Bristol, Avon). 2006;21:89–98.
117. Nakamura H, Kakurai S. Relationship between the medial longitudinal arch movement and the pattern of rearfoot motion during stance phase of walking. J Phys Ther Sci. 2003;15:13–8.
118. Lee SY, Hertel J, Lee SC. Rearfoot eversion has indirect effects on plantar fascia tension by changing the amount of arch collapse. Foot. 2010;20(2):64–70.
119. Wearing SC, Smeathers JE, Sullivan PM, Yates B, Urry SR, Dubois P. Plantar fasciitis: are pain and fascial thickness associated with arch shape and loading? Phys Ther. 2007;87:1002–8.
120. Maganaris CN, Narici MV, Almekinders LC, et al. Biomechanics and pathophysiology of overuse tendon injuries: ideas on insertional tendinopathy. Sports Med. 2004;34:1005–17.
121. Almekinders LC, Weinhold PS, Maffulli N. Compression etiology in tendinopathy. Clin Sports Med. 2003;22:703–10.
122. Lyman J, Weinhold PS, Almekinders LC. Strain behavior of the distal achilles tendon: implications for insertional Achilles tendinopathy. Am J Sports Med. 2004;32:457–61.
123. Chang R, Rodrigues PA, VanEmmerik RE, Hammil J. Multi-segment foot kinematics and ground reaction forces during gait of individuals with plantar fasciitis. J Biomech. 2014;47:2571–7.

124. Karas MA, Hoy DJ. Compensatory midfoot dorsiflexion in the individual with heelcord tightness: implications for orthotic device designs. J Prosthet Orthot. 2002;14(2):82–93.
125. Bolivar YA, Munuera PV, Padillo JP. Relationship between tightness of the posterior muscles of the lower limb and plantar fasciitis. Foot Ankle Int. 2013;34:42–8.
126. Birtane M, Tuna H. The evaluation of plantar pressure distribution in obese and non-obese adults. Clin Biomech. 2004;19(10):1055–9.
127. Gravante G, Russo G, Pomara F, Ridola C. Comparison of ground reaction forces between obese and control young adults during quiet standing on a baropodometric platform. Clin Biomech. 2003;18(8):780–2.
128. Ozdemir H, Yilmaz E, Murat A, Karakurt L, Poyraz AK, Ogur E. Sonographic evaluation of plantar fasciitis and relation to body mass index. Eur J Radiol. 2005;54:443–7.
129. van Leeuwen KD, Rogers J, Winzenberg T, vanMiddelkoop M. Higher body mass index is associated with plantar fasciopathy/'plantar fasciitis': systematic review and meta-analysis of various clinical and imaging risk factors. Br J Sports Med. 2016;50:972_81.
130. McMillan AM, Landorf KB, Barrett JT, Menz HB, Bird AR. Diagnostic imaging for chronic plantar heel pain: a systematic review and meta-analysis. J Foot Ankle Res. 2009;2:32.
131. Mohseni-Bandpei MA, Nakhaee M, Mousavi ME, et al. Application of ultrasound in the assessment of plantar fascia in patients with plantar fasciitis: a systematic review. Ultrasound Med Biol. 2014;40:1737–54.
132. Menz HB, Thomas MJ, Marshall M, Rathod-Mistry T, Hall A, Chesterton LS, Peat GM, Roddy E. Coexistence of plantar calcaneal spurs and plantar fascial thickening in individuals with plantar heel pain. Rheumatology. 2019;58:237–45.
133. Kirkpatrick J, Yassaie O, Mirjalili SA. The plantar calcaneal spur: a review of anatomy, histology, etiology and key associations. J Anat. 2017;230:743_51.
134. Waclawski R, Beach J, Milne A, Yacyshyn E, Dryden DM. Systematic review: plantar fasciitis and prolonged weight bearing. Occup Med. 2015;65:97–106.
135. Werner RA, Gell N, Hartigan A, Wiggerman N, Keyserling WM. Risk factors for plantar fasciitis among assembly plant workers. PM R. 2010;2:110–6.
136. Wearing S. Rethinking the treatment algorithm for plantar fasciitis. Podiatry Today. 2019;32(11):32–9.
137. Mokone GG, Schwellnus MP, Noakes TD, Collins M. The COL5A1 gene and Achilles tendon pathology. Scand J Med Sci Sports. 2006;16(1):19–26.
138. Posthumus M, September AV, O'Cuinneagain D, van der Merwe W, Schwellnus MP. The COL5A1 gene is associated with increased risk of anterior cruciate ligament ruptures in female participants. Am J Sports Med. 2009;37(11):2234–40.
139. September AWV, Schwellus MP, Collins M. Tendon and ligament injuries: the genetic component. Br J Sports Med. 2007;41:241–6.
140. Wearing SC, Hooper SL. Mechanisms underlying the development of plantar fasciitis. Dansk Sportsmedicin. 2010;12(4):10–4.
141. Soslowsky LJ, Thomopoulos S, Tun S, et al. Overuse activity injures the supraspinatus tendon in an animal model: a histologic and biomechanical study. J Shoulder Elb Surg. 2000;9(2):79–84.
142. Archamault JM, Hart DA, Herzog W. Response of rabbit Achilles tendon to chronic repetitive loading. Connect Tissue Res. 2001;42(1):13–23.
143. Carpenter JE, Flanagan CL, Thomopoulos S, Yian EH, Soslowsky LJ. The effects of overuse combined with intrinsic or extrinsic alterations in an animal model of rotator cuff tendinosis. Am J Sports Med. 1998;26(6):801–7.
144. Wearing SC, Smeathers JE, Yates B, Urry SR, Dubois P. Bulk compressive properties of the heel fat pad during walking: a pilot investigation in plantar heel pain. Clin Biomech. 2009;24(4):397–402.
145. Bojsen-Moller F, Flagstad KE. Plantar aponeurosis and internal architecture of the ball of the foot. J Anat. 1976;121:599–611.
146. Baxter DE, Thigpen CM. Heel pain-operative results. Foot Ankle. 1984;5(1):16.
147. Kumai T, Benjamin M. Heel spur formation and the subcalcaneal enthesis of the plantar fascia. J Rheumatol. 2002;29:1957–64.

Index

D. H. Richie Jr, *Pathomechanics of Common Foot Disorders*,
https://doi.org/10.1007/978-3-030-54201-6

MIX
Papier aus verantwortungsvollen Quellen
Paper from responsible sources
FSC® C105338

If you have any concerns about our products,
you can contact us on
ProductSafety@springernature.com

In case Publisher is established outside the EU,
the EU authorized representative is:
Springer Nature Customer Service Center GmbH
Europaplatz 3, 69115 Heidelberg, Germany

Printed by Libri Plureos GmbH
in Hamburg, Germany